The Second Jurassic Dinosaur Rush

The Second Jurassic Dinosaur Rush

Museums and Paleontology in America at the Turn of the Twentieth Century

PAUL D. BRINKMAN

The University of Chicago Press   Chicago and London

PAUL D. BRINKMAN is a research curator at the North Carolina Museum of Natural Science in Raleigh.

The University of Chicago Press, Chicago 60637
The University of Chicago Press, Ltd., London

Printed in the United States of America

18 17 16 15 14 13 12 11 10 1 2 3 4 5

ISBN-13: 978-0-226-07472-6 (cloth)
ISBN-10: 0-226-07472-2 (cloth)

Library of Congress Cataloging-in-Publication Data

Brinkman, Paul D.
The second jurassic dinosaur rush : museums and paleontology in America at the turn of the twentieth century / Paul D. Brinkman.
p. cm.
Includes bibliographical references and index.
ISBN-13: 978-0-226-07472-6 (cloth : alk. paper)
ISBN-10: 0-226-07472-2 (cloth : alk. paper) 1. Paleontology—United States—History—19th century. 2. Dinosaurs—Study and teaching—United States—History—19th century. 3. Natural history museums—United States—History—19th century. I. Title.
QE705.U6B75 2010
560'.176075—dc22

2009046342

♾ The paper used in this publication meets the minimum requirements of the American National Standard for Information Sciences—Permanence of Paper for Printed Library Materials, ANSI Z39.48-1992.

For my father

Men of science are often carried away by personal ambition and by a fierce spirit of competition with their rivals, as well as by a sense of scientific power, to employ methods which are wholly unworthy of the true spirit of American scientific research. —HENRY FAIRFIELD OSBORN, 1931

Contents

Illustrations

Acknowledgments

This book began long ago when I was in the Program in the History of Science and Technology (HST) at the University of Minnesota. I am deeply grateful to Sally Gregory Kohlstedt, John Beatty, Alan Shapiro, Ronald Rainger, and Olivier Rieppel for the invaluable guidance they provided. I am likewise grateful to the HST faculty, students, and staff, past and present, who provided a vital network of support and encouragement. A few deserve special mention, including Mary Anne Andrei, Susan Rensing, Margot Iverson, David Sepkoski, Richard Bellon, Michel Janssen, Barbara Eastwold, Bob Seidel, and Piers Hale. I am also grateful to the program and the university for generous financial support, including a writing fellowship.

In addition to a PhD in the history of science, I have many years of experience as a field, lab, and collections technician working with vertebrate fossils. I worked for many years at the Field Museum, where I first got interested in the life and career of Elmer Riggs, one of the key people whose story I tell in this book. The Field Museum provided a second institutional home for me. The staff of the museum library was especially accommodating, providing access to critical research materials, office space, and invaluable professional support. Ben Williams made these arrangements possible. Christine Giannoni made them palatable. The Geology Department, likewise, has been generous with access to important materials. Many friends, including staff and fellow graduate students, at the museum's Friday evening seminar series contributed to this project in ways that are hard to explain, or remember. Finally, the Field Museum Scholarship

Committee awarded me with the Lester Armour Graduate Student Fellowship, which made my residence in Chicago possible.

Innumerable staff members—too many to list individually—at dozens of institutions have helped me with critical access to special collections. Those that merit special mention because of the magnitude of the help they provided include Armand Esai, Jerice Barrios, and Bill Simpson of the Field Museum; Bernadette Callery and Betty Hill of the Carnegie Museum; and Susan Bell of the American Museum.

All of my family and friends contributed to this project in important ways. My mother gave me a copy of Url Lanham's *The Bone Hunters*, and then encouraged me to go to graduate school and write my own book. My brother Chris provided a year of free lodging. Brothers Joel and Lorne made several of my research trips possible by providing transportation and a place to stay. I also want to thank Bill Hammer, the Munson family, and all the staff, students, and other participants in the Augustana college field fossil vertebrate course, for giving me the opportunity to dig every summer; the faculty, staff, and students of the vertebrate paleontology community in Raleigh, North Carolina; and, in no particular order, John Foster, John McIntosh, Brent Breithaupt, Janet Voight, Carol Urness, Rob Cowie, Dave Clarke, John Flynn, Lance Grande, Tom Rea, Betty Shor, Michael Kohl,Tom Brown, Karsten Lawson, Evan Roberts, Lori Belk, Dan Paulsen, Allison Smith, Steve Emrick, Chris Regis, Beth Erdey, Cathy Dowd, Andrew Leman, and the family of Elmer Riggs.

Finally, I'd like to thank my editor, Christie Henry, and the rest of the staff at the University of Chicago Press for their dedication and hard work.

Introduction

The second Jurassic dinosaur rush was a fierce contest to find, collect, and exhibit the world's first and (ideally) largest sauropod dinosaur. It began in the mid-1890s and spilled into the twentieth century, as the institutional setting for American vertebrate paleontology shifted from private collections to urban museums funded by large-scale philanthropy, including the American Museum in New York, the Carnegie Museum in Pittsburgh, and the Field Columbian Museum in Chicago. The competition among these museum paleontologists to find more and better-quality Jurassic dinosaurs fostered the development of new and improved techniques for excavating, packing, and handling fossils in the field and for cleaning and mounting fossils for display. At the same time, the vast accumulation of new specimens resulted in an important period of Jurassic dinosaur research, as a new generation of paleontologists sought to revise and improve upon the work of Edward Drinker Cope and Othniel Charles Marsh, the bitter antagonists of the *first* Jurassic dinosaur rush in the 1870s and '80s. Henry Fairfield Osborn, founder and first curator of the Department of Vertebrate Paleontology at the American Museum, emphasized that early twentieth-century vertebrate paleontology was a collegial and cooperative venture. But Osborn and his rivals were every bit as competitive, petty, and proprietary as their infamous nineteenth-century predecessors. This book tells their story.

In many ways the second Jurassic dinosaur rush was more important than the first rush, and it had a greater and more lasting impact on science and society. The status of dinosaurs soared from prehistoric relic to cultural phenomenon, from arcane scientific term to household word, in the wake of the second rush. Once an obscure group of terrible lizards named and described from fragmentary remains discovered first in Great Britain and later in continental Europe, dinosaurs became a pop-culture marvel only after an even more obscure group of American field paleontologists helped make them so in the late nineteenth and early twentieth centuries. To be sure, dinosaurs achieved some renown in the 1850s, when several life-sized, squat, cement models of *Iguanodon*, *Megalosaurus*, and other prehistoric monsters went on display on the grounds of the Crystal Palace, in Sydenham, south of London. An account of a rowdy dinner party hosted by paleontologist Richard Owen inside the body of *Iguanodon* appeared on the front page of the *Illustrated London News*.[1] In the United States, dinosaurs garnered additional fame in 1868, when the first mounted skeleton of *Hadrosaurus* reared up, kangaroo-like, at Philadelphia's Academy of Natural Sciences. Later, they achieved a certain notoriety when they were the hotly contested trophies of the *first* Jurassic dinosaur rush, an unseemly struggle between Cope and Marsh, two American paleontologists who raced to collect, describe, and name them in the last quarter of the nineteenth century. American Jurassic dinosaurs created a sizable stir in the scientific community during the first rush, but they remained relatively unknown outside a small circle of geologists.

The second rush began modestly, and more or less cooperatively, when a few small university parties collected Jurassic dinosaurs in southeastern Wyoming starting in 1895. A party of professional collectors, with more ample means, from New York's American Museum of Natural History raised the ante when they entered the field in 1897. Meeting the extraordinary expectations of H. F. Osborn, their intensely ambitious, but largely deskbound superior, was their chief motivation. Impressing Osborn could mean a chance at permanent museum employment and professional advancement at the American Museum. But the competition began in earnest only in 1899, when more professionals from the Carnegie Museum in Pittsburgh and Chicago's Field Columbian Museum joined in the search for Jurassic dinosaurs in the same region and opened productive quarries only a few miles from the main American Museum locality. A mercurial lepidopterist named William Jacob Holland, who was Osborn's equal in blazing ambition, was the director of the Carnegie Museum. Holland hired a cadre of talented paleontologists and spurred them to succeed. He aimed to please steel baron Andrew Carnegie, the museum's benefactor and namesake, who took a personal interest in acquiring an exhibit-quality dinosaur for Pittsburgh. The Chicago effort, by contrast, was decidedly more blue collar. It suffered for years from inadequate funding and a lack of reliable institutional support. Elmer Samuel Riggs, Chicago's representative in the field, was a lowly assistant curator with limited means and virtually no influence with museum administrators. Nor did any of Chicago's museum patrons, including founder Marshall Field, take any particular interest in dinosaurs. Osborn's army of subordinates, Holland's hirelings, Riggs and his several cut-rate colleagues, as well as many others competed to collect Jurassic dinosaurs for several consecutive field seasons, although this book is largely a story of the competition among the museums in New York, Pittsburgh, and Chicago. Their contest, which began in Wyoming, spread quickly to other states, including Colorado, Utah, South Dakota, and Montana, as rival parties scoured the West for Jurassic exposures, seeking out new and better dinosaur localities. By 1905, all three museums had closed their Jurassic field operations and moved on to collect other fossil faunas from other geological periods.[2]

The outcome of this field campaign was a wealth of Jurassic dinosaur specimens in unprecedented numbers and kinds and of unprecedented quality for America's well-heeled natural history museums. Paleontologists cleaned and restored the specimens; studied, described, and named them; then mounted many of them for display in lifelike poses, which thrilled and awed museum visitors and fed the vanity of their millionaire patrons. Sauropod dinosaurs from the American Jurassic were the largest land animals then known, far larger than their Cretaceous counterparts, and the contest to acquire them and put them on exhibit fueled a popular fascination for gigantic things. Mounted dinosaurs have become so ubiquitous in the aftermath of the second Jurassic dinosaur rush that it is easy to assume that they have been a staple of natural history museum exhibits for centuries. Yet Philadelphia's *Hadrosaurus* and a few duplicate casts in other museums were the only American examples prior to 1895. Museum displays mounted in the early twentieth century and the media attention they attracted ushered dinosaurs into the consciousness of the American public for the first time. And though their popularity has waxed and waned over the intervening decades, dinosaurs have remained a cultural fixture ever since. Newspaper coverage provides a crude measure of their growing popularity. A search for "dinosaur" in the *Washington Post*, for example, from 1877 through 1895, turns up nothing. But after 1896, the word appears with ever increasing frequency.[3] It was the beginning of modern *dinomania*.[4]

Yet we know remarkably little about the history of the second Jurassic dinosaur rush. Instead, scholars have focused their attention on the earlier dinosaur collecting contest between Cope and Marsh. Dozens of books on their notorious exploits have been written, and new ones appear in print every few years. John S. McIntosh pointed out this disparity in 1990 in a short article entitled "The Second Jurassic Dinosaur Rush." McIntosh broke new ground by defining and briefly describing this phenomenon, identifying its chief protagonists, and summarizing their major accomplishments. But his short article did little to address the broader social and scientific context of the second rush. Since that time, several important contributions on various aspects of the second Jurassic dinosaur rush have appeared, including Ronald Rainger's history of Osborn's vertebrate paleontology program at the American Museum and Tom Rea's history of the discovery and later celebrity of "Dippy," the mounted *Diplodocus* at the Carnegie Museum.[5] Even so, the first rush remains far better known than its successor.

The present work is an attempt to treat the second Jurassic dinosaur rush with the book-length narrative history it deserves. I adopt a microhistorical approach to the subject. Drawing liberally upon a rich archival record, I endeavor to cover the three major museums and all the personnel involved as intimately and as thoroughly as the documentary evidence will allow, but always with an eye toward telling an interesting and coherent story about scientific practice. Only through a microhistorical approach is it possible to convey the urgency and the intensely competitive nature of the contest for fossils. It is also easier to relate the complicated and often volatile working relationships that developed between museum patrons, administrators, paleontologists, and their support staff. The structure of these relationships was rigidly hierarchical. The men at the top maintained command and control of their subordinates in the field with a steady stream of instructional letters and cables. Often these communications showed an appalling insensitivity, bordering on ignorance, to the hardships inherent in dinosaur fieldwork. Deskbound paleontologists spurred their collectors relentlessly because field success was the foundation for research and exhibit development back at the museum. Not surprisingly, credit for success settled disproportionately on the men at the top of the museum hierarchy who held the purse strings. It seldom trickled down to the low-ranking men at the bottom who did the overwhelming bulk of the hard, skilled labor. The unequal distribution of credit caused much resentment, sometimes leading to turnover and long-lasting grudges.

Building representative collections of fossil vertebrates of all kinds, and from all times and places, was the general scientific goal of all three museum programs in vertebrate paleontology. Meeting this goal was the motive behind dozens of fossil collecting expeditions. The second Jurassic dinosaur rush had its own unique set of motives, however. A race to collect and exhibit the first and largest sauropod dinosaur drove the second rush. Magnificent paper restorations of dinosaurs published by Marsh in the 1880s and '90s were the most likely culprits for catalyzing the race. Marsh provided a scientific motive, also, as many of the paleontologists involved in the race were working explicitly to revise his seminal work on North American dinosaurs.

But dinosaurs do not spring fully and perfectly formed from the ground. Nor is the acquisition, reconstruction, and scientific interpretation of dinosaurs an entirely unproblematic enterprise. The present account, which details the day-to-day practices of dinosaur paleontology and explains how and why Jurassic dinosaur field campaigns were planned and carried out by America's museum paleontologists, offers a look at what it was to do vertebrate paleontology in the United States in the late nineteenth and early twentieth centuries.

ONE

Scientists Wage Bitter Warfare

The *first* American Jurassic dinosaur rush began in April 1877, when Arthur Lakes and Henry C. Beckwith unearthed gigantic bones from a now-famous hogback ridge near Morrison, Colorado (see figure 1). Lakes sent samples to two prominent Eastern paleontologists: Othniel Charles Marsh, a brooding, of American solitary Yale professor, and Edward Drinker Cope, a brilliant and combative Quaker from Philadelphia. The two pillars of vertebrate paleontology, Cope and Marsh were already bitter rivals. Indeed, Marsh showed interest in the fossils of Lakes only when he learned that his competitor had received some samples, also. Cope wrote to Lakes and offered to hire him as a collector, but he was too late. Lakes had already agreed to work for Marsh. Worse still, he asked Cope to forward the fossils he had sent him to Marsh in New Haven. Marsh published his first announcement of Lakes's discovery in the July 1 issue of the *American Journal of Science*. Later in the month, he received a letter from "Harlow and Edwards" (William Harlow Reed and William Edward Carlin) disclosing the existence of another locality for large fossil vertebrates somewhere near Laramie, Wyoming. (They initially kept their real names and the exact location of their discovery, Como Bluff, secret from Marsh.) When another letter from Wyoming warned about rival collectors nosing around at the site, Marsh worked quickly to secure the new locality and its mysterious discoverers for his own purposes. Meanwhile, Oramel W. Lucas alerted Cope to a site near Cañon City, Colorado, one hundred miles

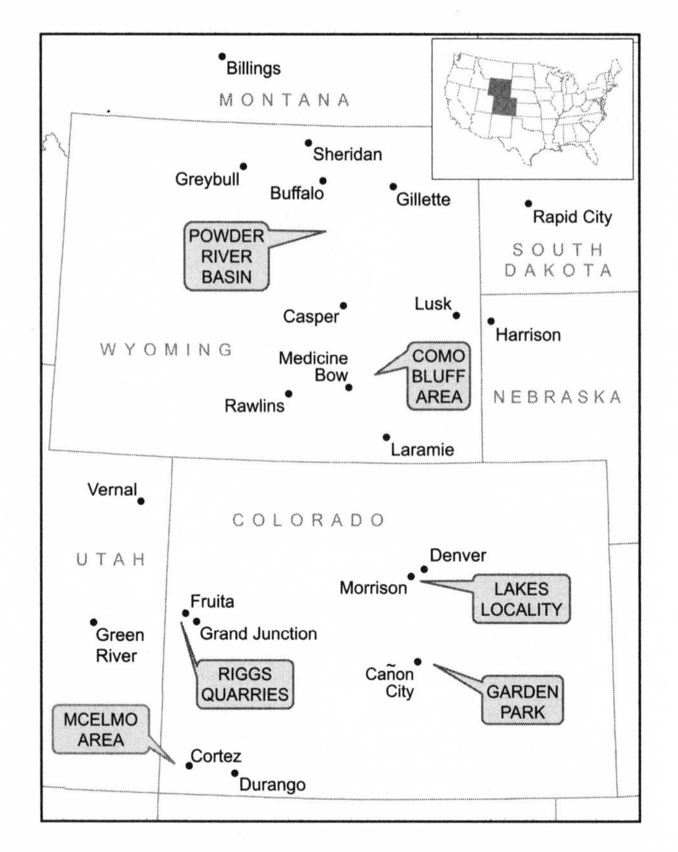

Figure 1. Map of Wyoming, Colorado, and parts of surrounding states showing the principal fossil localities worked during the second Jurassic dinosaur rush. Map drawn by Cathryn Dowd.

south of Morrison, where more enormous fossils had been found. Cope hired Lucas to dig them up and send them to Philadelphia for study and description. In an article dated August 23, he published his own notice of the discovery of gigantic fossil reptiles. With that, the first race for American Jurassic dinosaurs was under way.[1]

The ensuing fossil feud between Cope and Marsh netted a wealth of new data that revolutionized the study of Jurassic dinosaurs. Before 1877, Jurassic dinosaurs were a very poorly known group, thanks in large part to a dearth of good fossils in Great Britain, Germany, France, and other traditional centers of paleontology. The Jurassic beds of the American West, on the other hand, showed themselves to be far more extensive and more fossiliferous than their European equivalents, and they yielded a superabundance of new dinosaurs, better preserved and often far more complete than anything found previously. The relative completeness of the specimens found in America provided, for the first time, a much clearer picture of dinosaur morphology, while the richness of the American fossil record opened a new window onto dinosaur diversity. New World fossils were an invaluable scientific resource that lent a considerable competitive advantage to any paleontologist who could acquire them.[2] Cope and Marsh hired collectors to exploit this resource whenever and wherever possible. They found spectacularly productive fossil localities and excavated thousands of important specimens, many destined to become types. Back East, Cope and Marsh described and named dozens of new genera and species of dinosaurs, often on the basis of fragments or isolated elements, but sometimes from relatively complete skeletons. Many of the most familiar dinosaurs are the products of the first American Jurassic dinosaur rush, including the bizarre, dorsal-plated *Stegosaurus*, the fearsome, meat-eating theropods *Ceratosaurus* and *Allosaurus*, and the gigantic, long-necked, and whip-tailed sauropods, *Camarasaurus*, *Diplodocus*, and *Apatosaurus* (aka *Brontosaurus*).[3]

Marsh was the clear winner in the race for Jurassic dinosaurs, but his victory made him the object of Cope's envy. Although Cope enjoyed a fair number of successes, his Yale rival took the lion's share of the finest fossils. Marsh's final tally of specimens is not precisely known, but his several Jurassic localities in Colorado and Wyoming yielded not less than 1,115 crates of (mostly) dinosaur bones. From that rich harvest, he described twenty-one new genera and forty-one new species.[4] Cope, who described about fifteen new species of Jurassic dinosaurs, Cope, was not one to accept defeat gracefully, however. With time, his resentment of Marsh's accomplishments festered into a thirst for revenge. In short, he was jealous, and he conspired with some shady characters to try to put an end to his troubles with Marsh. "Either he or I must go under," he wrote to a colleague.[5] Marsh, for his part, could afford to be magnanimous in his relations with Cope, but he chose not to be. He was once overheard muttering about his archrival, "Godamnit! I wish the Lord would take him!"[6]

Unfriendly rivalry exploded into public scandal in January 1890, when Cope and Marsh traded insults in the pages of the *New York Herald*. A headline that read "Scientists Wage Bitter Warfare" was the opening salvo in an ugly press campaign initiated by Cope and contrived to bring down Marsh and his chief accomplice, John Wesley Powell, director of the U.S. Geological Survey. To make his case, Cope mucked liberally from a desk drawer full of what he called "Marshiana"—a decades-old collection of notes on his archrival's alleged transgressions and errors. Marsh, too, had kept a careful record of Cope's most humiliating miscues, which he gleefully recounted in his published rebuttal. The result was an embarrassment for American science, replete with unsavory accusations of plagiarism, incompetence, dishonesty, fraud, theft, trespassing, and misappropriation of government funds. A number of Marsh's scientific assistants were implicated in the scandal. Many prominent American scientists also were drawn unwittingly into the fray. Others took sides, including Cope's Princeton allies William Berryman Scott and Henry Fairfield Osborn. When the reading public quickly lost interest in the strange scientific feud, the story disappeared, but not before Cope and Marsh were both permanently tarnished by the scandal. Marsh suffered the most when, as part of the fallout, he lost his lucrative federal funding in 1892. Forced to scale back his field and lab operations, he recalled his field-workers and continued his work at the museum in New Haven with more limited means.[7]

Jurassic dinosaur collecting, however, had already ground to a halt. The great size and weight of Jurassic dinosaur bones rendered the effort to excavate, transport, and prepare them very costly and labor-intensive. For economy's sake, Cope and Marsh had directed their collectors to concentrate on the very best, the biggest, or the newest materials. Productivity eventually reached a point of diminishing returns, which brought an end to collecting before the localities were exhausted. Marsh transferred Lakes to Como Bluff in 1879, and thus abandoned the Morrison locality after two short seasons. Cope's collectors quit their Cañon City locality in 1883. A local rancher named Marshall P. Felch then opened two nearby quarries for Marsh in 1877, and worked them again from 1883 to 1888. Como Bluff proved to be Marsh's longest lasting Jurassic endeavor. Starting in 1877, a long series of field-workers mined this area continuously on the Yale professor's behalf. When Marsh's collectors finally abandoned the field in 1889, they left it open to any takers.[8]

Several able paleontologists stepped into the breach. Wilbur Clinton Knight, professor of geology and mining at the University of Wyoming in Laramie seized the initiative in 1895–97, when he and Reed, one of Marsh's former collectors, and codiscoverer of the Como Bluff locality, made a local collection of Jurassic dinosaurs and marine reptiles. Reed had been collecting these specimens independently for years, and offering them for sale to any number of paleontologists, including Marsh and Osborn. Knight hired him to collect fossil vertebrates and develop exhibits exclusively for the university, which he proceeded to do with abandon. By the late 1890s, the Jurassic dinosaur collection at the University of Wyoming was probably the second largest in the world—after Marsh's storied collection at Yale. The sheer size of this collection, which weighed over eighty tons, prompted university officials to build a new science building to house and display it.[9] Knight, meanwhile, inspired to unprecedented flights of fossil vertebrate generosity by the Jurassic riches available in southeastern Wyoming, began devising a means to spread the wealth more broadly.

Samuel Wendell Williston was an ally of Knight and Reed in the Jurassic of Wyoming. Williston earned an MD in 1880 and a PhD in entomology in 1885, both from Yale. He also had an abiding interest in fossil vertebrates. But Marsh, for whom he was working as a field and lab assistant, jealously guarded access to the precious Yale collections, including the Jurassic dinosaurs that Williston had had a hand in collecting at Morrison, Cañon City, and Como Bluff. Frustrated by Marsh's despotism, Williston quit the museum in 1885, but remained at Yale to teach anatomy. Cope then dragged him into his feud with Marsh when he published excerpts of a letter Williston had written that was exceedingly critical of his former boss. New Haven was now not nearly big enough for the two of them. When an opportunity to teach geology opened in April 1890 at the University of Kansas (KU) in Lawrence, Williston eagerly accepted. At KU he rekindled his interest in paleontology, beginning with the abundant marine reptile fauna of the western Kansas Cretaceous. He also made the acquaintance of Knight, who had similar research interests. In 1895, Knight invited Williston to bring a group of his best students to collect Jurassic dinosaurs in a new locality near Lusk, Wyoming . Williston was still a little bitter about his experience at Yale, and he viewed Knight's new locality as an opportunity to make inroads into Marsh's Jurassic dinosaur monopoly. He wrote to Osborn, predicting, "I propose to go into the Jurassic and I trust will make it interesting for [Marsh]. I learn . . . that he . . . proposes now to write a book. . . . I shall await the results with interest, and may, perhaps, if I get the material from a brand new locality that I am after, undo some of his work."[10] But the new locality proved relatively unproductive. Williston and his enthusiastic students cleaned it out completely in only a few short days.[11]

Dean of the Terrible Lizards

By the 1890s, Marsh reigned supreme as the dean of American dinosaur paleontology. His Yale successor, Charles Beecher, claimed that "Marsh stood as the sole possessor of an acute and comprehensive knowledge of [the dinosaurs], one of the most wonderful and difficult groups known."[12] In November 1896, Marsh staked his claim to dinosaur supremacy with the publication of his hefty and lavishly illustrated monograph, The Dinosaurs of North America. A second contribution, "Vertebrate Fossils [of the Denver Basin]," which appeared shortly thereafter, bolstered this claim. These works were the first of a kind, presenting a comprehensive account of dinosaur structure and classification, and summarizing the results of a long series of Marsh's own papers, which had been issued scattershot over the previous quarter century. Marsh's biographers have pointed out the significance of the former volume, calling it "the foundation of dinosaur knowledge" and "a testament to [Marsh's] unsurpassed knowledge of [dinosaurs]."[13]

These books also served as a clarion call for Marsh's rivals. Cope, who was grappling with an agonizing ailment that would end his life in April 1897, at age fifty-six, was in no shape to carry on with his infamous feud. Legions of paleontologists of the next generation, however, including Williston, Knight, Reed, and especially Osborn, were eager to take shots at Marsh's prized monographs. Marsh had kept his fossils under lock and key at Yale for years, pending the outcome of his meticulous studies. Now his results were available in print, and rivals could gain access to the collections through his printed descriptions and detailed figures.[14] The books provided a baseline set of facts and ideas that could be tested in the field and confirmed, or, better still, contradicted by new and better specimens. Marsh and his work were now subject to revision. Many field paleontologists would take *The Dinosaurs of North America* west for ready reference during the second Jurassic dinosaur rush. Others, wishing they had, would write to their respective museums to have the indispensable volume sent out. That they all referred incessantly to Marsh's work was an unmistakable indication of its importance, but it was also a sign that they felt it could be bested.

One of Marsh's most important contributions to dinosaur paleontology was the publication of a series of complete, lifelike drawings of reconstructed skeletons that served to illustrate the probable former appearance of a number of dinosaurs, including *Brontosaurus* (Figure 2), *Stegosaurus* (Figure 3), *Ceratosaurus* (Figure 4), *Camptosaurus*, and *Laosaurus*, all from the American Jurassic. By their very nature, skeletal reconstructions were somewhat speculative—Marsh thought of them as working hypotheses.[15] Marsh's paper reconstructions were particularly important visual aids in the late nineteenth century, before mounted dinosaur exhibits proliferated widely in American museums. Indeed, they likely played a crucial part in inspiring museum paleontologists, administrators, and benefactors alike that exhibit-quality dinosaur skeletons displayed in lifelike poses would appeal to the American public.

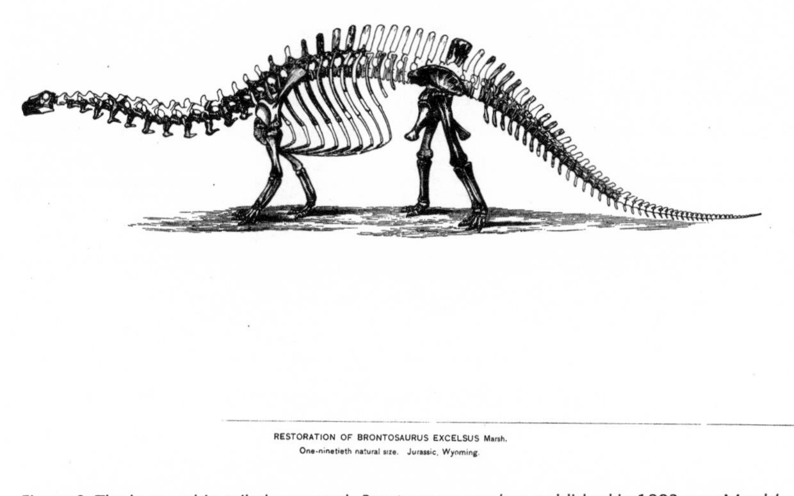

Figure 2. The huge, whip-tailed sauropod, *Brontosaurus excelsus*, published in 1883, was Marsh's first and most famous dinosaur restoration. This slightly modified reconstruction—including several extra vertebrae—comes from Marsh, "Dinosaurs."

Marsh's work set a high-water mark in the brief tradition of fossil reconstructions. The earliest known dinosaur reconstructions—some dating to 1838—were crude and based on scant fossil evidence. So too were the boxy, life-size models executed by Benjamin Waterhouse Hawkins and displayed on the grounds of the Crystal Palace, near London, beginning in 1854. Richard Owen oversaw their design, and thus they resembled the "elephantine lizards" he envisioned when he coined the term dinosaur in 1842. Despite their shortcomings, the Crystal Palace models were immensely popular. In 1868, Hawkins, along with Cope and his mentor, Joseph Leidy, mounted one of America's first dinosaurs, *Hadrosaurus foulkii*, at the Academy of Natural Sciences in Philadelphia. It, too, was an immensely popular exhibit. *Hadrosaurus* was the world's first and only mounted dinosaur fossil until 1883, when paleontologist Louis Dollo and his chief preparator Louis De Pauw exhibited first one and then a series of fully mounted *Iguanodon* skeletons in Brussels, Belgium.[16]

Marsh, along with many of his contemporaries, was skeptical of the value of mounted fossil vertebrates, and he mocked the Crystal Palace models as an "injustice" to dinosaurs. He declined to mount any of his own invaluable dinosaur skeletons for display. A cautious scholar, he worried particularly that a prematurely mounted restoration would be prone to error, and that such errors, once established, would be impossible to dislodge from the public mind. Besides, paper reconstructions would be much easier to manage for the gigantic Jurassic dinosaurs found in the American West. Cope, however, struck first when he displayed a life-size paper reconstruction of his *Camarasaurus supremus* at a meeting of the American Philosophical Society in December 1877. He showed it again in Europe in 1878, where it excited tremendous interest. Despite its popular reception, he never published this illustration in his lifetime. Marsh, by contrast, brought out a paper menagerie of no less than twelve complete dinosaur reconstructions, beginning in 1883 with *Brontosaurus excelsus*, in the *American Journal of Science*. Late in 1895, he compiled these reconstructions on a pair of large plates, privately printed and issued as separates. They also made an encore appearance in *The Dinosaurs of North America*, which circulated widely among American paleontologists the following year. Marsh's stunning reconstructions were duplicated repeatedly in textbooks, encyclopedias, newspapers, and other popular publications. They served to familiarize working scientists and the general public with the fantastic size and grotesque skeletal appearance of America's Jurassic dinosaurs.[17]

Marsh's publications played a direct part in sparking the second Jurassic dinosaur rush. In 1890, Andrew Carnegie, the steelmaking millionaire, showed Marsh's reconstruction of *Brontosaurus excelsus* to a committee that was considering his proposal to build a museum in Pittsburgh. Carnegie explained that he wanted a mounted sauropod dinosaur to serve

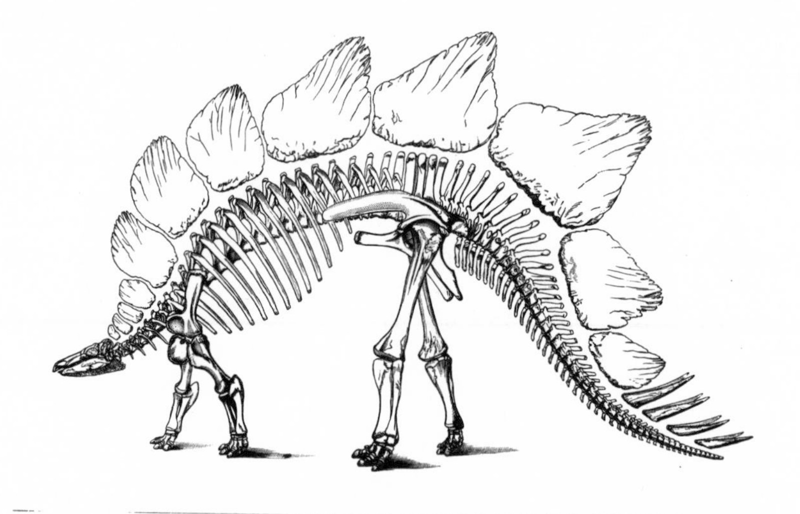

RESTORATION OF STEGOSAURUS UNGULATUS Marsh.
One-thirtieth natural size. Jurassic, Wyoming.

Figure 3. The dorsal-plated *Stegosaurus ungulatus* was one of the strangest dinosaurs from Marsh's Jurassic menagerie. From Marsh, "Dinosaurs."

RESTORATION OF CERATOSAURUS NASICORNIS Marsh.
One-thirtieth natural size. Jurassic, Colorado.

Figure 4. *Ceratosaurus nasicornis* was a large, meat-eating theropod dinosaur from the American Jurassic. From Marsh, "Dinosaurs."

as the new museum's centerpiece. This idea "knocked them flat–every one," he boasted in a letter. Five years later, Carnegie had his museum, but still no dinosaur. He invited Marsh to the dedication ceremony in 1895, and then goaded him from the podium to donate a dinosaur to Pittsburgh, but all in vain. In December 1898, Marsh's *Brontosaurus* appeared prominently on the front page of the *New York Journal and Advertiser*. The banner headline of the accompanying story read, "Most Colossal Animal on Earth Just Found out West." According to legend, the story cut short Carnegie's breakfast, and inspired him to write the $10,000 check to Director Holland that touched off the Carnegie Museum's participation in the second Jurassic dinosaur rush. In fact, he had already begun making an effort to acquire that very dinosaur.[18] Likewise, at Chicago's Field Columbian Museum, established in 1893, Geology Curator Oliver C. Farrington, a Yale graduate, wanted to mount a Jurassic dinosaur. To drum up support for this venture, he cited the superlative dimensions of the sauropod dinosaur *Atlantosaurus*, which he had taken from Marsh's "Fossil Vertebrates [of the Denver Basin]."[19] In New York, at the American Museum of Natural History, work was already under way on building an exhibition series of mounted fossil mammals. Jacob L. Wortman, who was nominally in charge of fossil vertebrate fieldwork at the museum, first advocated for the acquisition of Jurassic dinosaurs for exhibition purposes in the summer of 1896, shortly after the distribution of Marsh's reconstructions as separates.[20] Osborn ignored Wortman's plea until the summer of 1897, the first field season following the publication of *The Dinosaurs of North America*. The influence of Marsh's monographs, and especially of his dramatic dinosaur reconstructions, is very likely responsible for drawing the attention of so many museum paleontologists to the potential exhibit value of American Jurassic dinosaurs at roughly the same time.

Museums and American Vertebrate Paleontology

Once the exclusive domain of private, entrepreneurial scientists such as Cope and Marsh, American vertebrate paleontology in the 1890s shifted gradually into large urban science museums established and funded lavishly by Gilded Age philanthropists. A complex suite of motives fueled the philanthropists' generosity. Some were vain and wanted merely to build a lasting monument to themselves. Some were also motivated by civic rivalry; if other great cities featured a prominent natural history museum, then theirs must have one as well. Most, however, felt a strong sense of noblesse oblige. They supported museums and other cultural institutions in order to provide for the moral and cultural uplift of the masses in the cities where they had made their fortunes. With some, the object of their charity genuinely was the betterment of the general populace. For others, institutions like the museum were a mechanism for maintaining the status quo. In the latter half of the nineteenth century, New York's wealthy elites expanded their philanthropic activities to encompass the support of natural science at institutions such as the American Museum of Natural History. They were not necessarily interested in promoting scientific research so much as they were in providing wholesome educational and recreational opportunities for underprivileged New Yorkers.[21] Andrew Carnegie at one time thought of his Pittsburgh museum as "one of the chief satisfactions of my life. [It] is my monument." He spelled out the reason for establishing the Carnegie Museum in his autobiography: "[I]t was the good of the people of Pittsburgh I had in view, among whom I had made my fortune, the unfounded suspicions [of self aggrandizement] of some . . . only quickened my desire to work their good by planting in their midst a potent influence for higher things."[22] In Chicago, lumber merchant Edward E. Ayer pried a crucial one million dollar contribution from retailer Marshall Field not only with an implied promise of immortality but also by urging that he had a rare chance to do a good turn for countless needy citizens. Ayer worked on the reluctant benefactor, pleading: "You can sell dry goods until Hell freezes over; you can sell it on the ice until that melts; and in twenty-five years you will be . . . absolutely forgotten. You have an opportunity . . . of being the educational host to untold millions of people . . . in the Mississippi Valley."[23] Field committed the requisite funding and his name was later affixed to the museum.[24]

Well-funded museums provided the ideal institutional setting for American vertebrate paleontology. A far-flung, specimen- and labor-intensive discipline that required collecting in a variety of places, vertebrate paleontology was very expensive to pursue systematically. In the 1870s, Cope and Marsh, who were both independently wealthy, drove their predecessor Leidy from the field with their profligate spending (and with their unseemly competition, which Leidy detested). Both went broke in the process. Marsh for many years enjoyed the additional advantage of a generous U.S. Geological Survey appropriation, which helped him outcompete his Philadelphia rival. But federal subsidies for such an impractical and esoteric discipline as vertebrate paleontology left Powell and his survey vulnerable to criticism from the enemies of government science. In 1892, using wasteful government spending on paleontology as a rallying point, a group of hostile legislators attacked the U.S. Geological Survey budget. In July, Congress gutted it, specifying an end to nearly all federal funding for paleontology. Powell sent Marsh a curt and humiliating telegram demanding his resignation.[25] When Marsh's federal funding dried up, it presaged a new era of museum-based paleontology, bought and paid for by some of the wealthiest men in America. Carnegie, for example, bragged in a letter to Marsh that he had $50,000 to spend every year, forever, in order to get a dinosaur. Similarly, at the Field Columbian Museum, which was backed by Field's millions, paleontology had, in theory, a practically limitless pool of money on which to draw.

The shift to the new museum setting revitalized and transformed American vertebrate paleontology in important ways. In its new institutional setting, paleontology benefited from new and often lavish sources of private support, which funded a surge of renewed activity, including the race to obtain exhibit-quality Jurassic dinosaurs. Only a sustained, methodical, and relentless field collecting campaign, and a new commitment to the proper care and handling of fossils by field-workers, could meet the demand for quality specimens. Likewise, the need for more and better contextual data required changes in field practices. In the fossil preparation laboratory, the high volume of work to be done, and the need to get it accomplished more efficiently, encouraged the development of new, labor-saving lab techniques. A new emphasis on public education and popularization in the museum context inspired a period of innovative new displays, such as mounted fossil vertebrate habitat groups, comparative exhibits, and full-color, fleshed-out reconstructions, to disseminate results to the visiting public. With respect to scientific contributions, museum paleontologists instigated a wave of Jurassic dinosaur revisionism, much of which, for better or worse, was aimed deliberately at bringing down Marsh's earlier body of work. Finally, intermuseum rivalry initially encouraged the continuation of the competitive, proprietary, and petty milieu of the Cope/Marsh era, but it eventually gave way to a modestly cooperative working arrangement among museum paleontologists.

The disciplinary context of these changes has been the subject of some debate. Ronald Rainger argues that, by the early twentieth century, American vertebrate paleontology was a peripheral field of scientific inquiry. Increasingly utilitarian federal sponsorship for science emphasized practical studies in mining, engineering, and soil science, at the expense of paleontology. Meanwhile, the rise of experimental biology and genetics chased paleontology out of most American university curricula. Experimental biology expanded rapidly in the early twentieth century, largely because of the utility of its approach to biological phenomena. By formulating and testing hypotheses, experimentalists generated exciting causal explanations for specific biological problems. Paleontology, on the other hand, could only provide speculative, untestable, historical explanations in such traditional fields as morphology, biogeography, and systematic biology. According to Rainger, these subjects held little interest for the new generation of American biologists. Similarly, Steven Conn characterizes vertebrate paleontology as "the last, most significant contribution made by . . . museums to turn-of-the-century science. Paleontology was the apotheosis of the older natural history and its last stand."[26]

Yet American vertebrate paleontology was not as moribund as these accounts suggest. Although marginal relative to other burgeoning subfields of biology or geology at those universities that continued to support it, vertebrate paleontology survived and continued to grow modestly in the academy. And at large urban natural history museums, it fared particularly well, especially around the turn of the twentieth century. Though few in number and relatively slow in rate of growth, American natural history museums, according to Peter J. Bowler, sustained "a steady level of activity" in systematic and evolutionary biology in the late nineteenth and early twentieth centuries. Vertebrate paleontology was a vital component to this work. Similarly, Lynn K. Nyhart argues that natural history, especially in new institutional settings such as museums and zoos, grew in size but declined in relative importance with the gradual expansion and diversification of biology. Rainger, too, concedes that a number of biologists maintained an interest in morphology and taxonomy, despite the exaggerated claims to the contrary by early twentieth-century experimentalists such as Osborn's Columbia University rival, Thomas Hunt Morgan. Indeed, Rainger argues elsewhere that one should think of American paleontology as a distinctive biological research program that maintained its commitment to morphological questions and methods through the early twentieth century, a period of dramatic changes in American biology. This conception allows for the expansion of experimental biology and genetics, but without asserting that paleontology was supplanted or replaced by the new biology.[27]

Problems inherent in the study of fossil vertebrates—rather than the declining interest of biologists—were largely responsible for the relatively slow rate of growth for American vertebrate paleontology. For example, the great cost of making and maintaining fossil collections precluded all but the best-funded institutions from fully supporting vertebrate paleontology. Furthermore, the small number, relatively small size, and cumbersome nature of fossil vertebrate collections imposed practical limits on the number of researchers who could have the requisite access to vertebrate fossils. So there is room for argument respecting the so-called marginal standing of vertebrate paleontology during the first few decades of the twentieth century. From some perspectives, vertebrate paleontology at the turn of the twentieth century could be described as an exciting and prosperous scientific discipline. Indeed, at the end of 1903 Osborn published a progress report that emphasized the expansion of American vertebrate paleontology and the dramatic growth of fossil collections. He wrote that "collections . . . are so extensive and multiplying so rapidly, that it is gratifying to report that the number of specialists engaged in the field, in museums and in research work, has rapidly increased."[28]

Vertebrate paleontology served a number of purposes in American natural history museums. Because many of these museums received a significant part of their funding from state or local governments, they accepted public education as part of their mandate. Administrators and curators alike took this responsibility very seriously. Farrington believed that the educational advantages of the museum "entitle it to a place in public estimation at least on par with that accorded the public library, and in some respects one equal to that given the public school."[29] Osborn felt that "arousing interest and spreading accurate information" among the general populace were two of the museum's chief responsibilities. "If these obligations are unfulfilled," he wrote, "the metropolitan museum fails in its purpose and deserves the withdrawal of public support."[30] Mounted fossil vertebrates proved useful for teaching visitors about science, especially evolutionary biology and historical geology. They were also believed to be invaluable for teaching non-English-speaking immigrants and the illiterate urban masses about morality, order, and proper behavior.[31] Species that misbehaved went extinct was the implicit message of vertebrate paleontology exhibits. Morris K. Jesup, president of the American Museum and a staunch supporter of Osborn's program, believed that the most important feature of evolutionary science was that it conveyed moral lessons to students. Mounted fossils also helped museums fulfill their missions by bringing record crowds inside to see them. Patrons of museums, most of whom had little or no real interest in paleontology per se, supported vertebrate paleontology, directly or indirectly, because mounted fossil vertebrates tickled their fancy. Carnegie is the perfect example. Moreover, according to Rainger, New York's wealthy elites supported vertebrate paleontology at the American Museum because they believed it tended to shore up the established social order. Vertebrate fossils became invaluable museum commodities. Often rare or unique, and sometimes spectacularly large, fossils made beautiful objects for display. They attracted both public interest and private support from philanthropists who were eager to make a particularly conspicuous contribution to the museum.[32]

Building elaborate displays of vertebrate fossils to meet the goals of major donors required large numbers of reasonably complete, exhibit-quality specimens. Some of these could be acquired by purchase or exchange, but the most efficient way to build a collection of fossil vertebrates was through prolonged fieldwork. The demand for higher-quality specimens drove museum paleontologists to devise improved methods for collecting, handling, preparing, and mounting fossils. No longer would it be acceptable simply to pluck weatherworn fragments from the surface, or to take up fossils in "potato fashion," roughly hewn from the earth like a common spud.[33]

Museums provided a multitude of new professional opportunities for paleontologists and support staff. Administrators like Osborn and Holland created these opportunities, and then chose the men who could fill them to mutual advantage. They controlled the funding, provided direction and oversight, and reaped the majority of the credit for work accomplished. From their charges, the men who performed most of the skilled labor, they expected unflagging loyalty and hard work. The men who staffed these new museum positions during the second Jurassic dinosaur rush came from two very different groups. Often, there was friction, even mutual hatred, between them. The first group comprised the men who had been cut loose by Cope or Marsh in the wake of the first Jurassic dinosaur rush. These were men with considerable skill and experience, but also some frustrated ambitions. With their reputations already made, they expected a certain amount of freedom of action and a more equitable distribution of credit. Several of them would have a very difficult time adjusting to the new order of museum paleontology, which continued to be very much like the old order under Marsh. The second group consisted of young, ambitious upstarts in paleontology, many of whom hailed from small universities in the Western states. These were men eager to take full advantage of the opportunities presented by the surge of activity in American museum paleontology. Osborn and Holland both had far better luck coping with individuals from the latter group, which produced men who tended to be more loyal and less independent. Although opportunities to publish, to work somewhat independently, and to enjoy credit for work accomplished were better for some museum paleontologists than they had been for the unfortunate souls who worked under Marsh, the structure of American vertebrate paleontology in the new museum setting continued to be rigidly hierarchical, with benefits going predominantly to those at the very top.[34]

When American vertebrate paleontology shifted into the museum setting, another contest to collect fossil vertebrates broke out. Building collections was an essential first step in establishing a viable research and exhibition program in vertebrate paleontology. As one paleontologist put it, "The best Museum will be the one that has the fossils!"[35] The most cost-effective means of acquiring fossils was to collect them in the field, which each museum program endeavored to do as rapidly as possible. Competition was especially fierce for Jurassic dinosaurs from 1899 through 1905. This new contest, however, was different in degree, if not in kind, from the first Jurassic dinosaur rush of the 1870s and '80s. The open antagonism and public feuding that had characterized the relationship of Cope and Marsh was a thing of the past. The contestants in the second Jurassic dinosaur rush would seldom sink as low as their notorious predecessors. But petty rivalry persisted in the new museum setting, where it assumed an understated and less public guise. Osborn liked to claim that American vertebrate paleontology in the post-Cope/Marsh era was a kinder, gentler science. "A division of subjects and the friendly co-operation of different institutions have been brought about," he wrote in

1903.[36] But this was not, strictly speaking, an accurate characterization. Friendly field encounters and occasional cooperative ventures among paleontologists notwithstanding, competition to collect fossils was still the norm for the first several years of the twentieth century.

Osborn and the American Museum

Henry Fairfield Osborn, who was the founder and first curator of the Department of Vertebrate Paleontology (DVP) at the American Museum, the archetypal program that other museums emulated, was one of the principal contestants in the race to collect fossils. Osborn was born in 1857 into a family that moved among the highest circles of New York's social and economic elite. His father was the president of the Illinois Central Railroad. His brother, Frederick, was a close friend of a young Theodore Roosevelt. The banker and financier J. Pierpont Morgan was his uncle. Osborn attended Princeton beginning in 1873—a maternal great uncle had been one of its founders. There he caught the science bug and befriended a like-minded classmate named William Berryman Scott. Inspired by an account of one of Marsh's Yale expeditions to the far West, Scott idly proposed that Princeton students should undertake to do the same and Osborn heartily agreed. The first Princeton Scientific Expedition went afield in the summer of 1877. In Colorado, they stopped at Morrison and visited Lakes, who had just begun collecting Jurassic dinosaurs there for Marsh. In the Bridger Basin in Wyoming they made their own collection of fossil mammals. When they returned, both worked hard to make the most of their experience. With a little helpful advice from Cope, who was initially suspicious of the Princeton pair but later became their staunch supporter, Scott and Osborn studied the fossils they collected and published their results. Princeton's president, James McCosh, encouraged both students to take graduate courses and to pursue an academic career. Osborn's father reluctantly submitted to his son's decision to become a scientist, rather than enter the family business, but he expected him to "build an empire in the pursuit of knowledge."[37]

Osborn caught an unexpected break in 1880, when the American Museum of Natural History offered him a position as geologist. Because Cope and Marsh were then at the height of their powers, he refused the offer, choosing instead to join Scott on the Princeton faculty.[38] There he developed a modestly successful research program devoted to fossil mammal morphology and systematics. Ten years later, he was eager for greater opportunities, and he reopened negotiations with the American

Museum. The museum's president, Morris K. Jesup, had recently resolved to launch a new program in vertebrate paleontology. Jesup wanted to lure Yale's paleontologist to New York, but Marsh declined. Other prominent paleontologists, including Cope, expressed interest in the position, but the trustees—a cross section of New York banking and business tycoons—were predisposed to favor Osborn. With a growing reputation as a paleomammologist, Osborn was certainly qualified for the post. More importantly, he was socially acceptable to the trustees who, along with Osborn, were all cut from the same cloth. Wealthy and well connected, Osborn could attract the requisite financial support to hire a crackerjack staff, to field multiple expeditions, to purchase important specimens and collections, to pay handsomely for timely fossil vertebrate locality information, and to collect, prepare, study, and exhibit a vast quantity of first-rate vertebrate fossils. Osborn was perfect. He had acceptable scientific credentials and sterling social connections. He joined the museum staff in October 1890 and received a curatorial appointment in April 1891.[39]

The timing of Osborn's move to the American Museum, just on the heels of the Cope/Marsh feud, is very suggestive. Perhaps he sensed that the feud would ultimately bring down both protagonists. A savvy reporter who interviewed Osborn in the days after the *Herald* scandal first broke seemed to sense this, too. He wrote, "Professor Osborne looks like a college senior and an athlete, but he stands very high in learned circles, and has been named as a possible successor to Professor Marsh, when that gentleman gets out of the way." Osborn was then quoted as saying that he was "glad the matter has at last come out. It will clear the atmosphere . . . and great good will be accomplished." The reporter, fishing for Osborn's meaning, continued to ask questions. "'It has been stated,' said I, not knowing I was treading on Professor Osborne's corns of ambition, 'that Professor Cope is the best equipped man in the country to fill Major Powell's position. Is that true?' 'Well, I don't know about that,'" Osborn replied mischievously. "'Perhaps he is. Professor Cope is a scientist of high attainments and one of the most powerful and original thinkers alive.'"[40] But Cope's research program, by early 1890, was already in tatters. Marsh's triumphs had demoralized him, while a disastrous succession of unlucky investments had ruined him financially. He was looking for paying work, and shopping his enormous fossil collections around in an effort to raise capital. Osborn even loaned him the money to keep his journal, the *American Naturalist*, afloat. Despite his precarious circumstances, Cope wanted to go forward publicly with his fight with Marsh, and Osborn encouraged and enabled him. Both were probably disappointed when Marsh seemed to suffer no immediate harm in the aftermath of the *Herald* scandal, and Cope soon gave up the fight. Marsh's comeuppance was in the offing, however, and Osborn played a part in bringing it about.

Osborn's long-term ambition was to establish institutional dominance in American vertebrate paleontology. Marsh's program at Yale was the finest in the country in the early 1890s, and Osborn sought to undermine it. He accomplished this first by aiding and abetting Cope in his feud, and later, by borrowing some of Cope's underhanded tactics. He criticized Marsh's work in print, for example. He also hired away several of Marsh's talented assistants, and enlisted some of Marsh's Western associates to switch allegiance to the American Museum. Worst of all, he successfully attacked Marsh's federal government funding. In June 1892, he wrote a letter to Senator William B. Allison, head of the Senate Appropriations Committee, disparaging the value of vertebrate paleontology relative to invertebrate paleontology in government science. The former Osborn regarded as "less practical and of a more purely scientific character." In short, he argued that Marsh and the U.S. Geological Survey were spending taxpayer money inappropriately on vertebrate paleontology.[41] Marsh lost his federal appropriation in the summer of 1892, just at a time when a calamitous national depression affected his personal finances. The financial pinch reduced Marsh's scientific empire to a much more modest condition, and severely limited his ability to pursue fieldwork.[42]

Beginning in 1892, Osborn sent American Museum field parties into several of Marsh's staple fossil localities. Marsh tried to block access to his best sites, but Osborn called on his influential connections to trump the effort. In 1897, he set the stage for the second Jurassic dinosaur rush when he sent his collectors to Marsh's locality at Como Bluff. Osborn could have entered the Jurassic merely to make a collection for the American Museum, but another more personal objective probably motivated him as well. He likely intended to gather the fossil evidence necessary to make a serious revision of Marsh's entire body of scientific work. The stated purpose of the 1897 expedition was to find and collect Jurassic mammals. Marsh had done some groundbreaking work in this area, and Osborn wanted to go over it again carefully, with the object of overturning his work. If Osborn's field-workers could also make a representative collection of dinosaurs at Como Bluff, then Marsh's opus, *The Dinosaurs of North America*, would also be subject to revision. Osborn was likely inspired in this direction by the letter from Williston, mentioned above, respecting Knight's new Jurassic dinosaur locality. Williston wanted a shot at Marsh, too, but Osborn was in a much better position to take one.

When Osborn visited his collectors at Como Bluff in June 1897, he kept a field notebook, which he filled with gossip about Marsh. Most of this gossip he obtained from Reed, a disgruntled former employee. According to Reed, for example, all of Marsh's Jurassic mammal specimens were collected under the auspices of the U.S. Geological Survey. They belonged, therefore, to the National Museum of Natural History. Reed claimed also that Marsh took pleasure in the hardships suffered by his field-workers. Furthermore, Marsh spent precious little time in the field during Reed's seven years in his employ. This might have been for the best, however, as the Yale paleontologist was no good at fieldwork. He got sick. He complained. And when he undertook to show Reed how to work out a beautiful dinosaur skull he struck it squarely in the center with the business end of his pick. "God—don't do it as I do it—but do it as I tell you to do it," was his sage advice. Osborn recorded these anecdotes in his notebook and thus inaugurated his own assortment of Marshiana, modeled after the notorious collection of his late mentor, Cope. Osborn's interest in collecting dirt on Marsh provides ample evidence of his antipathy toward the man. It also suggests a continuation of the Cope/Marsh feud even beyond the former's grave. Indeed, Osborn seemed determined to carry the banner for Cope, and he only let up after Marsh died unexpectedly in March 1899.[43]

Marsh's death was another boon for Osborn's program at the American Museum. To begin with, it marked the end of Marsh's reign as America's foremost paleontologist, and it left a vacancy for a fossil vertebrate specialist at the U.S. Geological Survey. Osborn ultimately ascended to this position in 1900, despite his formerly hostile attitude toward the federal government's role in fostering vertebrate paleontology. Marsh's successor at Yale wanted to retain control of his predecessor's unfinished, Survey-funded monographs on the various groups of dinosaurs and fossil mammals. Osborn, however, lobbied successfully for their control, and apparently won some government support to complete them. Access to Marsh's specimens was essential for these projects, so Osborn obtained permission from Yale officials to send a staff of assistants and collaborators to undertake this work. (It is not clear whether Osborn investigated any of these specimens himself.) Access to part of Marsh's collection became easier for any qualified student of fossil vertebrates when the National Museum finally pressed its claim to a rightful share of the bounty after Marsh's death—another victory for Osborn.[44]

Succeeding Marsh as U.S. Geological Survey paleontologist, and inheriting the lion's share of his unfinished work, may seem like an odd fate for Osborn, who was one of the Yale professor's harshest critics. But there are stranger and subtler ways in which Osborn assimilated his former rival. For example, Osborn started using some of Marsh's favorite expressions in his correspondence. He referred to any particularly rich fossil deposit as "Holy ground," for instance, often crediting Marsh when he used the phrase. He also advised some of his collectors not "to go duck hunting with a brass band," Marsh's favorite maxim for maintaining a low profile in the field. Many of his work habits and attitudes also became remarkably Marsh-like. He preferred not to do his own fieldwork, for example, but was free with dispensing fieldwork advice. Also, more than one of Osborn's collectors complained (or implied) that he did not fully appreciate the hardships they suffered in the field. Finally, Osborn had a habit of delegating large portions of the research and even some of the writing on his various projects to his staff of subordinates, often without giving sufficient credit. Like Marsh, Osborn was secretive about his movements, possessive of his localities, and envious of the successes of his rivals. This was especially true during the height of the second Jurassic dinosaur rush.

TWO

Osborn's Sorry Valentine

On Valentine's Day, 1896, Osborn accepted the unwelcome resignation of Olof August Peterson, his skilled field assistant. Peterson had worked for Osborn's department at the American Museum of Natural History for five years and was an expert fossil hunter. Fed up with his menial position, Peterson elected to accompany his brother-in-law, John Bell Hatcher, an ingenious fossil collector and geologist, on a Princeton expedition to Patagonia. Peterson was so disaffected with working for the DVP that he agreed to go to Patagonia for next to nothing—merely for the privilege of keeping half the animal skins he could capture and dress during the expedition. His sudden departure left Osborn without a reliable field hand. But what was most exasperating for Osborn was the apparent hostility Hatcher showed in ensnaring his employee. In 1890, while he was still negotiating for his own position in New York, Osborn had tried to hire Hatcher, then Marsh's chief collector, for the American Museum. He was angrily rebuffed when Hatcher learned that Osborn was not yet officially an American Museum employee. (Hatcher believed that Osborn had been misrepresenting the facts.) Wealthy and extremely well connected, Osborn was unaccustomed to rejection or failure. Yet when Peterson and Hatcher embarked for South America on February 29, they scuttled many of Osborn's grand plans.[1]

Osborn's ambition for the DVP was to build and lead the finest vertebrate paleontology program in the world. This would require a comprehensive collection of fossils and a small army of subordinate laborers to do the work of routine scientific support that Osborn was unable or unwilling to do himself. The most cost-effective means of acquiring fossils was by collecting them in the field. But Osborn was no field-worker.[2] Instead, he attracted a coterie of low-ranking, low-paid fossil collectors who supplied his program with specimens and who had to depend on him personally for patronage. To fill these ranks, he hired loyal, hard-working fossil enthusiasts, many from the rural West, who were grateful for the opportunity to work for him.

Jacob L. Wortman, who had worked as a collector for Cope for many years, was Osborn's first hire. Wortman would serve as assistant curator and field foreman, a position of some standing officially, although this was largely symbolic. Osborn hired Peterson, who, along with Hatcher, had also worked as a field assistant for Marsh, to help Wortman in the field. Beginning in the summer of 1891, he sent Wortman and Peterson on annual collecting expeditions to the fossil localities of his choosing. The plan for the summer of 1896 was to return to the San Juan Basin of northwestern New Mexico to search the Eocene Wasatch beds for an exhibit-quality skeleton of the primitive, large-bodied fossil mammal *Coryphodon*. The DVP already possessed a rich collection of Eocene fossil mammals acquired on several previous expeditions—some, also, had been obtained by purchase. Little of this material was exhibition worthy, however. So Osborn detailed Wortman to find and collect something good enough to put on display.[3] It was a difficult proposition in the best of circumstances. Now that Peterson was Patagonia bound, Wortman would need experienced help to make it happen.

Finding Good Help

To fill the unexpected vacancy, Osborn and Wortman first looked in-house at Walter Granger, who worked in the museum's Taxidermy Department. Granger was then twenty-three years old. An avid hunter, trapper, and outdoorsman, he had learned taxidermy as a teenager, practicing on specimens he collected in the Green Mountains of his native Vermont. In 1890, he joined the museum staff as a part-time assistant to the building superintendent. It was an opportune time to join the museum, which was then poised for a burst of new research activity, and he soon came to the attention of the scientific staff. In 1894, ornithologist Frank M. Chapman sent him out to collect extant mammals and birds on the Great Plains in company with Wortman's fossil collecting expedition. He joined the DVP party again the following season, but this time he spent half his time hunting for fossils. Granger, who was gregarious, extroverted, and self-effacing, made an excellent impression on Wortman. And he enjoyed fossil vertebrate fieldwork immensely. Chapman supported Granger's transfer to the DVP, which took place on March 1, 1896.[4]

More help was needed, however. Wortman had been favorably impressed by a group of student field-workers he met while collecting fossil mammals in the White River badlands in the summer of 1894. These young men were participating in a University of Kansas paleontological expedition led by Samuel Wendell Williston, a respected colleague of Wortman's. Recalling this chance encounter, Wortman queried Williston about his students.[5] Williston responded with a letter enthusiastically recommending Barnum Brown as a field assistant, proclaiming,

> Brown has been with me on two expeditions, and is the best man in the field that I ever had. He is very energetic, has great powers of endurance, walking thirty miles a day without fatigue, is very methodical in all his habits, and thoroughly honest. He has good ability as a student also and has been a student with me in anatomy, geology and paleontology. He practically relieved me of all care in my last expedition [to Wyoming in 1895], looking after te'm [team], provisions, outfit, etc. I can not say enough in his praise.[6]

Williston's endorsement landed Brown his first position as a professional fossil vertebrate collector.

Thrilled with the opportunity to take a paying museum job, Brown abandoned his studies and left Lawrence on April 2, 1896, heading ultimately for southern Colorado to rendezvous with the DVP party. As luck would have it, Brown's first expedition with the DVP was plagued with problems, including a frustrating lack of fossils. The novice field-workers, on the other hand, made a good first impression. Wortman wrote to Osborn that "Brown and Granger are turning out first class 'Bug hunters' and I am so far not disappointed in them in any way."[7] Osborn was unhappy with the dearth of fossils, but pleased by Wortman's reports about the new assistants. In fact, he consistently expressed the highest confidence in Brown's potential as a collector in his correspondence with Wortman. For example, Osborn wrote: "I am glad that Brown and Granger are turning out so well, and think you have a couple of reliable men. I have no doubt you will make a first-class man out of Brown"; and "I am very much pleased to hear that Brown and Granger are doing so well.

It shows the truth of the old adage—that 'there are plenty of fish in the sea.' I trust you will find Brown a loyal fellow, who will stand by us."[8]

Meanwhile, another one of Williston's KU students, Elmer Samuel Riggs, with only a few weeks remaining until his graduation, was facing an uncertain future. Riggs also aspired to be a paleontologist, and he must have envied Brown's good fortune in landing a paying position with Osborn. Williston would not be taking a party into the field that summer, so Riggs needed to find something constructive to do. With the idea of pursuing a PhD, Riggs wrote to Osborn in March 1896 to inquire after opportunities at the American Museum and Columbia University (where Osborn was a faculty member). He also asked for a position as field-worker or lab assistant for the summer.[9] Osborn responded favorably a week later with a letter describing the opportunities for graduate work in paleontology in New York. Having recently formulated a plan to make a comprehensive collection of the fossil horses of North America, Osborn asked Riggs about the possibility of sending him into the Miocene beds in Wyoming's Hat Creek Basin to pursue independent fieldwork. However, a special appropriation to meet the costs of this collection would have to be secured first through Osborn's connections. Osborn mentioned having spoken to Wortman, and reported that the latter had promised, if possible, to send Riggs into the field for at least part of the summer.[10] Riggs replied enthusiastically that he was "thoroughly convinced that the American Museum is the place to pursue the course of study in Mammalian Paleontology that I have laid out for myself." By way of convincing Osborn that he was up to the challenge of performing independent fieldwork, Riggs wrote, "I was born and bred a country lad and have been accustomed to taking life as I find it. . . . I assure you that I will do my best to prove that your confidence is not misplaced."[11]

Osborn had asked Riggs to have Williston write a letter of recommendation. As he had for Brown, Williston rose to the occasion with a very favorable letter, writing,

> Of all of [my students] Brown and Riggs were my especial favorites. . . . I could trust them implicitly, they were zealous, intelligent, and capable of extraordinary exertions. Mr. Riggs has pursued studies in Geology (20 weeks) under me, Vertebrate Anatomy (20 weeks) and Paleontology (20 weeks)[.] During the present year he has been doing special work with me in paleontology and now has a good paper on Dinictis in press. He is very careful and patient in handling specimens, has picked up all the information about collecting that he could from Wortman, Hatcher and myself, and could find more specimens than any member of my parties. . . . I am warmly interested in all their

welfares, because they have worked so faithfully for me and I shall be really glad if you are able to send Mr. Riggs into the field.[12]

Williston was careful not to compare his students directly in this letter, except for one important detail: "Mr. Riggs is more of a student than Mr. Brown; that is he is quicker in apprehension."[13] He placed far more emphasis on Riggs's academic ability than he had for Brown, suggesting, perhaps, that he had higher ambitions for the former.

Williston's paternal support won Osborn over. Osborn wrote to Wortman at Chama, New Mexico, claiming, "I feel confident that [Riggs] would be able to undertake our Horse work and do it up in good shape."[14] Wortman, who was supposed to have authority over DVP fieldwork, was less impressed. He replied: "In regard to Riggs I have an important plan to unfold to you which I think would be a decided improvement on the one you suggested. . . . Regarding the horse story there will be plenty of time next season and both Brown and Riggs can take a flyer at it. I am afraid Riggs does not know enough about it to start him out alone without previous training." As an alternative, he recommended sending Riggs and Brown into Marsh's famous Jurassic dinosaur locality at Como Bluff, Wyoming, later in the season.[15] Presumably, Wortman himself would there and then provide the necessary training and supervision.

Wortman, whose position in the DVP could be a difficult one, undoubtedly felt threatened by the idea of sending other collectors into the field independently. Osborn's second choice (after Hatcher) to fill the position of assistant curator and head of fieldwork, Wortman had a sometimes troubled relationship with his superior. Wortman started working for Osborn in 1891, and was feeling professionally underappreciated at least as early as 1894.[16] Older, better educated, and more experienced than the other collectors, he nevertheless lacked the social and scientific status of Osborn. He therefore occupied a tenuous middle-management position between the DVP's stable of ambitious and increasingly experienced collectors, and Osborn, its undisputed leader. Osborn's willingness to send collectors to do fieldwork independently must have eroded Wortman's sense of self-importance. In his field correspondence with Osborn, Wortman defied this tendency with insinuations of his own indispensability. For example, he repeatedly made reference to his previous expeditions under Cope, or with Peterson, as if to remind Osborn that he had essential Western collecting experience and know-how that his office-bound superior lacked. An excellent example of this type of remark is this: "I will be compelled to stay with the party to start them in at the quarry. *I am the only one who knows the exact place"* (emphasis added).[17] Problems of status continued to bother Wortman in subsequent field seasons.

Meanwhile, Riggs waited in Lawrence with growing impatience while Osborn and Wortman haggled over plans. Wortman continued to advocate for a separation of forces, with one party to focus on fossil mammals, and the other on Jurassic dinosaurs (but both under his general supervision). Osborn, however, now preferred that the party concentrate on finding sufficient remains to mount an exhibit-quality *Coryphodon*. In the end, he decided that the entire party, with the addition of Riggs, would relocate to Wyoming and search for fossil mammals in the early Tertiary horizons of the Wind River and Big Horn basins. Accordingly, Riggs met the American Museum party in Casper, Wyoming, in the first week of July 1896. Wortman telegraphed Osborn with some final objections about the plans. This telegram occasioned a misunderstanding about Riggs's role in the party that Wortman hastened to correct: "When I wrote you or rather telegraphed you that I did not think the plan was for the best, I meant solely in taking the whole outfit into the Big Horn. It had no reference whatever to the employment of Riggs which I think was a capital stroke. He is a good man, a good worker and will without doubt develop into a first class collector."[18]

From Casper, the DVP party proceeded overland to the Big Horn Basin. A six-week search of the local exposures yielded numerous *Coryphodon* remains. Yet good specimens were scarce and difficult to come by and the weather was miserably hot. "The boys are in good health and spirits but are getting rather sick of the job," Wortman reported.[19] They disbanded in Casper on September 20 once Wortman was confident that the collection was sufficient to meet the expedition objective of mounting a complete, exhibit-quality skeleton.[20]

Riggs and Brown earned a measly $50 per month (plus transportation and expenses) for their work with the American Museum. Riggs had borrowed $50 in advance on his wages from Osborn, however, so that at the end of the expedition, he netted only $83.33.[21] Returning to Lawrence as a college graduate with a very small amount of capital, he was faced with the necessity of finding work. Williston still had some money remaining from his previous appropriation for fieldwork. He put Riggs to work for a year as an assistant in the University of Kansas preparation lab. Strangely, there is no record that Riggs made any effort to join the American Museum's next fossil hunting expedition in 1897. In a letter to Wortman, Brown claimed mysteriously that "I think that [Riggs] will hardly be with us next year though I have not asked him," and he was right.[22]

Another Opportunity for Brown

Brown returned to KU after the American Museum expedition to continue his schooling. A lackluster student, he whiled away the winter months in Lawrence, pining for a chance to return to professional fieldwork. Hoping to secure a permanent position as a paleontological assistant with the DVP, he initiated a letter campaign with Wortman (and Granger), expounding on the excellent opportunities to collect Jurassic dinosaurs in Wyoming and confidently offering his expert services:

"[T]here are . . . plenty of fossils, good roads to haul over, which would be a big thing in handling those bones[,] and good camps [with] plenty of wood and Mt. water. The deposits I examined are about thirty miles west of Laramie, in clay and lime stone. Reed and Knight have worked this bed considerable, though I could take two or three men with teams and take out tons of material in a short time if that was necessary. Reed assures me however that he can take me to a much richer deposit, especially in mammal jaws about fifty miles north of Laramie.[23]

Osborn's need of skilled assistants to do fieldwork and fossil preparation in the DVP was still critical. Throughout the spring and summer of 1896, Wortman had praised Brown's field abilities effusively in his letters to Osborn, writing: "Granger and Brown are developing into first class 'bone diggers' "; "Brown is catching on splendidly"; and "the new men of the party have developed into first class assistants."[24] When Osborn approached Wortman to discuss plans for collecting fossils in the summer of 1897, Brown undoubtedly received a very warm endorsement from the field foreman.

Osborn wrote to Brown on March 24, 1897, with another offer of summer employment as a collector. (There is no record that Wortman played any direct role in these negotiations between Osborn and Brown.) An opportunity to continue working at the museum part-time in the winter as a fossil preparator, while finishing up his bachelor's degree at Columbia University, also was available. Brown was to go to Como Bluff, Wyoming, to make, if possible, a "complete representative collection" of the Jurassic mammal fauna that had been found there in abundance by Marsh's collectors in the 1870s and 1880s. Osborn expected Brown to discuss his plans with Williston, who was familiar with both the locality and the desired specimens, but he otherwise preferred to keep his interest in Como Bluff a secret.[25]

Deeply grateful for the opportunity, Brown conferred with Williston at once, and then sped an enthusiastic letter of acceptance to Osborn, his new patron, plying him with youthful optimism. Brown had learned as much as possible about the situation at Como Bluff from veteran fieldworkers Williston and Reed. According to Williston, the fossil mammal quarry, though abandoned and partially buried on Marsh's orders, was never exhausted. A mile south of the railroad depot at Aurora, Wyoming, on the steep northern slope of Como Bluff, Marsh's historic quarry could be relocated easily and reopened at a modest expense. From Reed, Brown learned of another locality nearby, north of the Medicine Bow River, where a similar outcrop with promising indications existed that had never been adequately prospected. According to Brown's letter, the opportunity for taking even better material was excellent, "for the methods used in preservation of specimens were primitive and the collectors were not careful workmen."[26] In short, the young collector believed the outlook was very good.

"As to the reptiles," Brown's letter to Osborn continued, "I think we can obtain any amount of material." He explained that at the end of the previous summer, immediately following the DVP expedition of 1896, he had worked together with Reed for a few days in a quarry west of Laramie. (Riggs likely worked with Reed and Brown as well.) There he found seemingly "inexhaustible" numbers of well-preserved dinosaur bones tangled in a disarticulated heap. He was careful to point out that Jurassic mammals and dinosaurs were, of necessity, distinct and separate goals, as mammal remains had not previously been found in close association with the much larger reptiles. Indeed, Brown's careful inspection of the dinosaur quarry in Reed's company yielded no trace of fossil mammals. Furthermore, the search for dinosaurs, which would likely be more geographically extensive as well as more labor intensive than any excavation for small mammals, would entail additional equipment and expenses including, at a minimum, a team of horses, a wagon and scraper, and a paid field assistant to help handle the heavy materials.[27]

Brown seems gently and adroitly to have coaxed Osborn to favor the dinosaurs. Osborn's initial letter extending the job offer to Brown omitted any mention of these animals. So when Brown broached the subject of collecting dinosaurs in his reply he was making an unsolicited pitch for his own particular interest in large-bodied fossil vertebrates. He was likely encouraged in this pursuit by the recollection of Wortman's idea of sending him (along with Riggs) to Como Bluff the previous summer—Wortman being very keen to make a collection of Jurassic dinosaurs. From his consultation with Brown, Williston suspected a confusion of motives for the expedition. In a letter to Osborn he wrote: "Brown . . . may, perhaps, be able to discover other beds, but it would not be wise for him to undertake the search at first. Teeth of mammals are never or rarely ever found in connection with the large bones of the reptiles, and it will be idle to search for them among such, unless the reptile bones are wanted more than the mammals."[28] Despite Williston's advice to the contrary, Brown sought Osborn's approval to purchase a team and wagon for prospecting; he also recommended hiring his KU classmate, Harold William Menke, as a field assistant. Both of these requests were ultimately granted, although neither was absolutely necessary to a modest expedition after the tiny teeth and jaw fragments of Jurassic mammals.

Who Set the Dinosaur Agenda?

As founder and curator of the DVP, Osborn has been given the credit for launching the American Museum's initiative into dinosaur collecting in 1897.[29] Osborn's own explanation of these events gives the misleading impression that he decided on the shift of emphasis from mammals to dinosaurs in a perfectly rational and deliberate way. According to Osborn, "From 1890 to 1897 . . . such substantial progress had been made [with fossil mammals] that I decided to push into the history of the Age of Reptiles."[30] In fact, this decision was reached in a very haphazard fashion. Moreover, others had championed the cause of dinosaurs long before Osborn, who was reluctant, at first, to deviate from the mammals. Collectors in the field, for example, often took an opportunistic approach to fossil hunting, particularly when the intended material goal of a given expedition proved elusive. In the summer of 1892, in the Laramie beds of southeastern Wyoming, the DVP party made a cursory search for dinosaurs while collecting Cretaceous mammal teeth. Fossil mammals, however, were a much higher priority on this expedition, no doubt to satisfy a directive from Osborn, and thus no dinosaurs were brought back to New York.[31] Wortman reported with obvious enthusiasm on the prospect of bagging some dinosaur remains on this expedition in his letters to Osborn. But the curator's replies to these letters have not survived. In 1893, Peterson and Wortman, on separate expeditions in northern and southern Utah, respectively, each made a brief reconnaissance for Jurassic dinosaurs, but neither collector obtained any specimens. Peterson, who had been detailed to explore the Uintah Basin for Eocene fossil mammals, learned about a promising Jurassic dinosaur prospect along the east bank of the Green River, near Vernal. But the specimen proved disappointing, and he gave up the search. No matter—Osborn was indifferent to Peterson's report about the dinosaur, instructing, "I want you to devote all your attention to the mammals."[32] Osborn could be single-minded about his dedication to fossil mammals, especially during the first few years of his tenure, before the size and quality of the DVP's fossil mammal collection gave him cause to reflect on the limits of its scope. Wortman's scientific interests, on the other hand, were more egalitarian. If Osborn never explicitly prohibited Wortman's interest in reptiles, it was only to indulge his collector's enthusiasm for vertebrate fossils of all kinds. Still, the fact remains that not a single noteworthy dinosaur specimen entered the DVP's collections before 1897.

Although the DVP did not commit fully to dinosaur collecting until 1897, Wortman was nevertheless the first to advocate openly for the shift of emphasis in 1896. That summer, in the San Juan Basin, the DVP party located some large Cretaceous dinosaur remains, but nothing of sufficient quality to justify a prolonged excavation. When Riggs was tapped to join the party in midsummer, Wortman and Osborn disagreed about how best to utilize his services. Wortman wanted to send Riggs and Brown to the Jurassic beds of southeastern Wyoming, which were then being worked successfully for marine reptiles and dinosaurs by Reed and Knight of the University of Wyoming. Osborn assumed, incorrectly, that Wortman wanted Riggs to collect Jurassic mammals.[33] Anxious to firm up his expedition plans, Wortman wrote to Osborn with a question and a clarification of his objectives: "Will we send Brown [and Riggs] into the Jurassic at Laramie . . . or what[?] Brown can work to much better advantage with Riggs helping him and there is not the slightest doubt but that they could make a good showing among the big reptiles there. We need something of this sort to help out our exhibition."[34] Riggs, for his part, was eager to work, but less than enthused about gigantic Jurassic dinosaurs. He confided to Osborn that "like yourself, I should be much more interested in mammals than in the dinosaurs, although Mr. Reed who is working the new beds reports an abundance of material."[35] Osborn decided ultimately against Wortman's plan.

Although he was a committed paleomammologist, Osborn's great ambition ultimately exceeded his class loyalty. He received his curatorial appointment as head of the American Museum's newly minted Department of Mammalian Paleontology in mid-April 1891. He immediately began an ambitious program to acquire fossil mammals in the field. In 1895, with the financial assistance of several wealthy trustees, the museum purchased Cope's massive collection of fossil mammals. Osborn even dipped into his own deep pockets to help raise the asking price for this magnificent collection. Replete with important type specimens and many more or less complete skeletons taken from a panorama of North

American localities and horizons, Cope's collection promoted Osborn and his program to the very front rank of mammalian paleontology. Now he could reasonably claim to have fulfilled his original contract with the museum to establish a leading research program devoted to North American fossil mammals. Thereafter, he began looking to other places to collect mammals, including Patagonia and South Africa, and to other groups of animals. That same year, he advocated a change of name for his program and the trustees approved. The new title, Department of Vertebrate Paleontology, better defined the scope of Osborn's mounting ambitions.[36]

Osborn did, at times, betray a burgeoning interest in fossil reptiles. In the spring of 1897, he swept through several Western museums collecting data on fossil reptiles from the likes of Williston, Knight, and George Baur, another disgruntled former assistant to Marsh. A notebook he kept on this tour suggests that his interest was fairly general, and not limited to any particular group or topic.[37] At the same time, however, he was writing and lecturing on the subject of mammalian descent from the reptiles. When his friend and mentor Cope died in April 1897, Osborn was, by his own admission, "very sad." But he was positively heartbroken about the undecided fate of Cope's remaining collections. His fervent wish was to get Cope's fossil reptiles for the American Museum and to co-opt Baur, then languishing at his lonely outpost at the University of Chicago, to curate it. "We could make the grandest department in the world," he declared, with a trace of megalomania.[38] (Osborn eventually landed the reptiles, but not the researcher, who died unexpectedly the following year.) In some instances, the extension of Osborn's reach into other fossil groups was serendipitous, as when William Diller Matthew, a new assistant in the DVP, went in 1894–95 to a North Carolina coal mine to extract some primitive mammals, and he returned instead with a magnificent specimen of a crocodile-like phytosaur. Later, in the summer of 1897, Osborn sent Matthew to the chalk beds of Kansas to collect marine mosasaurs.[39] This latter expedition was the DVP's first unambiguous and deliberate incursion out of the mammals and into the fossil reptiles.

With respect to the first Como Bluff expedition in 1897, it is impossible to state with confidence how Osborn initially felt about collecting dinosaurs because no record of any direct statement he might have made on this question survives. Two things are certain, however. First, Brown never had a clear mandate from Osborn to collect dinosaurs until after he had already found one during the latter's visit to the field camp in early June. In a letter dated May 8, for example, which reported on the frustrating conditions at the mammal quarry, Brown made a plea for permission to prospect for new quarries, especially those rich in dinosaur remains: "There are certainly other mammal beds [at Como Bluff] and plenty of reptile material. Shall I not collect everything? . . . Shall I not open up reptile quarries where there are good prospects or do you want me to confine my attention to mammals exclusively[?]"[40] Second, although Brown did confine his work to the mammal quarry (at least for the time being, and presumably following his patron's instructions), Osborn permitted him to purchase a team and wagon, and to hire Menke, who joined the expedition from Lawrence in mid-May. Both of these steps were of somewhat dubious value in quarrying for mammals, but were necessary preconditions for prospecting and handling dinosaur bones.

Breaking Ground at Como Bluff

Arriving in Laramie in early May 1897 to consult with Reed, Brown learned that the situation at Como Bluff was not as sunny as he had previously reported to Osborn, and he wrote to him immediately with a revised assessment. "I learn from [Reed] that things are in very discouraging condition up there. . . . I feel pretty badly mixed up in the situation, that I have misinformed you somewhat," he wrote apologetically.[41] In actual fact, Marsh's fossil mammal quarry was four miles—not one—from the train station at Aurora, Wyoming, where there was no mail service and nowhere to board. A complete camp outfit with tent and cooking equipment would now be necessary to work at this location efficiently. A team and wagon would also be useful for resupply (and for prospecting). These few items alone would cost Osborn almost $120, up front. But the worst news concerned the quarry itself. According to Reed, another one of Marsh's collectors had worked the quarry after his departure, expanding it back into the bluff until the back cut in the bank exceeded twenty feet in height. This cut had caved in since active work at the quarry had stopped, and all this waste would have to be removed. Moreover, the amount of overburden resting upon the bone layer increased rapidly the deeper you dug into the bank. Thus, an enormous and ever-increasing amount of laborious stripping would be required merely to expose the layer where fossil mammals had once been found. Reed's opinion was that good material yet remained at this locality. "The question is," Brown asked, "will it pay to work [the quarry] under those conditions[?]"[42]

Evidently, Osborn could not or did not take Brown's not-so-subtle suggestion. It seems clear from his letters that Brown wanted to forego

Marsh's mammal quarry in favor of prospecting for new dinosaur leads. Possibly he expected that Osborn would take immediately to the dinosaurs if and when a reasonably good specimen could be located. In any case, Osborn would not give up his Jurassic mammal agenda quite so easily, and he sent Brown and Menke to work at the old quarry.

With team, wagon, equipment, and provisions, Brown and Menke left Laramie in mid-May and pulled their wagon in the direction of Aurora. They set up camp along Spring Creek, only a short distance from Como Bluff. Reed helped them locate the mammal quarry, which was completely caved in, exactly as promised. Using a team and scraper, Brown and Menke set about skimming the debris off the top of the mammal layer. Heavy rains, however, soon turned the quarry into a muddy morass, too slick for horses' hooves. They had to dig a deep drain, and in so doing they struck some bone fragments and crocodile teeth, but no mammals.[43]

Osborn, meanwhile, was on a whirlwind Western trip, part vacation, part fact-finding expedition, and part public relations tour. Time permitting, he also planned to do a modicum of fieldwork. He stopped in Chicago to commune with Baur, see his colleague's collections, and visit the recently opened Field Columbian Museum. He likewise stopped in Lawrence to see Williston and his University of Kansas collections. There he also met Williston's "stalwart six-foot" students. (Riggs, then working as Williston's museum assistant, was almost certainly among this group.) In Denver, Osborn met with several members of the Colorado Scientific Society. One member gave him information about fossil mammals in the Huerfano beds, in southeastern Colorado, which Osborn shortly intended to visit with Wortman. Another, Marsh's old collector Arthur Lakes, had information on an entirely new fossil mammal locality near Akron, Colorado. Osborn made arrangements to go there with him for an exploratory tour. But Osborn's tentative fieldwork plans were thrown into turmoil when he accepted an invitation to accompany a raucous party of railroad executives on a narrow-gauge tour of Colorado in his friend Edward T. Jeffery's private railcar. (Jeffery, formerly of New York, and now president of the Denver & Rio Grande Railroad, did many important favors for Osborn and his collectors in Colorado.) Wortman had to make the short expedition with Lakes, which netted little, and then spent a day or more organizing a team, wagon, and complete camp outfit, by himself, for the Huerfano trip. After several fine days of luxury railroading, Osborn cozied up to the depot in La Veta, Colorado, in a special train, arranged by Jeffery, where Wortman met him with the wagon. He endured three trying days of fieldwork in the Huerfano beds, and then split for civilization.[44]

Osborn next spent an "extremely satisfying" day in Denver hustling new finds and new localities for the DVP. He made a follow-up visit to Lakes, who, from 1877 through 1880, had collected Jurassic dinosaurs for Marsh. Lakes assured Osborn that any future fossil vertebrate discoveries he made would be sent to the American Museum. Osborn was very pleased with himself, and he wrote excitedly to his wife that "my visit has attracted wide attention—all over Colorado—and it will be the means of bringing all the new discoveries and 'finds' our way instead of to [Marsh,] who has hitherto reaped everything and stored all away miser-like in the cellars of the Yale Museum. I believe you will now realize how important it is for me to come out myself."[45]

After brief visits with relatives in Cheyenne, and with Knight and Reed in Laramie, Osborn headed for the DVP field camp at Como Bluff, arriving June 1. He met Brown for the first time at the depot in Aurora, and then they rode in the wagon together four miles east to camp, where he met Menke. Both, he thought, were "capital fellows." Osborn brought groceries, including chocolate and other "delicacies" from Laramie, in order to lubricate his introductions with the men, so they undoubtedly liked him, as well. He spent his first afternoon examining the quarry with Brown. He found, to his chagrin, that the mammal layer was still covered with four to six feet of debris. Brown, however, somehow impressed his guest with the amount of dirt removed, for Osborn expressed himself as "extremely pleased with the work they have already done." The entire party spent a very cold June 2 digging in the mammal quarry. Osborn, swinging his Marsh pick like the other bone diggers, blistered his hands. They worked steadily until chased off the bluff and into their tents by an ominous black cloud that thundered for three hours and dumped a wet blanket of snow and hail on their progress. After the storm passed, they ventured back out into the cold air to work, Osborn donning a white sweater and overcoat to fend off the chill.

Immediately opposite camp, at the foot of the bluff, Brown picked up some broken fragments of bone. Badly weathered fragments of dinosaur bones littered the base of the bluff in many places—Brown, perhaps, was drawn to these particular fragments by the fresh appearance of their breaks, or broken fragments; fresh breaks indicates something recently broken and therefore better preserved. Together with Osborn, he climbed the slope and found a promising looking dinosaur femur protruding from the surface (see figure 5). With shovel and picks they followed their lead into the bank and found that it articulated with a large tibia. Osborn

Figure 5. Barnum Brown (left) and Henry Fairfield Osborn (right) at Como Bluff, Wyoming, June 2, 1897. Brown's prospect, later identified as a *Diplodocus*, was the DVP's first dinosaur discovery. Image #17808. Photograph by H. W. Menke. American Museum of Natural History Library.

was elated. Back in camp, while Brown cooked dinner, he penned a hasty letter home to his wife, proclaiming, "You need not pity me because I am happy as a bone hunter can be."[46]

Osborn departed the next day, leaving Brown and Menke with instructions to develop the new dinosaur prospect to determine its extent. In the meantime, they were not to neglect entirely the fossil mammals. But the new dinosaur quarry quickly became Brown's ruling passion. "Please pardon my seeming negligence in writing for Cope quarry is a veritable gold mine and I have been in bones up to my eyes," he wrote to Osborn with obvious relish. "Wish you were here to enjoy the pleasure of taking out those beautiful black, perfect bones."[47] (Before his departure, Osborn had named the new quarry after his late mentor, Cope. He also decided that it would be DVP policy to name quarries after "distinguished paleontologists [including] Cope, Leidy, Owen," and others, but not Marsh. This policy was later abandoned.) After two weeks of furious pick and shovel work, Brown and Menke exposed numerous caudal vertebrae tailing to the left of the femur, all more or less articulated. The bones were complete and fairly well preserved, although riddled with cracks. To consolidate the bones, Brown coated them with a thick application of gum Arabic as soon as they were exposed. Expenses were running high, and Brown wrote to Osborn, asking for an advance on a portion of the July budget, if possible. Lumber for crating the bones would be expensive, he warned. As for the mammal quarry, he confessed that they had devoted only two work days there since first locating the dinosaur. Naturally, Osborn was delighted by the news about the dinosaur. But he continued to pressure Brown to give some attention to the mammals also. He sent him a small hand lens, a matrix sample, and even a beautiful and exceedingly fragile Jurassic mammal jaw by way of encouragement.[48]

Wortman, meanwhile, finished up fieldwork in the Huerfano beds in early June and made his way to Pueblo, Colorado, to decompress. There he received an unexpected tip about a large dinosaur skeleton to be had west of town and about thirty miles east of one of Cope's classic dinosaur localities at Cañon City. Judging from the fragments he was shown, Wortman thought the specimen might be an example of the rare, dorsal-plated *Stegosaurus*. On June 18, he decided to go scout the prospect for himself. There he found numerous scattered remains of at least one large Jurassic dinosaur, but it was badly weathered. Meanwhile, Osborn had written a glowing account of Brown's specimen and sent it to Wortman from the field. The Colorado specimen must have seemed unworthy by comparison, so Wortman headed north to join Brown and Menke at Como Bluff.[49]

Wortman arrived at the dinosaur quarry on June 23 and was immediately taken with Brown's magnificent specimen. Twenty-three caudal vertebrae, including seventeen with their corresponding chevrons attached, were, by then, in sight. A portion of the pelvis was also exposed at the point where Brown and Osborn had first opened up the prospect. Wortman concluded that the vertebral column took a sudden turn at the pelvis straight into the bluff. Accordingly, the party began to make a cut into the bank at that point, and immediately they discovered five or six ribs resting, apparently, in their natural positions. Ten feet of dorsal vertebrae and a few more ribs turned up in the new cut over the next four days. They also discovered three additional caudals at the far left end of the dig. Wortman's expectations for a complete, exhibit-quality skeleton—with the exception of one hind limb and the extreme tip of the tail—were now very high. He suggested that Adam Hermann, the DVP's talented preparator and fossil mount-maker, "may start figuring on a method of mounting the monster."[50]

From a careful in situ examination of the structure of the pelvis, Wortman was convinced that the specimen was a remarkably complete and excellently preserved example of Marsh's *Morosaurus grandis* and he asked Osborn to send out his copy of the Yale professor's latest book on the dinosaurs in order to familiarize himself with the structure of these animals. The bones themselves were "little crushed and in a good state of preservation generally," Wortman reported to Osborn. But he was worried about the extensive network of cracks and what this implied in terms of care and handling of the specimen. "It will be decidedly for the best for me to stay and help Brown [recover the skeleton]. While his ideas are good and he is very careful I think on such a specimen as this we ought to have the benefit of all the experience at our command," he advised.[51] No doubt he was also worried about his ambiguous role on the expedition.

THREE

Wortman Assumes Charge

From the moment they joined forces at Como Bluff in the summer of 1897 there was friction between Wortman and Brown. As nominal head of DVP fieldwork, Wortman came to Wyoming expecting to assume control of Brown's dinosaur excavation, despite the latter's priority of discovery. Brown, however, had a different understanding about their respective roles at the dig. His commission to work in the Jurassic beds had come entirely through Osborn—Wortman had played no part in their negotiations. As far as Brown was concerned, he was working directly for the DVP curator, rather than under the field foreman. Meanwhile, Osborn had done nothing proactively to clarify the relationships of his field-workers, possibly because he had not anticipated any trouble. There was no open conflict of authority at first. Wortman declined to assert his leadership in any absolute way, while Brown simply refused to act submissively. Instead, Wortman made subtle efforts to usurp control of the camp while feeding Osborn with a steady diet of cautiously critical pronouncements on Brown's abilities. Brown, meanwhile, resisted Wortman's modest efforts to assume control and countered his attempts to take credit for the work accomplished.[1] In this way, tensions between the men accrued all summer.

A longtime proponent of the exhibit value of dinosaurs, Wortman was particularly eager to take full advantage of the promising new foothold the DVP had established in the Jurassic. Once Brown's specimen was recovered, he planned to prospect additional Jurassic exposures to the north and west of Como Bluff. He worried about the imminent arrival of collectors from the University of Wyoming. "Knight and Reed will soon be here to work in the country where we intend to prospect and they have an awkward habit of going around and 'locating' everything in sight," he warned Osborn. "My opinion is that the prospects are good to get a splendid collection of Jurassic material and I think it would be well to concentrate our force upon it while the opportunity lasts." He advocated the pooling of resources in Wyoming, and wanted Granger and his outfit, then collecting fossil mammals in Nebraska, to join him in the hunt for dinosaurs.[2]

Wortman also contrived an ambitious plan to dispense with much of the heaviest, hardest, and most time-consuming work at the quarry, and he shared it with Osborn. His idea was to use a pump and hose to direct a high-pressure stream of water onto the bedrock where the bones were exposed, thus excavating the bluff more quickly, and with far less manual labor. Hydraulic mining was a common practice in the West and was used profitably in Alaska during the Klondike Gold Rush (1896–99), but would it work for fossils? According to Wortman, "This clay under the influence of a stream of water melts like sugar and I am sure the thing is perfectly practical." He estimated that material and equipment would cost between $1,500 and $2,000. Brown also urged the hydraulic experiment on Osborn. But his letter of advocacy apportioned credit for the idea differently than Wortman had. He wrote, "Dr. Wortman has fell in with my scheme to hydraulic this bluff and I believe it is very feasible." Osborn's reaction to the plan was decidedly hostile—he balked at the idea of hosing down the specimens with water. But Wortman was certain that water posed no serious danger to the fossils, which were much harder than the surrounding matrix. As proof he offered some distal-most caudal vertebrae recently recovered from a gully—these bones had washed out beautifully in the rain. The initial investment was large, he conceded, but he expected that it would "pay handsome returns." With luck, the new technique would work just as well at other localities with comparable results. In Wortman's view, the best reason to develop this technique was for the competitive advantage it would convey. "[W]e would in the course of a few years have as good if not better [a] collection than Marsh," he argued, assuming that the mere mention of Osborn's hated rival in this context would inspire him to action. But Osborn was never taken with the hydraulic scheme, and he put a permanent kink in it.[3]

Reed dropped in for a visit on July 1. He was making his way north to scout some Jurassic exposures in the Freezeout Mountains, precisely where the DVP party intended to begin prospecting at their first opportunity. His visit occasioned a slight change of plan, which Wortman explained in a letter to Osborn. "In view of the fact that [Reed] is in the 'holy ground,' in all probability 'locating finds' as fast as he can," Wortman wrote, "we have determined to take out everything up to the bank, store [the fossil bones] in the station [at Aurora] and 'make a sneak' for the new ground." Apparently, the irony of racing to the Freezeouts to preempt the University of Wyoming party, when it was Reed and Knight who had generously provided intelligence on the local Jurassic beds in the first place, was lost on Wortman. In any case, work on the cut was suspended temporarily while the party worked together to recover the exposed bones. They began at the end of the tail, which terminated abruptly at a small gully, taking up the bones in manageable blocks of three or four vertebrae partly encased in soft, crumbly matrix. (The well-preserved distal caudals mentioned above were recovered from this gully.) Closer to the pelvis, the vertebrae were much larger and more complex. The jointed clay matrix and the insidious web of cracks in the bones rendered them vulnerable to breakage and very difficult to remove intact. To overcome this difficulty, the party developed an effective new collecting technique. Here Wortman's experience and ingenuity proved invaluable. With characteristic self-satisfaction, Wortman clarified this point for Osborn, crowing: "I am . . . very glad that I happened along just as I did for I fear that Brown's skill would have been put to a severe test in taking up this specimen."[4]

Before the summer of 1897, one common method for reinforcing fossils for shipment from the field had been the use of light strips of cloth wound around the specimen and stuck on with a coating of flour paste. Inexpensive and easy to obtain in bulk, flour was an almost ideal adhesive for fieldwork. Brown intended to apply this technique, which he had learned on previous expeditions under Williston, on the new dinosaur. Wortman was certain that this would have ended in disaster, however, as the flour paste bandage was simply not strong enough to consolidate heavy dinosaur bones. In years past, Wortman had sometimes resorted to the use of plaster of Paris poured directly over the surface of friable specimens, and he decided to adapt this technique to the dinosaur. Accordingly, the specimen was left in situ while the party excavated as much as possible of the surrounding rock, leaving the bones exposed in relief on a stone pedestal. The party then coated the exposed surfaces with plaster. Next, they tunneled under the specimen and attempted to apply the wet plaster to the underside, but they could not get it to adhere. Wortman then spread a burlap sack on the ground and directed his incredulous colleagues to pour wet plaster over it and then drag the mass through

Figure 6. Barnum Brown applies plaster to a block from his *Diplodocus* prospect at Como Bluff. The plaster jacket, often reinforced with boards and strips of rawhide, protected delicate fossils from damage during transit. Members of the DVP field party perfected the plaster jacket technique in the summer of 1897. It then spread quickly to other museums through the swapping of field personnel. Image #17811. American Museum of Natural History Library.

the tunnel. With a man gripping each end of the sack, the wet plaster was drawn up tightly against the underside of the specimen and held there until the plaster cured. But when they tried to remove the sack it held fast to the plaster, which had oozed through the open mesh of the burlap. Thereafter, burlap strips soaked in wet plaster and applied latticelike and allowed to set up on the fragile fossils provided the perfect field bandage (see figure 6). For extremely heavy or unwieldy specimens, the party applied a reinforcement of boards fixed to the plaster jacket with additional burlap bandages. Another innovation was to wrap long strips of wet rawhide around the entire bundle, nail them to the boards, and then allow them to dry and shrink. By this means, the DVP party perfected the handling, shipping, and long-term storage of large, fragile fossil vertebrates.[5] This indispensable technique spread rapidly to other institutions through the swapping of personnel.

Although Reed's visit inspired Wortman's competitive instinct to prospect, Knight's subsequent visit on July 9 had the opposite effect. Knight dropped in on a Sunday, the DVP party's traditional day of rest.

Nevertheless, he and Wortman went out walking along the bluff and found another prospect between Brown's first dinosaur specimen—now identified as *Diplodocus*—and the mammal quarry where Brown and Menke had first begun to excavate. A few days digging at the new prospect showed it to be a *Brontosaurus*. Always optimistic about new discoveries, Wortman thought the chances were excellent for getting another reasonably complete skeleton. He was now confident that there was work enough at Como Bluff to occupy the whole party through the end of the season. Too much work, in fact, as he made another urgent plea for bringing Granger and his outfit to Wyoming. Osborn apparently raised no objection to the joining of forces, so Granger left his Nebraska field site in mid-July, expecting to arrive at Como Bluff before the end of the month. Granger was anxious to "sink a pick" into the new dinosaur. W. D. Matthew, who had been collecting marine reptiles in the chalk beds of Kansas, also decided to join Wortman's party briefly at Como Bluff. He arrived around July 18 and remained for two weeks, providing valuable assistance in the quarry. With yet another favorable prospect to develop, Wortman postponed the exploratory trip to the Freezeouts indefinitely. He was still eager to see the new ground for himself, however, and hoped to make the trip before the end of the season. Meanwhile, Knight assured him that there were plenty of bones for everybody.[6]

As the new dinosaur prospect showed increasing promise, Wortman's faith in the completeness of Brown's *Diplodocus* began to wane. By July 20, he had given up on it completely, without having found the forelimbs, neck, or skull. Work on the mammal quarry then resumed in earnest. Another one of Brown's early prospects, a smaller dinosaur not yet identified, also occupied a few diggers from time to time. Due to the extremely jumbled nature of the deposit, progress in the *Brontosaurus* quarry was made only grudgingly. Still, the skeleton was showing very well, including limb bones, spinal column, and pelvis. Brown, apparently, stayed with the *Diplodocus*, working it by himself, and his careful prospecting was rewarded by the discovery of some additional bones, including significant parts of the pelvis. Undercutting the bank a bit too aggressively, Brown one day accidentally dropped a section of rock on one of the ilia, doing slight damage but affecting long delay.[7]

Matthew left at the end of July, a few days after Granger arrived from Nebraska with his outfit. Along with Granger came Albert "Bill" Thomson, a young man from South Dakota who would serve the expedition as camp assistant and cook, as well as collector. Matthew departed just as a series of furious rainstorms ensued, filling the arroyo next to the camp with a muddy torrent of water five feet deep. Muddy conditions in the quarries forced a temporary halt to the digging. In the interim, they built a large cookhouse twelve by fourteen feet where Thomson set up shop, and where men and material could be kept reasonably dry. Wortman had been hopeful that the two large skeletons could be extracted rapidly and made ready for shipping, but the bad weather occasioned another long delay in their work.[8]

Wortman's Discontent

Back at the museum, Osborn was alarmed by an unexpected newspaper article in the *New York Herald* about the DVP's dinosaur operation in Wyoming. Not wishing to draw any attention to his party's recent successes in the Jurassic, he wrote to Wortman immediately with a gag order. But news about the dinosaurs would prove hard to suppress. Unavoidable leaks sprung from any number of sources, including curious cowboys and railroad workers, ranchers who loaned the party horses and equipment, or citizens in Medicine Bow, the town where they purchased food supplies and services.

Osborn was already making tentative plans for the next field season. He was even thinking of sending collectors into the Southern Hemisphere within the next few months to take advantage of the austral summer. Pathbreaking work on mammal origins was then being done on the fossil mammals of Patagonia and on the mammal-like reptiles of South Africa. Osborn felt it was a matter of great importance to make representative collections in these places for the DVP. Wortman agreed, and recommended Brown for the trip to South Africa. As for Patagonia, Wortman wanted to go there himself. He had been preparing for such an expedition for more than a year, reading the relevant literature, studying Spanish, and publishing a paper on a possible North American origin for the edentates—a group of animals, including sloths and armadillos, that were confined mostly to South America. But Wortman also stressed the importance of making a good Jurassic dinosaur collection: "This country (here) should occupy our attention for several years with a good strong party and I can promise from what I have already seen that the results will be more than satisfactory."[9] Brown, for his part, was eager to go on either southern expedition, and he was willing to go immediately—even if it meant giving up a scholarship that Osborn had arranged for him to complete his undergraduate studies at Columbia.[10]

As the 1897 field season wore on into August, Wortman's enthusiasm began to flag. Sensitive and moody at the best of times, he grew increasingly discontented with the daily grind of quarrying. Osborn may have hoped to bolster his spirits with a letter containing photographs and descriptions of some mounted fossil vertebrates recently installed in an exhibit hall back in New York. But Wortman's reaction was surprisingly defeatist:

> Undoubtedly our hall looks well and the collection is worthy of all the severe hardships we have been through in getting it together but I cannot help but think often that it is *mighty little appreciated by those in authority*. So far as I am individually concerned I must confess that I don't like the outlook for the future. I can't keep up this kind of life indefinitely and I long to see the end hove in sight.[11]

Osborn read Wortman's remarks as a personal rebuke, and felt unjustly stung. This particular criticism must have seemed strangely familiar to him, given Peterson's resignation for similar reasons early in 1896 and Wortman's similar protests from the previous field season. Perhaps Osborn took a moment to reflect on the grounds for his colleague's complaints, and the possibility that he might somehow be at fault. It would have cut him to the quick, given that this fault was one of the many things that Osborn despised about Marsh.[12] But Osborn was a proud man of strong convictions. He sent his field foreman a defensive reply, and Wortman equivocated to avoid a conflict. Wortman wrote back insisting that he had not meant to imply in any way that Osborn failed to appreciate the results or the hardships suffered to achieve them, but this is almost certainly a false recantation. Moreover, he continued to sound an alarm of surrender, warning Osborn that he was on the verge of quitting:

> I do not know at what moment I will have to get out . . . and it is only with the greatest effort sometimes that I can keep the fires of my enthusiasm burning. This feeling of uncertainty is a disorganizer and many times within the past two years I have been sorely tempted to throw the whole thing up and call it a failure. However something may turn up to change the complexion of things and it has been in the hope of such that I have held on.[13]

Here, Wortman may have been trying to force Osborn's hand, although it is not perfectly clear what kind of changes he anticipated. The fundamental problem might have been about his salary, as even Osborn felt that Wortman was underpaid.[14] More likely, Wortman was making a bid for greater authority and responsibility in the DVP. He likely wanted to be more like the curator, with more time for research and less physical labor.

When he wrote "I certainly could not wish for anything better than that you were in sufficient authority [at the museum] to guide its scientific policy," he was obviously contemplating a promotion for Osborn to a higher administrative office.[15] This accomplished, he would then be free to slide into Osborn's vacated position as curator of the DVP. If this was Wortman's wish, it is difficult to understand why he resisted the curator's efforts to give the other DVP field-workers more independence, unless he feared that Osborn was trying to squeeze him out.

The mounting tension between Wortman and Brown boiled over a few days later. Apparently, the incident that finally set Wortman off was a claim he had heard that Brown had told Menke that the two of them were not subject to his directions. Wortman said nothing directly to Brown, but sent an urgent letter to Osborn explaining the situation, as he saw it, and demanding an appropriate adjustment by the curator:

Brown seems to be of the opinion that we are running two separate expeditions here, that he is in charge of one and I am in charge of the other. No conflict of authority or any unpleasantness has yet arisen simply because I have not asserted or attempted to exercise any authority but the occasion may arise at any moment. I do not know what you have told Brown or what arrangements you have made with him nor have you written me that I can recall that I was to assume charge of the expedition upon my arrival here. I have taken it for granted that I was to do so and be the responsible head of the concern. I have furthermore taken it for granted that Brown knows his place. If he don't[,] I think it is time that he should be informed. . . . I think it is decidedly for the best to have these matters clearly and distinctly understood at the very outset. . . . If I am in charge of the field work I should be so not only in name but in fact. . . . I don't know where the fault lies exactly but the remedy is to rest the requisite amount of authority in a responsible head and respect it. I hope you will attend to the matter without delay.[16]

It is not clear what, if anything, Osborn did to make amends. He must have done something to placate Wortman, at the very least, who wrote to thank him for his prompt action in settling the question of authority in the field. Wortman's satisfaction was short-lived, however, and he continued to take exception to Brown's attitude and behavior in camp. In the same letter to Osborn, he complained,

I am much surprised and not a little disgusted with Brown's actions . . . [H]e is a dangerous man to place in charge of anything. His knowledge of Paleontology and field work are too limited and his skill insufficient. . . . I am compelled to withdraw my recommendation that he be sent either to South Africa or South America. I say this not from any personal feelings in the premises but from an earnest conviction that it will be for the best interests of the Museum.[17]

A "good square friendly talk" that Wortman had with Brown later in the season failed to clear the air. What it did accomplish was to strengthen Wortman's resolve to demand of Osborn a clear delineation of authority in the field. "The whole difficulty seems to lie in the lack of authority in my possession," Wortman explained to Osborn. "[Brown] apparently thinks, even now, that my authority is only temporary and of little consequence anyway. . . . If I come into the field next year I must insist upon a very plain and distinct understanding being reached before we start."[18] Brown left the expedition less than a week later, and by then his differences with Wortman were irreconcilable. It seems that Wortman had heard a rumor about Brown that he was well prepared to believe, and on which he pinned the blame for their troubles. To Osborn he explained:

Brown left camp before we loaded the car. I took it for granted that he would hurry on to New York so as to get located before College opened. Instead . . . he had to hang around a woman for 3 or 4 days in Medicine Bow who seems to have been responsible for all the difficulty of which I have spoken. He has not only made a fool of himself and squandered his substance on this creature but he has caused a great deal of unfavorable comment and made the party ridiculous. I am completely disgusted with his actions and I doubt very much whether it will be expedient to keep him at any price. At all events I don't want him with me any more.[19]

In spite of the testy situation in camp, the 1897 field season closed successfully in early autumn. By the first of October, the dinosaurs were on their way east. Osborn's uncle, J. Pierpont Morgan, provided a freight car to carry the fossils from Aurora to Manhattan. Brown's *Diplodocus*, only partially crated, filled one end of the car and the *Brontosaurus* occupied the other. Separating them was a large quantity of bulk sediment from the mammal quarry boxed for shipment and awaiting careful sorting in the lab—Osborn remained hopeful of finding tiny Jurassic mammal specimens therein. On October 2, the remaining collectors headed north for a weeklong prospecting tour in the Freezeout Mountains and along the Little Medicine Bow River. Free at last from the pick and shovel work, Wortman was excited to examine other Jurassic exposures, and anxious to locate the dinosaur strikes that would bring the party back to dig them out the following field season.[20]

Reforming and Rewriting Marsh

Menke volunteered to stay behind, setting up a permanent camp in an old abandoned section house at Aurora. He passed the cold and lonely winter of 1897–98 there, protecting the DVP party's field site from claim jumpers and taking care of the horses and outfit. For New Year's Day, Osborn sent him a welcome box of cigars. When the battleship USS *Maine* exploded in Havana harbor on February 15, Menke read about it in the bundles of old New York newspapers and magazines Osborn sent out. Whenever snow covered the outcrops along Como Bluff, Menke spent his days in the cabin tediously sorting through a sizable pile of fossil mammal sediment looking for teeth and jaw fragments. Unfortunately, it proved to be almost completely barren. By April, he had found only three or four teeth and a single jaw fragment, probably reptilian. Nor did he find any other spots along Como Bluff where the conditions seemed favorable for obtaining more sediment. He also found the time to indulge his bird-watching hobby. A skilled amateur photographer, Menke snapped some ingenious pictures of the local wildlife from his cabin window. When the weather was clear, Wortman expected him to scout the area for new Jurassic dinosaur prospects. In this endeavor, Menke's luck was somewhat better. He found a few leads along the banks of Rock Creek and several more in the Freezeout Mountains. The most interesting discovery of the winter season, however, was a large selection of Marsh's old letters to his collectors, which Menke recovered from a pile of rubbish. The letters, which dated from the 1880s, contained some useful intelligence on the status of several of the quarries at Como Bluff, including Quarry Nine, the source of so many of Marsh's fossil mammals. Other letters included details about Marsh's notorious payroll practices and his habit for ordering his workers to rebury abandoned quarries. Osborn wanted these letters forwarded to him, as they would make invaluable additions to his growing collection of "Marshiana." And although he assured Menke that he would make no unfair use of the letters, he nevertheless insisted that his employee refrain from mentioning them to anyone else.[21]

Riggs, meanwhile, did no fieldwork in the summer of 1897. He was working as a fossil preparator in Williston's lab that summer when news arrived that he had been awarded a fellowship to continue his graduate studies in paleontology under William Berryman Scott, a dear friend and a former college classmate of Osborn's, at Princeton University. But an unhappy semester spent in New Jersey soured Riggs on life along the East Coast. Longing for an opportunity to head back west to collect fossils, he arranged an interview in New York with Wortman and Osborn. There, Wortman assured him that he could count on summer employment in 1898 with the DVP field party at Como Bluff. But Osborn, citing budget constraints, wrote to him later rescinding the offer. Feeling deeply dejected, Riggs wrote to Osborn and Wortman both with a desperate bargain: three months of summer work at half pay.[22] They accepted.

Osborn could afford to be choosy about his collectors. Wortman, his talented field foreman, seemed more contented with his DVP position in the sedentary winter than he had the previous summer. Younger collectors such as Granger, already a member of the permanent staff, and Brown, Menke, and Thomson, all regular seasonal collectors, were showing great promise. Other graduate students, including James Gidley at Princeton and Alban Stewart at the University of Kansas, had expressed a willingness to work as seasonal collectors for the DVP. Some young men volunteered to work for Osborn merely for the sake of the experience. But if Osborn was then feeling confident about his labor supply, the first sign that all was not perfectly well arrived in the mail in the early spring. A letter from a curator at the Field Columbian Museum in Chicago solicited Osborn's opinion on Brown's abilities as a collector and preparator of vertebrate fossils. Brown, who was then looking for a permanent museum position in vertebrate paleontology, had been writing to other institutions offering his services. Osborn waited a week before sending a positive but subdued endorsement. It seems clear from his reply that he hoped to retain Brown's services for the DVP. A few days later, a telegram arrived from Riggs requesting Osborn's consent to cancel their prior arrangement for summer employment. "Chicago offer seems promising," Riggs explained.[23] Brown was safe for the moment, but Riggs was lost to an upstart vertebrate paleontology program at the Field Columbian Museum.

With Wortman slated to take charge of the DVP field party at Como Bluff for the 1898 field season, Osborn had to scramble to find a congenial post for Brown. Osborn was committed to keeping him on the payroll, but reluctant to send him out to work again with Wortman. Mounting a second, smaller expedition to collect fossil mammals at another locality was one workable alternative, but a separate expedition would create a serious strain on the total budget for fieldwork. In a letter to Osborn, Wortman made a feeble stab at magnanimity, but his real feelings with respect to Brown's employment were painfully obvious:

> I don't want you to feel that you are not at liberty to send [Brown] to join our party. It is true he is not the most agreeable fellow in the world and would in all probability make

things unpleasant in camp, but I think we can get along with him. He must however know his place or get out. I will leave it with you to do just as you think proper and best.[24]

Osborn thought best to send Brown and Matthew to Kansas, thereby siphoning off some of the money that would otherwise have been available to the party at Como Bluff. Wortman was not pleased. Later in the summer when money was scarce, he laid the blame for the budgetary strain at Brown's doorstep and recommended that the young collector should be "unceremoniously bounced" from the payroll.[25]

The most immediate objective for a return expedition to Wyoming in 1898 was to secure enough additional fossil material to complete and mount for exhibition the partial *Brontosaurus* skeleton acquired in 1897. Possibly the smaller *Diplodocus* skeleton could be completed, as well. Wortman, who, along with Knight, found the larger of the two specimens, was very proud of his field accomplishments. He described the two skeletons as "by far, the most complete and perfect of their kind that have ever been collected." Yet he conceded that they were "not sufficiently complete to mount as entire skeletons in their proper anatomical positions." In characteristically optimistic fashion, he believed that another season's work at Como Bluff would supply all the missing parts of both dinosaurs.[26] Osborn, too, signed on to the dinosaur exhibition project. "We should bend all our energies this year to secure the remainder of the Camarasaur skeleton," he wrote Granger.[27]

Another motive for the DVP's work at Como Bluff, at once more scientific and more personal, emerged later in the spring, after Osborn had made a quick study of the available literature on American Jurassic dinosaurs, most of it authored by Marsh, his nemesis. He concluded rather hastily that Marsh's work on the dinosaurs was defective and was poised for a complete revision. Osborn's DVP, with its superior methods, was the perfect place to undertake this work. In particular, he believed that most of Marsh's best-known sauropod genera, including *Atlantosaurus*, *Apatosaurus*, and *Brontosaurus*, should be collapsed into synonymy with his late mentor Cope's *Camarasaurus*. Moreover, Osborn concluded that Marsh's famous restoration of *Brontosaurus excelsus*, of which Marsh was justifiably proud, was incorrect in many details, especially in including too few vertebrae in the body and the tail. (Osborn directed his staff to secure the specimens necessary to resolve these questions conclusively, although the collectors, of course, were powerless to influence the type or quality of specimens preserved in the rocks and exposed by erosion.) More time, better specimens, and sober reflection would show that Osborn was wrong on all these points. Undoubtedly his scientific judgment was clouded by the ancient grudge he bore toward Marsh. At this time, however, Osborn was positively giddy about the prospect of revising Marsh's work, and it served as a great motivator.[28] To Menke he confessed, "I started out in these beds with the idea of simply making a collection, not expecting to find much that is new. Now I have the idea that we shall make grand additions to our knowledge of these animals, and that Professor Marsh's work is as faulty among the reptiles as it has proved to be among the mammals, and must be reformed and rewritten from top to bottom."[29]

Acres of Bones

Granger left New York and headed directly for Como Bluff, arriving April 20, 1898. Menke had located several prospects in the vicinity of the section house at Aurora. Together they looked those prospects over when the weather was favorable. Later they intended to explore further in the Freezeout Mountains. Just before his departure, Granger had heard a rumor that Riggs would be coming to Como Bluff to collect for the Field Columbian Museum. Worried about the possibility of competition, Granger wrote to Osborn for advice. "If we are to have any opposition here in the field," he suggested, "we will be obliged to work out our good prospects at once and label them."[30] Osborn replied somewhat reassuringly, while at the same time echoing Granger's competitive instinct: "Riggs told me that his plans were to collect in South Dakota, but if he intends to enter the Jurassic you will have to open up all the prospects somewhat, and label them distinctly, so that there cannot be any doubt as to future ownership."[31]

Wortman took an indirect route to Wyoming. He left the museum in the company of a volunteer named Norman Grant about April 10. They headed first for St. Louis to investigate a tip about a deposit of mammoth bones. Next they went to Long Pine, Nebraska, where they rendezvoused with Thomson, who brought a team and wagon down from their winter quarters in South Dakota. While Thomson and Grant headed west with the wagon, Wortman steamed ahead by rail to Hay Springs, in a vain attempt to hunt up the remains of an alleged fossil hominid. Osborn was very excited about this fossil, and he urged Wortman to make every effort to procure it for the American Museum. Unfortunately, no more fragments of the elusive fossil turned up. Thomson and Grant met Wortman at Hay Springs with the wagon, and the three then proceeded west. The early May weather was abysmal, with snow and rain nearly every day. At Harrison, Nebraska, the party was compelled to tie up for two days to wait out a blizzard. Each night of the journey they had to find a livery stable for their horses and shelter for themselves, as the weather was too severe for camping out. Expenses ran very high. They followed the tracks of the Fremont, Elkhorn and Missouri Valley Railroad west from Harrison as far as Casper, Wyoming, and then cut south through the Shirley Basin, heading in the direction of Como Bluff. Wortman took note of the extensive Jurassic exposures along the southern and western slopes of the Laramie Mountains, but a brief examination of these beds for dinosaurs turned up no interesting leads. At last, after a long and most uncomfortable trip, Wortman, Thomson, and Grant arrived on May 12 in Aurora.[32]

Granger and Menke, meanwhile, had also been suffering from bad weather. Often it was too cold to mix plaster or too snowy to prospect. When confined to their quarters, they passed the time reading old issues of *Harper's* and the *New York Herald*, sent out for their amusement by Osborn. If the train stopped to take on water at Aurora, which it often did, they also could get the dailies from Denver. With these instruments they followed the progress of the Spanish-American War, which was declared on April 25, 1898. Despite the foul weather, they were sometimes able to work a few of Menke's local discoveries and prospect for new ones. They acquired a few more bits and pieces of sauropod dinosaurs in this way. In May they began work on a probable *Stegosaurus*. Osborn was extremely interested in this rare find, as he sensed another opportunity to amend Marsh's mistakes. Nothing significant came of this specimen, however. Once the rest of the party arrived in mid-May, they made short work of the remaining prospects.[33]

By early June, all the leads at Como Bluff had given out, and Wortman wanted to move to more promising ground. He was particularly impatient to get into the Freezeout Mountains, where most of Menke's best prospects had been found over the winter. From Aurora they hauled twelve miles north to a site along the Little Medicine Bow River, where another anticline brought extensive Jurassic exposures to the surface. They made camp at a spot on the river called Nine Mile Crossing. From this point they explored some Jurassic beds to the west, finding many prospects, but nothing particularly valuable. Just as they were breaking camp again, someone located a fairly promising-looking *Brontosaurus*, but there was insufficient time to explore it further. Wortman, who was normally a very thorough field-worker, must have been eager to push on. They moved upstream about three miles, to a place Wortman called Medicine Bend, and opened up one of Wortman's better prospects that had been discovered the previous summer. Here they found an abundance of well-preserved bones of *Stegosaurus*, *Morosaurus*, and a small unidentified dinosaur. Wortman was enthused. He estimated that it might take more than a month to extract all the bones at this locality.[34]

Another astonishing discovery made less than two miles to the northeast promised an embarrassment of dinosaur riches to be reaped over the next several seasons. Wortman referred to this other new locality as "the famous 'bone patch,'" and described it in an effusive letter to Osborn: "There are fragments enough . . . upon the surface to load a train. You have heard me speak of this place frequently during the last winter. I had concluded that there was no show for any more bones in that spot owing to the fact that the slope of the hill corresponds to the southerly dip of the stratum." In other words, Wortman initially feared that the bone-bearing layer had been shaved off the slope by erosion, leaving a relatively worthless stubble of broken fragments planted across the surface. Some years before, another collector had excavated a number of prospect holes at this locality and had apparently left several wagonloads of weathered bones lying abandoned in a heap. Wortman read this as an indication that there was no good fossil material to be had. But a closer inspection of the site showed him to be spectacularly wrong. "We went to investigate the place a little more carefully than we had done," Wortman continued,

> when to our intense surprise and delight we found that the slope of the bed is considerably greater than that of the hill[,] which in a few feet brings the bones down into good solid clay. . . . [The] bones are as 'fine as silk' and as black as ink. . . . [A]fter looking over the situation with the utmost care it is my opinion that there are from one to five acres of bones, covered by a layer [of overburden] ranging from one to five feet in thickness. . . . [T]he bone bearing layer . . . is about two and one half feet [thick]. I think so much of it that I will move the outfit to the spot and open it up immediately. If it is what I think . . . it is certainly the greatest bone bed in existence.[35]

Some years previously, on this same lonely slope, someone had accumulated a great quantity of weathered bone fragments to build the foundation of a crude shelter, now in prehistoric ruins. In honor of this curious circumstance, Wortman dubbed the new locality Bone Cabin Quarry (see figure 7).[36]

There has been some controversy about the question of when Bone Cabin Quarry was first discovered and by whom. The prospect holes mentioned by Wortman, and later plotted on a quarry map published in 1904, demonstrate that at least one fossil collector had worked this same

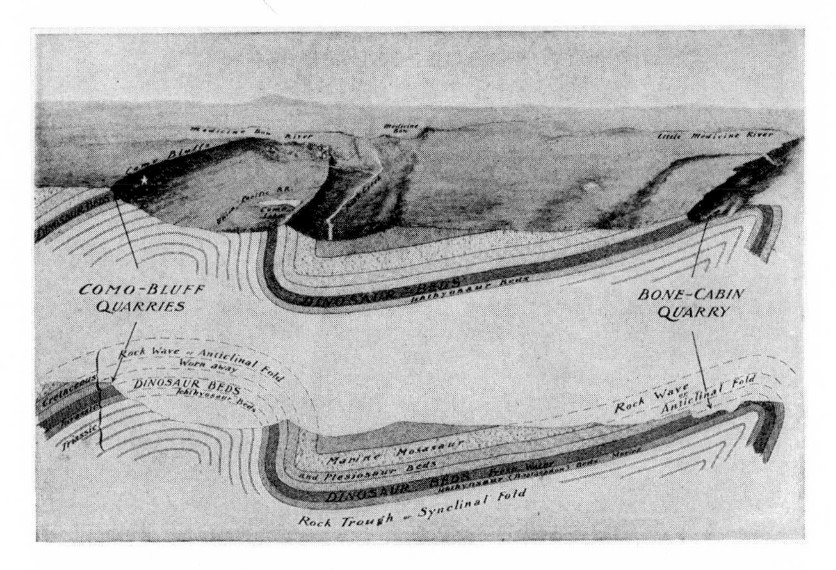

Figure 7. Diagram showing the geological relationship between the Jurassic dinosaur quarries along Como Bluff, on the Como anticline, and at Bone Cabin Quarry, on the Medicine anticline. From Osborn, "Fossil Wonders."

spot, however briefly, sometime before 1898.[37] Frank Williston (Samuel Wendell Williston's younger brother), who worked for Marsh in the 1870s and Cope in the early 1880s, is the most likely candidate.[38] Reed is another possibility.[39] He made a surprise visit to the new quarry the day after the DVP party first began to work it. Apparently he came intending to set up shop for himself, but was surprised and disappointed to find that the DVP party had gotten the jump on him. His visit indicates, at the very least, that he knew about the locality and suspected something about its possibilities. Nevertheless, it remained for the American Museum party, under Wortman's leadership, to discover and exploit the true potential of Bone Cabin Quarry, beginning in 1898.

Yet some published accounts credit Granger with the discovery of this locality at the end of the 1897 field season when he and other members of the DVP party made a one-week exploratory tour of the Jurassic beds north of Como Bluff. According to one recent account of these events, Granger was riding out on the Laramie Plains when he encountered one or more Mexican sheepherders who indicated a curious ridge some miles to the north. Granger rode off in search of the site. Once there, he climbed the ridge and spied numberless dark brown cobblestones, some in a jumbled heap, others strewn across the surface indiscriminately. On closer inspection, however, he realized, no doubt with great excitement and awe, that what appeared to be stones were, in fact, the badly weathered fragments of dinosaur bones eroding out of the bed beneath his feet. There was no time to develop the prospect, so he noted the locality, snapped a photograph, and moved on.[40]

A preliminary reconnaissance by Granger in 1897 is not inconsistent with the best-demonstrated facts, but does it constitute a discovery? Granger was an exceedingly modest man, and he never recorded any claims about the discovery of Bone Cabin Quarry. Nor does any of the archival evidence give him the credit. It is clear from Wortman's letter to Osborn, and from the DVP party's plan of attack for the 1898 field season, that the great potential of Granger's 1897 reconnaissance was not realized until the following June. Wortman reminded Osborn that they had spoken about this locality over the winter, but that he thought it would prove unproductive, which suggests that someone in the party other than Wortman had visited the site and recommended a more thorough search. But after more than a month spent exploring a number of indifferent prospects along Como Bluff and near Aurora, Wortman was most eager to relocate to the Freezeout Mountains. Someone urged him to take the trouble to investigate this site, despite his skepticism. Given the later claims of discovery made on his behalf, Granger seems to be the most likely party.

Wortman moved camp to the new locality within the next few days, probably on June 12. Using a team and scraper, the DVP party opened up a space about twelve feet square and uncovered a great profusion of bones (see figure 8). "[T]here are . . . so many [specimens] I would not know where to begin to describe them. . . . I never saw bones thicker in all my life," Wortman wrote. So thickly, in fact, were the bones commingled that it proved difficult to collect one without displacing or sometimes damaging another one. Some specimens had to be sacrificed in order to salvage others. In other cases, when the bones were found disarticulated, it could be difficult to tell which elements belonged together. Wortman adopted the strategy of photographing the quarry frequently to record the original positions of the bones. Menke's skill as a field photographer now proved invaluable. For clarity's sake, Wortman sometimes supplemented Menke's photographs with rough sketches of the bones lying in situ. These photographs and diagrams would prove extremely useful in reconstructing the animals back in the museum.[41]

After the near miss with Reed, Wortman circled the wagons; security and secrecy, he now felt, were paramount. Though he could have used more expert help to work the quarry to its fullest potential, he was reluc-

Figure 8. Walter Granger contemplates the confusing jumble of specimens at Bone Cabin Quarry in 1898. Image #17838. American Museum of Natural History Library.

tant to take on additional collectors for fear that news of the quarry would spread quickly, attracting rival institutions to the new DVP locality before they could grab all the finest material. Matthew's party, for instance, which included Brown and veteran freelance collector Handel T. Martin, was then having very poor luck hunting fossil mammals in Kansas and was available to haul to Wyoming on short notice, but Wortman did not want their help. He had no reservations about Matthew, whom he considered a faithful and agreeable colleague, but he was flatly opposed to letting Brown or Martin learn about the riches to be had at Bone Cabin Quarry. He feared that Brown in particular was likely to desert the DVP party at the first opportunity and offer his services to another institution. Martin posed a similar risk. To Osborn, Wortman explained:

> We have a splendid field here which I doubt not will furnish us all the Jurassic bones we will want and I trust you will be exceedingly careful not to make any false move to embarrass us. Marsh's maxim 'Don't go duck hunting with a brass band' is a mighty good one in practice and if we know our own interests we will keep shady until we have what we want. I have instructed the boys to keep mighty quiet about the quarry. . . . I tell everybody that our success is very poor so far this season.[42]

Instead of taking on more expert help, Wortman hired a common laborer to do the stripping, shoveling, and hauling, while he, Granger, and Menke attended to the detail work on the delicate bones. (N. Grant, who disappeared from the DVP records, had apparently quit the expedition by this time.) Wortman was very lucky in his choice, finding a Danish man named Peter C. Kaisen, then working as a section foreman for the Union Pacific Railroad. Kaisen was smart, eager, and a prodigious worker. He stripped off enough overburden in just a few days to keep the entire party busy excavating for weeks. Wortman soon put him to work on the bones. With Kaisen's help, the DVP party opened up a facing of more than thirty feet, to a depth of about four feet, with bones showing in the quarry bottom nearly every foot of the way. The sediment was gray sandstone and soft blue clay. It was pliant enough to pick apart with an awl, and work progressed rapidly through June. By early July, as Teddy Roosevelt's Rough Riders were charging up Kettle Hill in Cuba, Wortman abandoned the lesser prospect at Medicine Bend—there was simply too much work to be done now at Bone Cabin Quarry. "I think I can see work ahead here for a number of years at this one spot," he predicted.[43]

Late in July, with no sign of a slowdown in the quarry's productivity, Wortman began to plan ahead for the winter months. He was adamant about keeping someone at the quarry over the winter to safeguard the museum's interests. It also would be advantageous to have someone at the quarry to keep it dry through covering it in bad weather and digging drainage ditches, and to do some additional sod breaking and stripping whenever possible. "We might take chances and let the quarry take care of itself," he explained to Osborn, "but there is a likelihood that some fellow would 'jump' our claim." Menke, it seems, was unwilling to spend another solitary winter in Wyoming. He wanted to go to New York to learn more about the craft of preparing and mounting fossil vertebrates, although Wortman declined to make him any promises about a job at the museum. Kaisen, on the other hand, was available to stay. Wortman wrote to Osborn seeking his permission to make arrangements for the hardworking Dane to remain on the payroll.[44]

In August, Wortman began to feel gloomy, just as he had the previous field season. To Osborn, who was then on an extended European museum tour, he complained of feeling played out and stale. "I must confess to a sore, tired, worn out feeling which has a tendency to take the edge off one's enthusiasm," he wrote. Although the quarry was still producing extravagantly, Wortman now reported on a disturbing pattern with respect to fossil preservation that was just beginning to emerge. Rich in well-preserved and reasonably complete limbs, pelvic and shoulder girdles, and long series of caudal vertebrae, the quarry as yet had failed to produce a single good series of cervical or dorsal vertebrae. Skulls, thus far, were also conspicuously absent. A few fragments of skull bones had turned up at one time, but they were in such a poor state of preservation that Wortman neglected to collect them. Needing a skull to complete a mounted sauropod dinosaur, and tiring rapidly of the heavy quarrying, Wortman promised to spend some time prospecting in the Freezeout Mountains for more complete, individual skeletons.[45] This promise proved impossible to keep, however.

Wortman closed up work in the quarry promptly on September 15. There was still a great mass of bones exposed in their most recent cutting, but insufficient time remained to take them all up. To the south and especially to the west sides of the quarry, bones appeared to be preserved in ever-greater numbers. Wortman was absolutely confident of a complete representation of several of the many different animals encountered in the quarry, although it could take several years to recover all the parts. They spent a few days carefully packing their specimens in about one hundred custom-built wooden crates. The work of hauling the crates to the depot (apparently at Medicine Bow rather than Aurora) was very heavy and time consuming. With an early start, they could make a single round-trip journey in the wagon in one day, although reloading the wagon in the evening for an early start often meant working until nine o'clock at night. Handling the crates was made easier by a hoisting apparatus sent out from the museum. A fossil haul that filled two boxcars, again provided free of charge by J. P. Morgan, was ready to be shipped to New York around September 30, 1898.

Osborn never wrote with explicit instructions regarding a winter caretaker, so Wortman acted upon his own devices. He left a scraper, a team of horses, and a large supply of feed behind. The entire party then collaborated on the construction of a small cabin and a stable with their surplus lumber. Thereafter, they disbanded for the season, leaving Kaisen to fend for himself against the bleak Wyoming winter.

FOUR

Most Colossal Animal on Earth

Osborn's fear about the rising competition for fossil vertebrates in the American West was well founded. By the late 1890s, he and his DVP field paleontologists had amassed an enviable collection of fossil mammals from scores of localities in the Western states. They had also made a very respectable beginning in collecting dinosaurs near Como Bluff, Wyoming (see figure 9). Under Osborn's leadership, the DVP had become the leading center for vertebrate paleontology in the United States. But success inspired imitation. Just before the turn of the twentieth century, new, privately funded, museum-based programs in vertebrate paleontology started up at the Carnegie Museum of Natural History in Pittsburgh and the Field Columbian Museum in Chicago. The investment of private money into multiple museum paleontology programs fueled a period of intense competitive activity. A scramble ensued among museum programs to claim the best field sites and to collect, prepare, and exhibit the finest specimens.

The competition for American Jurassic dinosaurs, in particular, heated up in the summer of 1899. The DVP's sizeable operation in Wyoming, begun modestly at Como Bluff in 1897 and enlarged at Bone Cabin Quarry in 1898, continued apace. Small parties from the University of Wyoming—led by W. C. Knight and veteran collector W. H. Reed, now an assistant geologist and museum curator—also continued to collect in the Jurassic beds nearby. The field became particularly crowded after an invitation from the Union Pa-

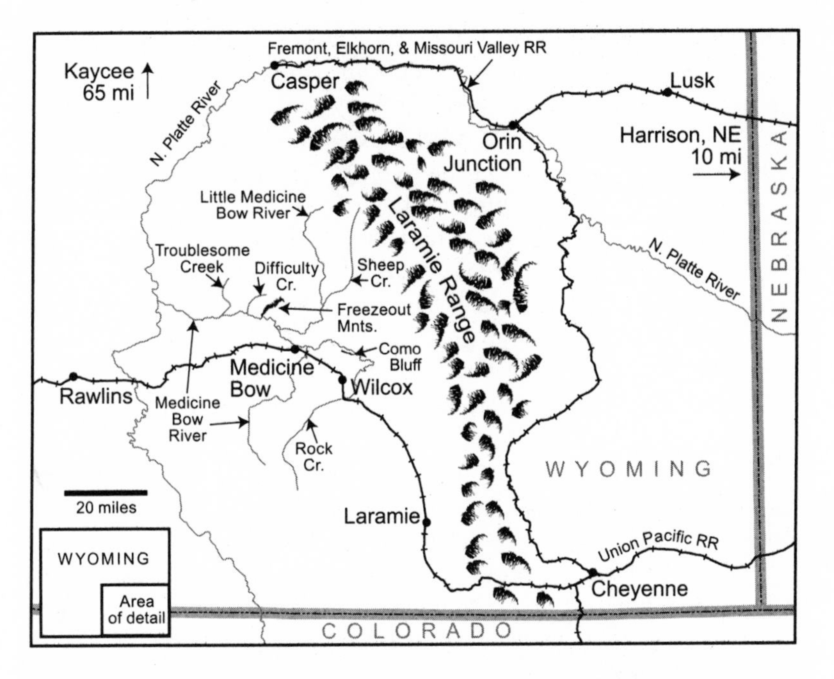

Figure 9. Map of southeastern Wyoming. Map drawn by Cathryn Dowd.

cific Railroad to participate in its Fossil Fields Expedition, complete with an offer of free rail transportation, flooded southeastern Wyoming with dilettante dinosaur collectors from scores of colleges, universities, and museums nationwide. It also attracted an experienced party from the University of Kansas, under S. W. Williston, who was deeply interested in Jurassic reptiles. Worse still, ambitious and relatively well-heeled parties from the new museum programs in Chicago and Pittsburgh crowded into the Jurassic beds near Medicine Bow, Wyoming, each within a day's ride of Osborn's principal locality. By the summer of 1899, the race to find, collect, and exhibit a Jurassic sauropod dinosaur was well under way.

Establishing Paleontology in Chicago

The founders intended the Field Columbian Museum to have a broad scope encompassing the diversity of materials acquired from exhibitors following the World's Columbian Exposition of 1893. A modest assortment of fossil vertebrates, casts, and reconstructions, obtained primarily from the Ward's Natural Science Establishment exhibit, was displayed at the museum from the outset. However, plans for a department of paleontology were indefinitely postponed while Director Frederick J. V. Skiff canvassed the American scientific community for a prominent paleontologist to serve as its curator. Meanwhile, stewardship over the museum's fossil vertebrates devolved on the Geology Department, under mineralogist Oliver Cummings Farrington, who held a PhD in geology from Marsh's Yale University. In the early 1890s, while Osborn was securing the institutional dominance of the DVP at the American Museum, Farrington channeled scarce departmental resources into hard-rock geology and meteoritics. A short-lived initiative into vertebrate paleontology by Oliver Perry Hay, an assistant curator in the museum's Zoology Department, ended abruptly when Hay was dismissed following a quarrel with Skiff in 1896.[1]

Farrington believed that the fossil vertebrate exhibit at the Field Columbian Museum failed lamentably to illustrate its field. Specimens on display suffered from incomplete or inaccurate labels inherited from the fair, others languished in storage, while whole faunas, periods, and localities remained entirely unrepresented in the museum's three paleontological exhibit halls. Lacking the time and training to launch a program in vertebrate paleontology himself, he lobbied Skiff repeatedly for money to hire a collector and a paleontological assistant. His motivation for promoting vertebrate paleontology at the museum had more to do with showmanship and popular education than with pursuing original scientific research, however. He held that valuable moral lessons for all classes of people could be taken from the examination of ancient life in the museum. In lectures to church groups and teachers, Farrington argued that the record of extinction and survival over the course of geologic time showed that forms of life characterized by "great size, strength, eccentricity, [or] selfishness [as exemplified by the] warring monsters of the early seas" have gone extinct, while "activity, agility, swiftness, intelligence, meekness, social organization, and a mastery over appetites and passions" characterized the survivors. Museum visitors who absorbed these lessons would become better citizens and better individuals. From the observation of gawking visitors at other museums, Farrington was also aware of the immense popular appeal of mounted fossil vertebrates, the larger the better. In an appeal to Skiff for funding to support an inaugural fossil mammal collecting expedition in 1898, Farrington argued that "finely mounted specimens of ancient mammals would probably too attract as much public attention as any kind of specimens that could be displayed." After several years of needling, Skiff at last adopted the cause of vertebrate paleontology. He shepherded Farrington's proposal through the museum's cumbersome approval process, netting $1,000 in seed money to mount a fossil collecting expedition in the spring and summer of 1898. Better still, funding was approved for Farrington to hire a skilled assistant.[2]

Farrington had his eye on Barnum Brown, who had written a letter to the Field Columbian Museum offering his services as an expert fossil hunter. But Osborn wanted Brown to stay in New York. Instead, Elmer Samuel Riggs, an experienced fossil collector and a biology fellow at Princeton, accompanied Farrington's expedition as an assistant. A graduate of the University of Kansas and a protégé of his former professor, Samuel Williston, Riggs was a fossil vertebrate specialist with a research interest in early fossil mammals. As a student, he worked under Williston in the field twice, in 1894 and 1895.[3] In the summer of 1896, he joined J. L. Wortman, W. Granger, and former KU classmate Brown on an American Museum expedition to Wyoming. Riggs was working toward a PhD in paleontology at Princeton when Farrington's offer—which promised a chance at permanent museum employment—arrived in the mail. In order to accept, Riggs had to cancel a less favorable arrangement he had made previously to collect that summer for Osborn at Bone Cabin Quarry. His field trial a success, he returned to the Field Columbian Museum in the fall as a fossil preparator and general assistant. He was promoted to assistant curator of paleontology on the last day of 1898.[4] His first priority was to launch an ambitious field campaign designed to build a respectable fossil vertebrate collection at the museum.

Riggs, however, was so preoccupied in the spring of 1899 with the multitude of responsibilities attending his new curatorial position that his proposal for summer fieldwork was not made ready and submitted for Skiff's consideration until May 26. Because of the tardiness of his proposal, Riggs lost creative control over his fossil collecting program and nearly missed the field season altogether. Following the rigid administrative hierarchy of the Field Columbian Museum, Riggs devised the plan and then passed it along to his supervisor, Farrington, who made the formal proposal to Skiff.[5] The director then sought approval for the project from Harlow Higinbotham, the president of the museum and chairman of its executive committee. Higinbotham, who had personally approved the museum's first fossil vertebrate collecting expedition the previous summer, and even fronted the money, rejected Riggs's proposal.

Riggs's original plan for the summer of 1899 reflected his research interest in fossil mammals and drew on his earlier experiences with the University of Kansas and the American Museum. The plan called first for a one-month trip to Long Island, Kansas, to collect late Miocene fossil rhinos from a well-known locality. Riggs had become familiar with this material in college, and, although he had never collected at the quarry himself, he knew that the chances of obtaining exhibit-quality specimens there were very good. The second objective, the Uinta Basin of northeastern Utah, was a much more risky proposition. Eocene fossils from the Uinta Basin were more ancient, but also comparatively rare, fragmentary, and ill preserved. With luck and hard work, Riggs could expect to make a representative collection of fossil mammals at this locality, but the odds against finding an exhibit-quality, mountable skeleton there were very long. On the American Museum expedition to Wyoming in the summer of 1896, Riggs learned firsthand how exasperating the search for Eocene mammals could be.[6]

As curator of the Geology Department and Riggs's immediate supervisor, it was Farrington's task to sell this plan to Skiff. In the 1890s, the imminent threat of extinction was used successfully by the museum's Zoology Department as a justification for undertaking expensive African fieldwork. In the Anthropology Department, curators cited the annihilation of American Indian cultures as one of their many motives for collecting ethnographic objects. Similarly, Farrington stressed the urgency of continuing to collect fossil vertebrates before a scourge of collectors from other museums had exhausted the known localities. But he erred in quoting Riggs's former professor Williston who described the Long Island quarry as "practicably inexhaustible." Farrington reminded Skiff that fossil collecting had been "auspiciously begun" the previous summer and argued that it was "eminently worthy of continuance." Expense should not be a problem: if railroad transportation for the collectors and their returning freight could be had for free, Farrington estimated the cost of the expedition would be $1,000—exactly the same amount that had been budgeted for fossil collecting in the summer of 1898. Apparently unimpressed by Farrington's plea for urgency, Skiff waited twelve days before submitting the proposal to Higinbotham. And it was not until June 22 that Higinbotham officially turned down the request, citing as his reasons "the advanced season, the expense, coupled with the fact that we can secure the objects desired at some future time just as well."[7]

Paleontology in Pittsburgh

Steel millionaire and philanthropist Andrew Carnegie was turning his attention to vertebrate paleontology at about the same time. In the autumn of 1898, he read a New York City newspaper article about a giant Wyoming dinosaur found by Reed and became determined to buy it for his new Pittsburgh museum. William J. Holland, a personal friend of Carnegie and director of the Carnegie Museum of Natural History, swung immediately into action. Pugnacious, overbearing, and prone to anger with his social inferiors, Holland was nevertheless eager to please his patron. When necessary, he could be an extraordinary bootlicker. Because Carnegie personally supported the development of vertebrate paleontology at his museum, money would be no object, although Holland at times had to coax his patron to spend it. Given an adequate budget to command, Holland promised his benefactor "the Prince of Lizards," and a collection of fossil vertebrates to rival or exceed the collections at Yale and the American Museum.[8]

A pastor by training and a practicing entomologist, Holland knew next to nothing about vertebrate paleontology, but he was supremely self-confident and a quick study. He first wrote to Reed to inquire about the availability of his dinosaur, and received a disappointing and unambiguous reply, reading (in part): "Dinosaur not for sale." However, a follow-up letter from Reed served to cloud the issue considerably. In it, Reed explained that while the dinosaur depicted in the article was owned by his employer, the University of Wyoming, another specimen described in the same article was still in the ground and belonged to him personally. "This specimen is for sale to the highest bidder when I get it out . . . we may be able to make a bargain," he wrote expectantly.[9]

Excited by the opportunity to make a quick deal, Holland wrote back immediately, hoping to get an honest assessment of the condition of Reed's sale specimen. His letter spells out precisely what the Carnegie Museum was after:

> I am anxious to know from you how perfect this skeleton is likely to prove. Do you think that it promises to be as perfect, for instance, as the skeleton of the Brontosaur which Prof. O. C. Marsh has at Yale? . . . I should like very much to obtain for this Museum a specimen as nearly perfect as possible of one of these huge saurians, and preferably the biggest specimen that has ever been discovered. I should dislike, however, to enter the arena as a competitor for a mere fragment of these remains, no matter how large they may be, because it is our wish to erect one of the largest specimens in our Museum Hall.[10]

All through the winter months Holland was eager to move forward with the dinosaur project, but he found himself caught between Carnegie, his paymaster, and Reed, the skilled craftsman he needed to bag the dinosaur for Pittsburgh. Neither would quote him a fixed price nor commit as rigidly as he had to the task at hand. He wrote repeatedly, asking for a frank estimate of the probable cost of excavation, but Reed was reluctant to provide one. Nor could he get Reed to quote him a salary (Reed, apparently, preferred that Holland extend him an offer). He also wrote to Carnegie with updates and to plead for his blessing: "What do you say? Shall I go out to Laramie and arrange with Professor Reed to disinter this monster and bring him east and restore him? I am ready to do anything which your judgment, under the circumstances, commends. . . . I take pride in serving you to the best of my ability in all such matters."[11] With the spring thaw fast approaching, Holland boarded a train bound for Wyoming, hoping to close the deal for the dinosaur.

Not surprisingly, officials at the University of Wyoming differed with their employee over the rightful ownership of the dinosaur prospect Reed was offering for sale and once again Holland was caught in the middle. Yet he refused to be beaten, repeatedly patronizing and antagonizing Wyoming officials in an effort to convince them to relinquish their claims to the dinosaur. He seemed to believe sincerely that the specimen belonged most appropriately at an out-of-state museum. His claim that it would benefit the university to have the dinosaur displayed in Pittsburgh rather than on the Laramie campus was received with deserved scorn and skepticism by university trustees. Meanwhile, Reed, tormented by the difficulty of serving one master while attempting to wheedle the best possible deal from another, offered little useful assistance. As for Carnegie, he was most concerned about his tenuous public image. The infamous labor dispute at his Homestead, Pennsylvania, steel plant in 1892, which left many dead and wounded, had been a public relations disaster. Often portrayed in the newspapers as a ruthless robber baron, Carnegie preferred to cultivate a legacy as a generous and enlightened philanthropist. He cabled Holland with instructions "not to antagonize the state of Wyoming." But it was already too late. Holland had hired an attorney, made land claims and purchases of his own, and had even outbid the university to retain Reed's services as a fossil collector, all the while resenting "that any of [the university officials] should have seemed to think that I was disposed to interfere in any way with their plans and purposes."[12]

The custody battle over Reed's dinosaur prospect ended in May 1899, during Holland's second trip to Wyoming, with no clear winner. Holland went west to attend personally to a survey and a legal claim to the land where the dinosaur had been discovered, but this task proved to be more physically demanding than he had expected. Although snow was still falling, spring flooding had destroyed all the bridges over the Little Medicine Bow River, thus separating Holland and his surveyor from their quarry. Undaunted, Holland rounded up some cowboys, requisitioned a cache of lumber and telegraph poles, and then supervised the construction of a rickety bridge just wide enough to accommodate a team and wagon. They crossed, then climbed into the Freezeout Mountains, about thirty miles north of Medicine Bow, and located the controversial specimen. Holland also found another specimen nearby. Like Carnegie, Holland's attorney urged conciliation and kindness. So Holland "put on the robes and mantle of peace." As he explained in a letter to Carnegie,

> The specimen which the Trustees refused to sell me, imagining that they had title to it, I have donated to them, being first, however, . . . sufficiently careful of the interests of the Museum to load up what [I] could . . . into the wagon which I took with me. . . . I only succeeded in bringing away some fragments of one bone, but these . . . made a heavy wagon load.

Holland hoped that by appearing magnanimous, and then dropping Carnegie's name with the governor and other prominent public figures of Wyoming, he could tap the necessary political pressure to force the trustees to give the dinosaur to the Pittsburgh museum:

> I believe that we shall ultimately get possession of our coveted monster. If we do not, it gratifies me . . . to report to you that I have myself located the remains of another beast of the same sort in the same locality. . . . I suspect that we shall find before long that it is not going to be . . . difficult . . . to get one of these huge saurians exceeding in size the largest specimens found by Prof. Marsh.[13]

Back in Pittsburgh, Holland was exceedingly busy with museum business, where making arrangements for his new Department of Vertebrate Paleontology was one of his highest priorities. A reconnaissance of the American Museum presented Holland with an unexpected opportunity. He traveled to New York on March 29, 1899, to confer with Carnegie about the dinosaur project and then spent the afternoon familiarizing himself with his DVP rivals. Here he met Wortman—then serving as Osborn's chief assistant—who impressed him as a skilled, experienced, and underemployed paleontologist. They had a frank discussion about Wyoming fieldwork. Wortman spoke kindly of Reed as a collector, but at the same time he encouraged Holland to recruit another man, one better educated and more competent to manage the scientific affairs of the Carnegie Museum's new paleontology program.[14] Before this meeting, Holland seemed content to head the program himself, with Reed as his principal field and lab assistant. Holland had supreme confidence in his own abilities and sufficient faith in his new assistant to proclaim in a letter to Carnegie that "Reed is a better man for this work than Osborn."[15] Wortman convinced him otherwise. In fact, the impression Wortman made was strong enough to inspire Holland to pursue his services as curator with the same determination with which the director was then pursuing Reed's dinosaur. Wortman, however, was then hopeful of an appointment at Yale as a successor to Marsh, who died unexpectedly on March 18, 1899. Finally, in May, after the Yale possibility dissipated due to a lack of funding, Holland extended a concrete offer. Wortman, dissatisfied by his ambiguous position between Osborn and the rest of the staff, accepted the job. And he took Arthur S. Coggeshall, one of Osborn's most skilled fossil preparators, with him.[16]

In hiring away Reed, Wortman, and Coggeshall from rival institutions, the Carnegie Museum acquired an ideal group to initiate vertebrate paleontology fieldwork. Holland wrote to Reed with the news, also letting him know that Wortman would be his new superior. He was proud of his coup, gushing, "I flatter myself that with Dr. Wortman's scientific ability and knowledge and your enthusiasm for work this institution . . . will make a record . . . superior to . . . any party . . . exploring and exhuming the fossil remains of Wyoming." Holland, like Osborn, was a very competitive and ambitious man.[17]

Less personally ambitious than some of his Eastern colleagues, and more genuinely interested in the advancement of science for its own sake (with little regard for the interests of any particular institution or person), Wilbur C. Knight was unfailingly generous in aiding other field paleontologists in Wyoming. Living in Laramie on the very frontier of field paleontology, Knight was isolated from his fellow naturalists and from other centers of learning. He had very limited means and too many responsibilities at the University of Wyoming.[18] Eager to see the fossil vertebrate resources of his state utilized, Knight, along with Edward L. Lomax, ticket agent for the Union Pacific Railroad, organized the "Wyoming Fossil Fields Expedition of July, 1899" for the benefit of colleges and universities nationwide. In a promotional brochure accompanying the invitation, Knight declared that "the fields are ample for all who wish to avail themselves of the opportunity to collect and create museums as large, if not larger than any that have been built up during the last quarter of a century."[19]

Getting Down to Business

With Wortman waiting for news from Yale and talking openly with Holland about his prospects at the Carnegie Museum, Osborn knew he would need to appoint a new field foreman. He chose Granger, a young and modest man, but a very capable field paleontologist who had been connected with the DVP as a collector since 1894. Granger departed for Medicine Bow to assume charge of field operations, arriving on April 29, 1899. Kaisen, who had safeguarded the field site through the long winter, was already at Bone Cabin Quarry, stripping away the overburden and drying out the quarry. Theirs was the first party to begin fieldwork that season, but they were not entirely alone, as Holland was then scouting in the Freezeouts with his surveyor. Granger knew they were out there. He wrote to Osborn, warning, "I do not anticipate any trouble from their party but I shall proceed at once to file our claims nevertheless." He staked out twenty acres for a placer claim, but never actually filed it, thus avoiding a considerable extra expense.[20]

Richard Swann Lull, assistant professor in the Department of Zoology and Entomology at the Massachusetts Agricultural College, and a future PhD student of Osborn's at Columbia University, volunteered to join the DVP party that summer without pay, "merely for the sake of what he can learn by contact with nature and the men who form the expedition."[21] Osborn, a great advocate of the benefits of volunteerism, gladly accepted. Lull was tall and gentlemanly, with a stiff, military bearing. He was also hard of hearing. He was personally motivated by a romantic notion of fieldwork, which he wrote about many years later: "The old-time expeditions were staged in the real West, at a time when lack of means of transportation . . . together with the very intimate contact every fossil hunter must have with his physical surroundings—with fatigue, heat and cold, hunger and thirst—made the search for the prehistoric a real adventure suited to red-blooded men."[22] Unfailingly courteous, his manner and appearance were aristocratic. Small wonder that Osborn was so taken with him. Osborn wrote to Lull describing the conditions of camp life. "The work will be . . . chiefly manual . . . upon six days of the week," he explained, "as our parties rest . . . on Sundays. The life is a vigorous and healthy one, and opportunities for the study of western geology and the direct contact with these wonderful deposits compensate for the hard labor. . . . Our parties are out strictly for business," Osborn warned, lest there be any misunderstanding about the nature of Lull's volunteer position, "everything is subservient to this."[23] And so that there would be no confusion on Granger's part, Osborn wrote him also, insisting that "you must not hesitate to give [Lull] his share of the rough quarrying."[24]

Osborn was concerned about Granger's fitness for his new leadership position. To ease his own anxiety, he shared with him some very specific advice about how to manage his charges. "You are not to treat [Lull] with unusual deference, as a Professor," Osborn instructed,

> but he is to serve exactly like all other members of the party, and in general it is very important for you now to quietly assume the leadership, and authority, in order to hold the party well together. This will be very important to your success this season. It is not necessary to be disagreeable or arbitrary, but it is necessary to think out before hand, exactly what work you propose each man to do, and then to assign it to him. If any man does not do his work, you are at liberty to discharge him. In case Professor Lull, I do not think you will have any trouble, if you do please write me about it.[25]

Renewed excavation at Bone Cabin Quarry began as soon as Granger made his appearance, although the full force of his party, as well as the bulk of their outfit, was yet to arrive. Dinosaur bones of all shapes and sizes started coming out of the ground every day, although nothing of special importance. The early May weather was still cold, but relatively dry. Reinforcements began to arrive in short order. Frederick A. Schneider, an apprentice collector, reached camp from New York "in good spirits" on May 8. Albert Thomson, the cook, brought a team and wagon and part of the museum outfit to camp on the 17th. Thomson's arrival overstretched the boarding capacity of "the shack," Kaisen's crude but warm wooden shelter, so that when Lull appeared three days later in the face of a howling spring blizzard, he found the party lodging in their tents.[26]

Rival parties were also assembling. Representing the University of Wyoming, Knight and a student assistant named Charles Whitney Gilmore, who had recently returned from a tour of duty in the army during the Spanish-American War, entered the Freezeout Mountains on or about May 5, 1899. Reed, now working for the Carnegie Museum, informed Holland of this fact in a letter, claiming, "I dont know what they are doing and I shal not interfere with them."[27] Granger also knew about the competition and he wrote to inform Osborn in New York. According to Granger, "Directly after [Holland's survey] party left Prof. Knight came out and began working there." Osborn understood Granger to mean that the University of Wyoming party was working at Reed's prospect.[28] Holland, however, was then still hopeful of acquiring possession of Reed's dinosaur for Pittsburgh, as Carnegie himself had "set his heart upon obtaining this particular specimen."[29] He was undoubtedly disturbed by news of Knight's movements. Wortman, too, was unhappy about the news. In a letter to his new boss, Holland, he wrote, "I am sorry to hear that Knight is up in the [Freezeouts] for that will mean I fear that he will 'camp out' on everything in sight there."[30] Knight was indeed searching for new dinosaur prospects, but not so selfishly as Wortman feared. When a severe snowstorm held Knight over at the DVP campsite, he and Granger had a long talk. "[Knight] tells me that he is sure there are more prospects near him and invited me to go up and look them over. He is apparently very friendly with our institution," Granger reported to Osborn.[31]

Reed, meanwhile, acting under instructions from Holland, scouted extensively among the Jurassic exposures north of Medicine Bow during the month of May 1899. He had some good luck, and wrote to Holland with the news. "I found some good bones in the Sheep Creek basin," he reported. "I picked up the float pieces and burried [*sic*] them and set up stone monuments fifty steps away to mark the places for future work."[32] Reed, too, was wary of the growing competition. After caching the specimens he resumed his explorations. Holland, however, wanted him to stay reasonably close to town so that he could link up with Wortman as soon as the latter arrived. But Coggeshall, Holland's new preparator, got there first on May 27. Reed met him promptly at the Medicine Bow station with the wagon. Wortman was running late. He stopped in Chicago, Omaha, and Laramie to purchase supplies and fix his arrangements for free railroad transportation, finally arriving in Medicine Bow sometime soon after his Carnegie Museum colleague. By the first week of June, all three collectors were actively seeking specimens in the field. Like Osborn, Wortman wanted to locate and stake a claim to the most promising specimens well in advance of the arrival of Knight's Fossil Fields Expedition horde. He left behind a blizzard of labels identifying all the Carnegie Museum prospects. To Holland he wrote, "In view of the contemplated influx of Bone Hunters in the near future Reed and I will make a trip to the head of the 'Troublesome [Creek]' and look over some promising ground in that locality. We may locate claims to hold quarries for the present at least."[33]

Back at his snug Manhattan office, Osborn fretted excessively about the competition gathering in Wyoming and he spurred his field party from afar to leave no stone unturned in the search for additional prospects. He thought his DVP crew could circumvent their rivals by laying claim to every promising fossil vertebrate they could locate. So, although they had all they could reasonably handle with the hard work of excavation at Bone Cabin Quarry, Osborn pressured them to prospect for new sites as much as possible. With the summer field season fast approaching, he wrote to Granger recommending action. "I think it is important for you to do some prospecting, as soon as you can," he urged, "that is after you get your quarry work well started."[34] Ten days later he was less ambiguous and much more emphatic:

> It is extremely important for you to begin to prospect, and if you find anything good mark them out for our party by placing an American Museum claim upon them. There is no doubt but by the first of June or soon afterwards, not only [Wortman's] party but others will be in the field, and it is very important for us not to be behind hand.[35]

In his next letter, written less than two weeks later, Osborn again urged, "I hope you will push your prospecting as vigorously as you can, and I trust that you will be able to locate something of great value during the present season."[36] Osborn seemed to feel a strong proprietary interest in Jurassic dinosaurs. His attitude toward rival collectors in the summer of 1899 is probably best expressed in this uncharitable reaction to the Fossil Fields Expedition:

> I think it would be well for us not '*to be duck hunting with brass guns,*' that is, to keep very quiet about our work and our methods, it seems a rather ungenerous way of acting towards other scientific men, but I cannot help thinking that this miscellaneous scrambling for Dinosaurs . . . will be a misfortune.[37]

Clearly, Osborn resented Knight's expedition. Yet the collectors from the Carnegie and Field Columbian Museum offered some far more serious competition. Wortman, the most experienced collector in the field that summer, posed the gravest threat. Bitter after his recent resignation, Wortman seethed with resentment at Osborn's autocratic style of leadership in the DVP. Osborn sensed his hostility while Wortman was still in New York and sent a panicky letter to Granger warning that his former field foreman was not "going to act in a friendly way" toward the American Museum. Wortman knew where all the American Museum prospects were located, and he was friendly—probably far friendlier than Osborn himself—with the members of the American Museum field party. He was especially close to Granger. Osborn worried particularly that Wortman, backed by Andrew Carnegie's millions, would try to hire away some of his best field assistants.[38]

Riggs Enters the Field

Harold William Menke, another one of Osborn's dinosaur collectors, had already moved to the Field Columbian Museum staff after the 1898 season. Like Riggs and Brown, Menke attended the University of Kansas in the 1890s, graduating with a bachelor's degree in 1897. At KU, he became friendly with Brown, who recommended him to Osborn for a position as field assistant. Brown described him as careful, hard-working, and intelligent, and he wanted his help handling heavy and fragile specimens. Menke worked for the American Museum for two seasons, 1897 and 1898, collecting dinosaurs at Como Bluff and Bone Cabin Quarry. In the intervening winter months, he braved the Wyoming cold, living in a cabin near the quarry, protecting it from would-be claim jumpers. Wortman, however, who had a troubled relationship with more than one DVP subordinate in the field, was not entirely satisfied with Menke's personality. "He is handy at many things," Wortman conceded, "a first class photographer and a fairly good bone man, but most awfully rough and uncouth." Probably the greatest strike against him was that he was a friend of Brown, whom Wortman despised. After the 1898 field season, Menke traveled to New York at his own risk and expense, hoping for a more comfortable way to pass the winter months. But, failing to land a remunerative position at the American Museum, he joined Riggs in Chicago in January 1899.[39]

Osborn was aware that when Menke defected to the Field Columbian Museum he possessed a detailed personal knowledge of many promising dinosaur localities in Wyoming—intelligence gathered at the behest of the American Museum, no less. Ever mindful of the competition, and wanting to protect his near monopoly on Jurassic dinosaurs, Osborn worried about what fieldwork, if any, Riggs was intending to do that summer for the Field Columbian Museum. Granger stopped to see Riggs in Chicago on his way out to Wyoming in the spring of 1899. Their discussion of field plans must have been somewhat circumspect, as Granger reported to Osborn that "from what I could gather from his remarks I think [Riggs] is intending to work in the Bridger or Uintah [basins] this season."[40]

Indeed, this is precisely what Riggs had proposed to undertake. But while he and Farrington were waiting for approval and funding for summer fieldwork, invitations to participate in the Fossil Fields Expedition arrived for each of them in the mail.[41] After Higinbotham rejected Riggs's original plan, Farrington worked quickly to submit a revision, much cheaper than the first plan, which would capitalize on the Union

Pacific's generosity. Farrington recommended that Riggs and Menke be sent to Wyoming to collect dinosaurs. He emphasized that the enormous size of the dinosaurs found in southeastern Wyoming would make "impressive and valuable objects for exhibition." The risks inherent in paleontological fieldwork would be mitigated, he implied, by relocating the "showings" that Menke had discovered, in the winter of 1897–98, while keeping watch over the American Museum's Como Bluff quarries. As an offer of free railroad transportation from Chicago, good for sixty days, was included with the invitation, expenses would be mitigated also. Banking on the museum administration's fondness for a good bargain, Farrington recommended that a miserly $350 be appropriated to cover the cost of the expedition and the museum approved.[42] Riggs would have his field season, but the poor precedent set by Farrington's low budget would be a source of regret in future summers.

Riggs and Menke arrived in Wyoming by July 14, 1899, one week ahead of the Fossil Fields Expedition, but six weeks and several months behind the Carnegie and American Museum parties, respectively (see figure 10). Ill-equipped for fieldwork, they stopped in Laramie and purchased basic field supplies from the W. H. Holliday Co., where Riggs opened an account. He bought mining shovels, picks and pick handles, an axe, a brush, a one-hundred-pound sack of cement plaster, another one-hundred-pound sack of plaster of Paris, wrapping paper, burlap, and a half dozen balls of twine. Gum arabic, which was sometimes painted on friable specimens to harden them up, was not available locally and had to be ordered later from Chicago. Picks, shovels, and brushes were used for removing overburden and exposing the fossils. Small specimens were wrapped in cheap brown paper and closed with twine, while large, fragile specimens were encased in rigid, protective casts made from strips of burlap soaked in wet plaster and spread like papier-mâché over the bones.[43] Riggs learned this latter technique, perfected by the DVP parties at Como Bluff and Bone Cabin Quarry during the previous two field seasons, directly from Menke. They forwarded their supplies by railroad to the town of Medicine Bow.

One of a few small settlements strung together by the railroad and serving a patchwork of distant ranches, Medicine Bow (or The Bow, as it was called locally) in 1899 was a rough frontier town later made famous in Owen Wister's 1902 novel *The Virginian*. It is located in a broad, sediment-filled basin surrounded by forested mountains. Skirting the mountains at the basin's edge and along many of the topographical high spots within the basin, long stretches of terrestrial sandstone and shale are exposed at the surface. These dinosaur-bearing strata are now referred

Figure 10. Elmer Samuel Riggs leads the Field Columbian Museum dress parade near Wyoming's Freezeout Mountains in 1899. Courtesy of the Field Museum, Negative #CSGEO3835.

to the Morrison Formation (Late Jurassic). In 1899, Medicine Bow was overrun by railroad workers who were regrading and straightening the Union Pacific tracks. Paleontologists also passed through in great numbers, arriving by train, buying up the town's groceries and hardware, and then departing for the fossil-bearing formations to the north. Because The Bow was the closest railroad town to the fossil beds, many used it as their local address and base of supplies.

Several fossil hunters recorded their grim impressions of the town. Holland, for instance, was particularly struck by the desolation. He remembered a depot, a water tank, a roundhouse, three saloons, about twenty houses, and abundant stockyards. Many were also impressed by the violence in town. Holland's landlady warned him ominously: "We have no law here." Arthur Coggeshall, Holland's new preparator and collector, reported "nightly shooting scrapes" in Medicine Bow between railroad workers and local cowboys. Granger noted the poor accommodations, claiming that "Medicine Bow . . . is a little bit the worst town I have ever seen . . . I would advise no one to spend a night here if he can help it." Ermine Cowles Case might have benefited from this advice. He arrived at The Bow late one night to join the University of Kansas party. Seeking out the only establishment in town that served meals and rented rooms, he found a bloody mattress in the street out front—apparently a man had shot himself the night before. Surprisingly, he was even more put off by the vermin and went to sleep in the railroad section house. There he found another man's valise, which he promptly placed outside, and then slept soundly behind a locked door.[44]

Although romantic notions of the Wild West have probably colored some of these memories, Medicine Bow could certainly be a violent place. In the late nineteenth century, The Bow was an outpost of civilization in a vast wilderness of sagebrush and bare stone, ideal both for finding fossils and for lying low. The Union Pacific Railroad provided easy access by ferrying capital and raw materials between the underdeveloped West and the urban East. These circumstances were as good for banditry as they were for paleontology, both of which flourished in such places, sometimes following strangely parallel courses. On June 2, 1899, for instance, Butch Cassidy's gang, the Wild Bunch, held up a Union Pacific train and dynamited its safe only a few miles from the DVP quarries at Como Bluff. Cassidy allegedly planned the robbery. While the gang was making its getaway, souvenir hunters scooped up relics of the infamous heist, now known as the Wilcox Train Robbery. Holland accepted a piece of the safe from his landlady, assuring her that it would be added to the Carnegie Museum collections. A few years later, with Pinkerton's detectives hot on their trail, Butch and the Sundance Kid fled the country, relocating to a lonely ranch in central Patagonia in 1901. After a few years of quiet respectability they reverted to a life of crime. Their reputations as dangerous gringo bandits made trouble for their fellow countrymen traveling in Patagonia, including a number of paleontologists who went there to plunder bedrock rather than banks. They were likely lured to remote Patagonia by coming across one of three articles that appeared in *National Geographic* around the turn of the century. The author of these articles was Princeton paleontologist J. B. Hatcher.[45]

Meanwhile, a mob of college geologists and their student assistants had been assembling on the campus of the University of Wyoming. After a suitable welcoming ceremony and many speeches, the main body of the Fossil Fields Expedition departed Laramie on the morning of July 21, 1899, heading northwest across the Laramie Plains. Riggs and Menke, along with some of the other more experienced participants, chose to operate independently. Following Menke's lead, the Field Columbian Museum party investigated several prospects with varying success. Eventually, they made camp in the Freezeout Mountains (see figure 11). Several other parties, all attracted to Wyoming by the Union Pacific invitation,

Figure 11. Field Columbian Museum camp scene in southeastern Wyoming, with tent and stove, 1899. Courtesy of the Field Museum, Negative #CSGEO3855.

were camped nearby, including a group of University of Kansas students led by Williston. Case, a paleontologist at the Wisconsin State Normal School at Milwaukee and a friend of Riggs from their undergraduate days, joined the KU party. Charles Schuchert, a fossil invertebrate specialist, headed another collecting party from the National Museum. There were also parties from the University of Minnesota, Augustana College, and a multitude of other schools. The University of Wyoming group included Gilmore, Knight's student and assistant. These several groups worked in and around the same area, often sharing resources.[46]

The Field Columbian Museum party enjoyed only moderate success over their first two weeks in the field, collecting isolated, fragmentary sauropod dinosaur material from at least three quarries, including a string of eighteen articulated *Diplodocus* caudal vertebrae found near Dyer's Ranch, at the base of the Freezeouts.[47] On July 29, Knight's Fossil Fields Expedition arrived with more than one hundred new collectors, plus their teamsters and cooks. They made camp, then scattered among the Jurassic outcrops, swinging pickaxes and collectively amassing several tons of dinosaur bones. They collected furiously for the better part of two days and then they were gone. Neither Riggs nor Menke seem to have recorded any reaction. Possibly Riggs missed all the action, as two of his routine reports to Skiff were written in Medicine Bow and dated July 29, 1899. Riggs's report on August 4 suggests a certain amount of unspecified satisfaction with the work accomplished in the field to date, but makes no mention of Knight's expedition.[48]

FIVE

An Overconfident Start

Like their Field Columbian Museum rivals, the Carnegie Museum party suffered from an inauspicious start to the 1899 field season. Wortman, their leader, arrived in Wyoming with almost boundless optimism. While he was still in Medicine Bow laying in supplies, the local postmaster informed him that his colleague, Reed, who had been fossil hunting in the Freezeout Mountains and along Sheep Creek, had already "succeeded in locating some *very promising prospects.*" Without yet having turned a single spade of earth, he wrote to Holland, confidently predicting that a railroad carload of fossil bones would be ready for shipment at a "comparatively early" date.[1] He then entered the field jubilant, with Reed and Arthur Coggeshall in tow.

As junior-most member of the Carnegie Museum party, many of the most disagreeable duties fell to Coggeshall, including the repeated handling of twenty hundred-pound sacks of plaster that Wortman, confident of the necessity, had purchased in Medicine Bow. At Holland's ramshackle bridge, for example, Wortman and Reed led the horses and then pushed the lightened wagon over the river, while Coggeshall carried the plaster across one sack at a time.[2]

They spent a week in the Freezeouts, busily chasing down Reed's long list of prospects, all of which proved disappointing. But Wortman's enthusiasm was still undimmed, largely because six new "good looking" prospects had come to light. Wortman ordered more plaster.[3] Reed, meanwhile, was also deferring success in his own report to Holland, writing, "We are doing fairly well but expect to do better in the near future. I think we are going to have a very prosperous summer."[4]

Work commenced immediately on a few of the most promising prospects, including a partial skeleton of *Brontosaurus* and another of *Diplodocus*. Wortman's mid-June letter to Holland reported indifferent prospecting results along Troublesome and Difficulty creeks. His mood, nevertheless, remained rosy. Regarding their work thus far, he wrote:

> Of the Brontosaur we already have out I should judge about 2 wagon loads and as soon as Bill [Reed] and I get at it we will snatch things out at a lively rate. . . . We have been hustling from daylight until dark ever since we struck the field and I feel well satisfied that you have made no mistake in employing Reed. He is earnest energetic and thoroughly interested in the welfare of our undertaking. He is an agreeable fellow to be with in camp and everything goes on pleasantly and smoothly. I have great expectations for the future of our department.[5]

With respect to the competition, he was positively triumphant:

> We have labeled our quarries and I do not apprehend any difficulty in holding them until we can get around to work them out. . . . We made a wise move in getting here early. There are many parties here already and they are getting pretty short on bones. The quarry of the American Museum is turning out badly and they are looking for other locations. Methinks they will get little. They lack experience.[6]

The tenor of Wortman's reports changed dramatically with his next installment, on June 28, 1899. First, he had some very bad news for Holland about their various prospects, every one of which had failed with "astonishing regularity." Each prospect in succession yielded only a few bones and then played out completely within a few feet. Wortman was growing increasingly frustrated. He wanted to make a final sweep of the area to the north, along Sheep Creek, where Reed had reported some interesting finds. His faith in Reed's prospects, however, was on the decline. Itching for success, he was suddenly eager to abandon the area altogether in favor of some other Jurassic exposures near the Black Hills. "I am more than anxious to get some good Dinosaurs this season," he declared, "and will not stop until we find them." To make matters worse, Wortman had similarly bad news about Holland's coveted prize. Wortman had taken the trouble to investigate Reed's initial find, the one that caught Carnegie's eye. It was a total bust. Only a single fragment of that celebrated dinosaur ever found its way to the Carnegie Museum—the piece collected by Holland himself in late April or early May. According to

Wortman, Wilbur Knight had removed Holland's posted claim and substituted one of his own bearing the date of May 11, 1899. He had then cut a trough through the prospect, fifteen feet long by ten feet wide, cleared out the bones, and shipped one wagonload to Laramie. Other bones and fragments were left piled up at the quarry. "There is evidence," Wortman claimed, "that [the fossils] were gouged out in the most primitive and unskillful manner."[7]

Holland was outraged by the news. Of course, he had never had any intention of letting the University of Wyoming take possession of this particular specimen, despite his earlier "generosity." But while he and his Laramie lawyer were jockeying for legal control of the dinosaur, Knight and his student assistant, Charles Gilmore, were busy carrying it off. It was a humiliating turn of events. Holland called it a "profound mortification," and responded to Wortman's report with a sweeping condemnation of his Wyoming colleagues, writing, "from the President of the University down I have a very poor opinion of the whole blooming outfit; they apparently do not know how to meet manly men in a manly way, but are as full of little narrow, petty jealousies as an egg is of meat."[8] He was so incensed that it made him oblivious to his own hypocrisy. To Stephen W. Downey, his lawyer, he fumed, "[Reed's dinosaur] is another added to the long list of valuable and interesting specimens which have been hewn into fragments by the hands of incompetent men, leaving nothing to science except a few broken bits of bone which utterly fail to tell the story which would have been told had slow and patient fingers been allowed to do their work."[9] He was disappointed, too, in the other failed prospects, and approved of Wortman's plan to move the party to the Black Hills, provided the Sheep Creek locality should prove as overgrazed as the Freezeouts.

Bone Cabin Quarry . . . and Beyond

At Bone Cabin Quarry, Pete Kaisen had done most of his winter stripping on the south and west sides of the quarry. Unfortunately, a full month of digging by the DVP party along the western strip turned up nothing good—the bone layer pinched out, and work on that quarter was abandoned in early June 1899. The late spring weather was still very pleasant for working, and the party logged approximately ten hours at the quarry each day. Bones continued to appear on the north side in fairly good numbers. The southeast corner also contained many bones, but at a depth that made taking them out very difficult and prohibitively expensive. By the first of June, Granger, Osborn's new field foreman, was convinced that his party would be able to work out the entire bone-bearing layer by fall. So, in accordance with Osborn's imperative, he decided to make a reconnaissance with A. Thomson, the teamster and cook, to locate additional prospects. Kaisen was placed in charge of the party in Granger's absence.[10]

Granger and Thomson left camp in the first few days of June with a light outfit. For a week or more they searched the nearby Jurassic exposures, first north of Sheep Creek, then in the Freezeout Mountains. They located two or three dinosaur prospects on this trip, but nothing of special significance. They also found a good locality for marine reptiles and another for fossil invertebrates, both near Sheep Creek. In the Freezeouts, Granger happened upon Wortman's party, which was then at work on their *Brontosaurus* prospect. At this time, Wortman still believed his specimen was a fairly good one, and he told Granger as much. Granger, in turn, told Wortman about the dim outlook at Bone Cabin Quarry, and also about his fruitless prospecting sweep to the north. (This conversation was the impetus for Wortman's triumphant letter to Holland of June 18.) Finally, they promised to respect each other's prospects, and then they parted ways.[11]

Granger decided to divide his force. Sometime before June 18, probably on his return journey from the Freezeouts, he established a branch camp near Nine Mile Crossing on the Little Medicine Bow River. Someone from the DVP party—probably Granger—had found something there the previous summer, but it did not then seem especially promising. Given the easy pickings at Bone Cabin Quarry, they left it undisturbed. Now, with pressure from Osborn mounting, it seemed an opportune time to develop the prospect. More important, Granger had just discussed the possible merits of this discovery with Wortman before anyone had had a chance to see much of it. Wortman might have considered it fair game. So, to make an ironclad claim to the prospect that Wortman would be bound to respect, Granger decided to open it up. Accordingly, he and several members of his party did some exploratory digging at the prospect and found some "very fair" vertebrae of *Brontosaurus* "somewhat broken up, but . . . quite hard and uncrushed." More bones led into the bank. The specimen looked like a keeper, and Granger wrote to Osborn with the good news.[12] Osborn received the news with pleasure and cautious optimism. He replied, "I am glad to hear . . . of your discovery of the big . . . Brontosaur . . ., I shall be very glad to hear whether this specimen represents an individual of the same size as ours, if it does, it is exactly what we need."[13] Plans were already afoot to display the big Como Bluff skeleton, acquired in 1897, at the American Museum in New York, but many bones were still lacking to complete a composite mount. Osborn hoped that Granger's discovery at Nine Mile Quarry would fill in the missing pieces.[14]

William Diller Matthew, recently promoted to associate curator (to replace the departed Wortman), came out from New York to join Granger's field party, arriving about June 22. A promising young geologist and paleontologist with relatively little fossil vertebrate field experience, Osborn nevertheless sent Matthew west to bolster the DVP party in the face of mounting competition. He went to work at the new Nine Mile Quarry with Granger, Kaisen, and (at times) Lull, the aristocratic Massachusetts professor. After ten days of digging, he wrote to Osborn with a detailed and very favorable report of the new dinosaur:

> The specimen is a *Brontosaur,* and corresponds very closely with the one from Aurora [Como Bluff] got in 1897. Parts preserved are all of a single individual. The best part is a connected series of four cervical and nine dorsal vertebrae with the left ribs in place and all or nearly all the right ribs displaced. There are also about a dozen caudals, one cervical, both coracoids and pubes, femur, parts of sacrum and ilia and various fragments of vertebrae ribs and limb bones exposed. There may be considerably more of the animal not yet exposed, but I do not look to find any connected series of vertebrae or completed limbs. One tooth is the only recognized fragment from the skull as yet.[15]

Matthew, who had a good nose for scientific problems, noted the peculiar preservation of the Nine Mile specimen, and speculated about the possible ecological implications:

> Carnivore *teeth* occur with the animal, and the floor on which it rests is littered with small unrecognizable fragments of bone. I suspect that in this and in most cases, the fragmentary condition of the skeleton is due not to currents or weathering, but to the carnivorous dinosaurs—which in this case seem to have left the bulky body intact but to have torn up and destroyed the extremities. The ilium and sacrum are broken up and other limb bones damaged, while the delicate vertebrae and ribs are perfect. The bone is very fine and the matrix excellent. Ends of limb bones and other cartilage-covered surfaces a little soft, but all other parts extremely hard, black and uncrushed.[16]

But the great value of Granger's discovery was as a display specimen. Matthew knew this was Osborn's fondest goal for the 1899 field season and he closed his letter with an optimistic appraisal of the dinosaur's exhibit potential:

This specimen assures us of the back and ribs for the mounted skeleton, and we may get the limbs under the bank, as I have seen no fragments of them. Cervicals I am afraid we shall have to reconstruct if at all from fragments of which I have already taken up a number. Sacrum may be reconstructible if we get more pieces of it. We are saving all recognizable fragments of importance as they may be usable to reconstruct some of the bones.[17]

In the last days of June, Wortman dropped in at the quarry for a visit with his DVP rivals. He was on his way to Medicine Bow from the Freezeouts, where his several Carnegie Museum prospects had all recently played out, and he shared his bad news with Matthew. They also discussed the itinerary of the Fossil Fields Expedition. Neither paleontologist harbored very kindly feelings for their fossil-hunting rivals. Matthew reported to Osborn that "[n]o-one seems to know much about [Knight's] great aggregation, as to where they will go or what they will do. Dr. Wortman thought they would go into the Big Horn [Basin], find next to nothing, and return disgusted. I hope it may be so." Granger suspected that Wortman's social call was, in actual fact, a prospecting trip aimed at the Nine Mile Crossing locality. "The Doctor may have come up to see what there was [at our prospect] but he did not let out anything of the sort if it was so."[18] Osborn was very pleased with the news, writing to Matthew that "Granger is to be greatly congratulated upon this find and we are fortunate we were not anticipated by our friend the Dr."[19] To Granger he wrote, "I am very much pleased with your discovery of [the] Brontosaur skeleton . . . if it is of the right size, it is a great hit, in fact, the very greatest you could have made. I shall look forward with great interest to seeing it, and in fact I am very anxious to see the quarry and get a better idea of the field."[20] Osborn, it seems, was on his way west to see the show for himself.

Hoping for Better Luck

On the evening of July 2, soon after his visit with the DVP party, Wortman and his Carnegie Museum colleagues established a camp at a grassy, well-watered spot on Sheep Creek, less than ten miles distant from Bone Cabin Quarry. From there, they could scout the surrounding countryside for more deposits of bone. The following morning, they wandered up the creek in the direction of a promising Jurassic outcrop. Reed, who had searched this area in May and marked a few prospects with rock cairns, probably acted as guide. First they found bony fragments strewn across the surface. Next, Wortman spied a row of vertebrae buried in the clay and partially hidden by sagebrush. Reed found another prospect nearby, even larger than the first.[21] They did some preliminary digging, trying to gauge the quality and exhibit potential of their discoveries. Wortman soon sent the good news to Director Holland, who was anxious for something with which to appease his patron, Mr. Carnegie. "I am very happy to report some good luck at last," Wortman beamed.

> We have two good prospects in sight. . . . The best one is a small Brontosaur of which femur, many vertebrae, pelvic and other limb bones are already in sight. The bones are in fine preservation and the prospect looks better every day. The second is a *very large* Brontosaur of which there are a large number of bones in sight with chances very favorable for many more. . . . It will make a superb mount.[22]

The Carnegie Museum party settled in at Sheep Creek for a long dig. Plaster was recalled from storage at the T. B. Ranch in the Freezeouts. Other supplies were laid in. Wortman ordered lumber and the party "built a flat-roofed, board-and-bat shack with a cook tent attached, pitched three big canvas wall tents for sleeping and a fourth for eating, and dug a well near

Figure 12. Camp Carnegie in 1899. © Carnegie Museum of Natural History.

the creek to ensure a good water supply."[23] They dubbed their new digs Camp Carnegie (see figure 12).

In time, they exposed more and more of their skeletons. It soon became clear that the "small Brontosaur" prospect was actually a *Diplodocus*, very well preserved. The "very large Brontosaur," on the other hand, was less complete, and far more friable than the other specimen. Taking this specimen out would require greater care and effort. Still, by July 19 Wortman was ecstatic. In a letter to Holland he bragged that

> [w]e have some mighty fine bones and a whole lot of them. I am almost afraid to say what I think we have but from present indications we will put in the best season I have ever had in the Bone Digging business. We have "a mighty good stagger" at a skeleton of *Diplodocus* which is a "rare bird indeed" and others almost too numerous to mention.[24]

Late in July 1899, Holland came out to see the specimen in person, bringing some young men from Pittsburgh, including George Mellor and Ira

Figure 13. Director William J. Holland (left, in armchair) visits Camp Carnegie in 1899 and takes a meal. Jacob Wortman is seated to Holland's left. © Carnegie Museum of Natural History.

Figure 14. The DVP field party visits their Carnegie Museum rivals at the Sheep Creek *Diplodocus* Quarry in 1899. From left to right: William Harlow Reed, Albert "Bill" Thomson, William Jacob Holland, Henry Fairfield Osborn, William Diller Matthew, Walter Granger, Jacob Wortman (squatting in foreground), and Richard Swann Lull. © Carnegie Museum of Natural History.

Schallenberger, to help with the labor and see the West. In Laramie, Holland gave a speech to the geologists assembled for the Fossil Fields Expedition. Then he and his associates headed for Medicine Bow. There, at the depot, Holland and Osborn met for the first time. Then Holland and his party loaded up with supplies and pulled for the Carnegie quarries at Sheep Creek. Holland was very happy with the novelty of camp life (see figure 13). He toured Sheep Creek Quarry with his small entourage, and posed for pictures with his rival Osborn (and his DVP party), who stopped by for a visit (see figures 14 and 15). Soon, however, he was stricken with a dangerous case of appendicitis. He raced back to Pittsburgh, in agony.[25]

Osborn Comes Calling

The American Museum had the largest and most expensive outfit in the field that summer and their expectations for success were correspondingly high. With Bone Cabin Quarry still yielding good returns, they were

Figure 15. Another group photograph showing the DVP field party and their Carnegie Museum counterparts. Standing from left to right: William Diller Matthew, Richard Swann Lull, Henry Fairfield Osborn, William Jacob Holland, Jacob Wortman, and William Harlow Reed (with black dog). Seated left to right: Walter Granger (with white dog), George Mellor, Ira Schallenberger, and Arthur Coggeshall. The man seated in the wagon is probably Holland's teamster. © Carnegie Museum of Natural History.

burdened with a great deal of very heavy labor. Meanwhile, Osborn continued to pressure Granger, the field foreman, to find and stake claims to additional prospects, hoping especially to collect a single, complete, and well-preserved sauropod dinosaur suitable for mounting. The difficulty of Granger's twofold mission was exacerbated by the awkwardness of his new leadership role within the party. Osborn was somewhat solicitous of Granger's troubling position, and he made a deliberate effort to see that Granger's influence over the staff was made as strong as possible. For example, general instructions, salaries, and promises of winter employment were all funneled from Osborn to the field crew through Granger. The DVP party was enjoying a very good field season thus far, yet when news of Wortman's Sheep Creek *Diplodocus* discovery reached them, they were thoroughly disheartened by their rival's success.[26]

In mid-July, Osborn arrived in Wyoming for a VIP tour of the American Museum's Jurassic dinosaur localities (see figure 16). Granger,

Matthew, and Lull shepherded him around the quarries for several days. Although Granger was nominal head of the DVP party, Osborn's visit left little doubt about who was really in charge. Fresh from a chance encounter with Holland in Medicine Bow, where he endured the news of the recent Carnegie Museum field triumphs and tendered his grudging congratulations, Osborn arrived at Bone Cabin Quarry eager to command and desperate for success. In a letter to his wife, written on July 21, 1899, Osborn described the situation in gloomy terms. "I reached here yesterday evening after an interesting ride across the plains, and my arrival is timely for I find matters somewhat disorganized," he related. "We have just taken out a very fine specimen, but have no more in sight, while Dr. Wortman after two months of very bad luck has made a ten strike, finding an unusually fine skeleton, just what we needed, in fact. I shall put revived life into the party and thoroughly reorganize their work."[27]

But things were not as bad as Osborn described and his stay in camp was brief. With Osborn's departure, work at the quarries settled back into a comfortable routine. Granger sent a polite thank-you letter to Osborn,

Figure 16. Henry Fairfield Osborn visits the DVP field party at Bone Cabin Quarry, 1899. From left to right: Frederick A. Schneider, Walter Granger (seated), Richard Swann Lull, Osborn (seated with dog), Albert Thomson, William Diller Matthew (seated with Osborn), and Peter Kaisen. Image #17906. American Museum of Natural History Library.

then on vacation in Colorado, claiming that "[w]e all enjoyed your visit very much and were sorry to have you go so soon." Adam Hermann, chief preparator for the DVP, arrived for a visit soon after Osborn left and stayed just one week. Camp life was a "trifle too rough for him."[28]

The work of stripping and excavating continued to net impressive returns at Bone Cabin Quarry, much better, in fact, than Granger had expected. At the southern end of the quarry, sufficient quality fossil material appeared to justify a full summer's work for the entire field crew. Kaisen found an interesting *Stegosaurus* prospect; another collector found a good *Morosaurus*. Results at the northern end of the quarry were even better. Diggers discovered a new outcropping there that led to a pocket containing some of the finest material yet discovered at the quarry. Granger was now confident that the fossils collected that summer would fill two railcars.[29] Matthew believed that there was ample material remaining in the quarry to work it for another full season.[30]

Finding additional Jurassic exposures was absolutely crucial to Osborn's peace of mind, however. So Granger continued to prospect the immediate area at every opportunity, but he never found anything particularly promising. Matthew also led an extended three-week prospecting trip. Many years later, he described the travails of fieldwork vividly:

> The day's work consisted in "looking out" as much as one could cover of the rock formations exposed in canyons and gullies, prowling over the weathered slopes, and climbing along the steeper cliffs, watching always for the peculiar colors and forms of weathered bone fragments, following up every trail of fragments to its source, and prospecting cautiously with a light pick or digging chisel to see what, if anything, is left in the rock. Such a prospect came often as a blessed relief after hours of climbing and scrambling had reduced one to a state of staggering weariness. If it was a valuable find, it might mean some hours or days of work to prospect and collect. More often . . . it was a minor find . . . and still more frequently nothing worth while. . . . Even at that it was an excuse for a little rest. . . . Toward sunset one must get back to camp, tired and discouraged by an unsuccessful day, or comparatively fresh and cheerful with a new discovery to report. . . . Then supper, a pipe, and to bed, and the same routine repeated the next day and the next.[31]

Matthew's expedition was also unsuccessful. Lull, like Matthew, had aspirations other than being a paleontological field hand; he accompanied the expedition to the Rattlesnake Mountains, west of Casper, together with Thomson, their teamster. They returned to Bone Cabin Quarry in early September with measured geological sections, rock samples, and a fine set of photographs, but no new dinosaur prospects.[32] News of these failures did nothing to mollify Osborn. Anxious to find more and better deposits, he decided to cast his net wider.

Sheep Creek Quarry

Back in Pittsburgh, Holland survived an emergency appendectomy and faced several long weeks of painful recovery. Although bedridden and packed in ice, he fired a sustained volley of unhelpful suggestions in letters to his paleontologist, Wortman, who was still toiling in the field. Holland wanted a perfect skeleton and he knew only just enough about vertebrate paleontology to make a real nuisance of himself:

> I am awfully glad to hear of the discovery of the scapula and the coracoid of the diplodocus. Yes, my dear fellow, the rest is all there, and do not you forget it. Dig out to the side and to the front, and you will find those little bones: they have not trailed very far away. There was not very much current in that pond where old Diplodocus lay down and died, but just enough to carry the legs to the tips of the toes out a little from the rest of the carcass.[33]

And again,

> I cannot see why one femur should be preserved, and so many of the ribs perfectly, a coracoid and scapula, and that at the same time the feet and legs should be wanting. These bones must be near by, and while it will perhaps involve considerable expense to get them, being scattered out and beyond the skeleton, I am quite sure they are there and that getting down as you say, to the bone level and working steadily, we are almost certain to strike it, and it is worth while expending a good deal of time and money in this instance to secure the whole thing.[34]

Wortman was a veteran of many fossil-hunting campaigns. Condescending and interfering instructions from a neophyte like Holland, however well intentioned, must have sorely tried his patience.

The work of excavating the two Sheep Creek specimens, meanwhile, was going very smoothly, despite several social calls by collectors from rival parties working nearby. "Our success in fact has been simply phenomenal and will not I fear be duplicated. I was talking with some of the Amer. Mus. party and they cannot get over the wonderful 'luck' as they call it that has attended our efforts," Wortman wrote to Holland.[35]

Figure 17. Resting in front of the board-and-bat shack at Camp Carnegie are, left to right, Paul Miller, Jacob Wortman, William Harlow Reed, and Reed's son Willie. Courtesy of the Carnegie Museum of Natural History, Pittsburgh, Pennsylvania.

Coggeshall and Willie Reed, the elder Reed's son, recently hired as inexpensive summer labor, worked on the *Brontosaurus*. This specimen was taken up and ready for shipment by the end of August. Wortman and the senior Reed worked on the *Diplodocus*. A great deal of this skeleton was found beautifully articulated, although many bones were missing. The Carnegie Museum party, supplemented by another locally hired laborer named Paul Miller, stripped away the overburden across a wide swath of ground searching for more bones. The pressure applied by Holland to find the entire skeleton was keenly felt by the entire party. All that turned up, however, was a fragment of the posterior end of the lower jaws. By mid-September, Wortman was tiring of the work, although still hopeful of recovering more of the skeleton the following season (see figure 17).[36]

Late Summer Success for Riggs and Menke

Late in August 1899, Riggs and Menke of the Field Columbian Museum inherited a quarry discovered first and worked profitably by the University of Kansas party and abandoned when their beloved old professor S. W. Williston returned to Lawrence. This locality, designated Quarry Six, was situated well up a steep hillside, about two hundred yards above the nearest space level enough to load and pull a horse-drawn wagon. With much labor, and the help of neighboring parties, Williston's student expedition had taken out "a large quantity of unusually good material," and Riggs was hopeful of continued productivity.[37] By August 27 his report to Skiff sounded more confident of success than had some of his earlier letters:

> We have just begun work on by far the most promising prospect of the season. A facing of twenty feet shows large bones throughout, and in a good state of preservation. The indications are that it will prove to be one of those large bone quarries such as the American Museum has been working during the past two years, and toward which the Carnegie Museum has been bending its energies this season.[38]

Later, after five more days of hard labor, Riggs was even more encouraged. Using dynamite and a horse-drawn scraper, the party had made a stripping seven by twelve feet in area and from three to six feet in depth through "the finest sort of indurated clay" down to a promising layer of dinosaur bones. They found pelvic bones, vertebrae, a very large scapula, and ribs "galore." The bones appeared to lead promisingly into the bank. In fact, a small tunnel had to be dug to retrieve the scapula (see figure 18). Time was running short, however. Their original rail passes were valid only through mid-September, but Riggs had sufficient funds from his original appropriation to remain in the field another two weeks. Accordingly, Riggs wrote to Farrington, curator of geology, enclosing the passes, hoping to get them extended until the first of October. In the letter he alternately pleaded and warned: "To leave those bones—the finest I have ever seen—is not to be thought of. We can not get them out in the time left us. Wortman and Reed have their eye on the place and would doubtless take possession as soon as we left."[39]

Other collectors also had their eyes on the Field Columbian Museum prospect. After seeing his charges safely back to Laramie, Knight returned to the Freezeout Mountains for a publicity visit in early September. Quarry Six was then yielding its finest returns. Riggs and Menke had just cleaned

Figure 18. A member of the Field Columbian Museum party, probably Elmer Riggs, tunnels for specimens at Quarry Six in the Freezeout Mountains. Courtesy of the Field Museum, Negative #CSGEO3830.

up their quarry to take a series of in situ photographs before taking out the bones when Knight's small party arrived. According to Riggs,

> Just as we had this display cleaned up for photographing, Knight came along with a party of seven geologists and newspaper men. Among them was Lester F. Ward, Dr. Schuchert and a correspondent of the N. Y. Tribune, Harper's, Frank Leslie's, etc. Knight's man, [Charles Gilmore], had little in sight so our quarry made all of the noise. We cornered the magazine man, gave him data, and promised photographs and finally took him down to dinner. As a result we are to have prominent mention in the Tribune, and later, an illustrated magazine write up. Ours were the only quarry views he could get so we have a corner in that way.[40]

Riggs got the extension he wanted, and worked hard to make the best use of his remaining time and money. The museum remitted Riggs's and Menke's salaries directly, and rail transportation was provided without charge, so Riggs was free to spend his appropriation locally on supplies and extra labor. With so much digging and packing to do, Riggs hired a local laborer, G. W. Patten, for nineteen day's work at about $1.00 per day. He also hired Z. H. Fales to provide meals for four weeks, hay for packing, and a team and wagon for hauling specimens from the Freeze-out Mountains to Medicine Bow.[41] In this way, the modest appropriation was rapidly depleted.

In early September, when the quarry seemed most promising, Riggs contemplated a possible return to Wyoming for the following field season. He reported to Farrington that "if the quarry holds up I shall put a mining claim upon it for another year." Less than a week later, however, the bubble had burst, and Riggs regretfully reported that "evidence of limits to the deposit begin to appear." By October 6, 1899, Riggs and Menke had quit the field, paid their debts, and were back in Chicago.[42]

The Carnegie party quit a little earlier. Wortman packed and shipped his specimens on September 27, and left the field soon thereafter. He planned to stop in Cheyenne to file a number of complicated claims on the land near Sheep Creek where they had collected their finest material, intending to return the following season. Coggeshall had already returned to the museum to begin the tedious process of preparing the bones for study and exhibition.[43]

Likewise, Granger and the American Museum party closed up shop for the season at Bone Cabin Quarry on September 25, reburying exposed bones to protect them from frost—and theft. His spoils filled ninety boxes on two freight cars. He returned to New York in the first week of October. Confident that another field season, at least, could be spent profitably in southeastern Wyoming, Granger left Bone Cabin Quarry fully expecting to return the following year.[44]

The Field Columbian Museum's dig at Quarry Six produced essentially the same Jurassic dinosaur fauna as the famously productive Bone Cabin Quarry, although it was not nearly as prolific. At the close of the field season, Riggs shipped more than five tons of material from Medicine Bow. The final tally numbered seventy-five dinosaur bones, including elements of the rear limb, shoulder, spine, and pelvis of *Brontosaurus*; the shoulder and spine of *Morosaurus*; the rear limb, shoulder, pelvis, eight ribs, and twenty-five tail vertebrae (eight in a series) of *Diplodocus*—all gigantic sauropod dinosaurs with long tails, long necks capped by small heads, and elephantine torsos and limbs. They also collected the rear limb of *Camptosaurus*, a duck-billed plant-eater, and the pelvis, rear limb, and foot of *Creosaurus*, one of the top predators of the American Jurassic. All of this material was new to the Field Columbian Museum collections. At least one record notes that the quarry yielded a single dorsal plate of the rare genus *Stegosaurus*, also, but this specimen was subsequently discarded. Riggs and Menke also collected ten Jurassic invertebrates and five recent mammal skulls (probably antelope, which they hunted for fresh meat). Finally, Menke, an exceptional field photographer, made ninety-nine exposures of quarry views, stratigraphy, and landscapes.[45] In aggregate, this was a very good return for a single field season, especially given the time, money, and manpower constraints under which the Field Columbian Museum party labored.

The American Museum party fared even better, however. In the summer of 1899, Granger and his field crew collected 131 dinosaur specimens from Bone Cabin Quarry, including some taxa new to the American Museum collections. Additionally, at Nine Mile Quarry nearby, the DVP party collected a large, articulated skeleton of *Brontosaurus*, approximately 50 percent complete. They also collected a fossil crocodile and an ichthyosaur (from the marine Jurassic beds). All of the Bone Cabin Quarry specimens were fairly fragmentary, but the *Brontosaurus* was reasonably complete. Osborn summarized the summer's results in a late September letter to a *Century Magazine* editor, claiming, "I suppose that our party secured as much material as all the other Museums put together, except the Carnegie Museum, which was very successful."[46] To his Princeton friend W. B. Scott he confided, "We have had a successful seasons work, not brilliant thus far."[47]

The Carnegie Museum party was indeed very successful. At their quarry at Sheep Creek, they collected a single specimen of *Diplodocus* that was more than 50 percent complete. It consisted of most of the spine from the neck back to the twelfth caudal vertebrae, most of the pelvis, the right shoulder bones, right femur, and eighteen ribs. The Carnegie party also collected partial skeletons of the dinosaurs *Brontosaurus*, *Morosaurus*, and *Stegosaurus*, as well as some marine reptiles and invertebrate fossils. Holland was less modest than Osborn in summarizing these results: "Of all the parties in the field this summer ours has proved the most successful."[48] And, in an irreverent letter to a friend, he claimed: "We obtained a quantity of material which would have made the mouths of Cope and Marsh water."[49]

Expanding the Search to Colorado

After his tour of the American Museum's Jurassic dinosaur localities, Osborn repaired to the Colorado Rockies for a mountain holiday. But dinosaurs remained his chief preoccupation throughout the summer of 1899. Back at the museum in September, he admitted in a letter to Granger that he was "very much disappointed" that a reconnaissance north of the Bone Cabin Quarry site late in the field season had turned up no promising dinosaur leads, and he resolved to search for other Jurassic exposures elsewhere.[50] Somehow, while vacationing in Colorado, he learned about a possible new dinosaur locality in the southwestern corner of that state.[51]

An informant had alleged, as Osborn related in a letter to his railroad friend E. T. Jeffery, that there was a "large and promising pile of bones about sixty miles [west of] Mancos [Colorado]."[52] But how reliable was this tip? Colorado was alive with rumors of large bones in the late 1890s. Rumors had come to the attention of USGS geologist Charles Whitman Cross, for example, but never with any reliable locality information. Nor, as far as Cross was aware, had any trace of fossil vertebrate remains been collected in the San Juan Basin area of southwestern Colorado. In June 1899, Cross alluded to these rumors when he wrote, "In the main McElmo Valley and in its various side canyons the clays, shales, and sandstones of the upper Gunnison [Morrison] are excellently exposed, and from current reports it appears that vertebrate evidence as to the age of the formation may there be found in abundance."[53] This is precisely where Osborn's tip placed the alleged dinosaur prospect.

At least one of Osborn's informants was Abraham Lincoln Fellows, resident hydrographer for the U.S. Geological Survey. Fellows had once lived in Cortez, Colorado, where he had served as county surveyor. He had information about a deposit of fossil vertebrates in southwestern Colorado and Osborn somehow found out that he had it. Osborn sent a letter of inquiry to Fellows, who, in reply, supplied abundant details about the deposit, providing the names of a number of Cortez residents who could lead Osborn's collector to the fossils.[54]

Osborn passed on this information to one of his young collectors, James Williams Gidley, who was then exploring the Tertiary beds of Texas for fossil horses. Osborn telegraphed on October 11, 1899, and instructed Gidley to head immediately for Mancos, Colorado, the railroad town nearest to the dinosaur prospect. Gidley, then a student at Princeton working as a seasonal collector for the DVP, had been collecting with a University of Kansas colleague named Alban Stewart. Osborn wanted Stewart left behind, most likely to keep the Mancos locality information from spreading to any of Stewart's KU paleontology connections. In his instruction to Gidley, Osborn urged secrecy and discretion:

If you find the locality, which I suppose to be a Dinosaur deposit, you had better investigate carefully and report immediately, where the exploration can be done next Spring, arranging with the nearest ranchman to protect the deposit until that time for a small consideration. . . . It will be well to do some prospecting also. Say absolutely nothing about it which can get into the newspapers.[55]

Gidley arrived in Mancos on October 21.[56] Exploring the area over the next several days, he found several indifferent dinosaurs prospects. His was a reconnaissance mission only, so he collected no fossils on this expedition. For secrecy's sake, no mention of his visit to McElmo Canyon appears in the American Museum's annual report for 1899. An unpublished summary of the expedition's results hints of disappointment: "The fossil reptile bones proved to be of little value."[57] Returning to the East Coast in early November, Gidley was the last of Osborn's collectors to abandon the field.

Aftermath

At the Carnegie Museum, despite their successful 1899 field season, all was not perfectly well with the vertebrate paleontology staff. Wortman, Reed, and Coggeshall returned from Wyoming in mid-October, ready to set up shop in Pittsburgh. They were joined in January by a new chief assistant curator, Olof August Peterson, formerly Wortman's disgruntled assistant at the American Museum and recently returned from the last of the Princeton Patagonian Expeditions. Osborn, smarting from Wortman's departure in May 1899 and needing more experienced field assistants, wanted to lure Peterson back to New York.[58] Getting him to come to Pittsburgh, instead, was another coup for Holland.

Wortman, meanwhile, was feeling restless. Once at the Carnegie Museum and coming into more prolonged contact with Holland, he began to lose confidence in the director's abilities as a scientist. And he deeply resented any interference from Holland in the management of his department, something he rightly felt far better qualified to lead. On January 30, 1900, Wortman showed his superior the proofs of an article he had written for *Science* describing the work of their new Department of Vertebrate Paleontology. According to Holland, there were one or two items in the article that needed to be altered, he claimed, in the interest of maintaining a harmonious work environment. But when Holland "courteously" proposed these changes, Wortman "became very angry; told me to 'go to hell' [and] covered me with uncomplimentary epithets."[59] Holland demanded his resignation on the spot. Exactly what Holland's objections were regarding the contents of Wortman's article is, more than a century later, impossible to determine precisely. Whatever the immediate cause of their flare-up, the real issue in the dissolution of their unequal partnership is that Wortman had a prickly disposition and a troubled work history that made him sensitive to the slightest censure, and Holland was a pushy, domineering egomaniac with no legitimate pedigree in vertebrate paleontology. Theirs was a relationship doomed to extinction.[60]

There was unexpected fallout at the Field Columbian Museum, as well. A *Cosmopolitan Magazine* article about the fossil resources of Wyoming, published in January 1900, did indeed feature Riggs, Menke, and their museum very prominently. At least three of Menke's field photographs (attributed erroneously to F. H. Menke) were reproduced in the article. Three other figures depicted their work at Quarry Six. Several views show men at work, at least one of whom is almost certainly Riggs. The text names them (surnames only) and celebrates their field exploits with some appropriately purple prose. About Riggs's specimens the *Cosmopolitan* writer reasoned: "At the Field Museum's quarry . . . the bones of five genera were exposed. . . . These . . . saurians may have died together in one vast culmination of animal ferocity, for one of the herbivores proves to have a suspiciously short tail."[61] Riggs, Menke, and their many superiors at the museum must have been delighted by the attention and the free publicity.

Unfortunately, the *Cosmopolitan* author took a few too many liberties with his story, and Riggs felt compelled to send a quick correction to *Science*. Riggs explained that "the erroneous impression has gone out in certain quarters that members of [the Field Columbian Museum party] were responsible for some of the [*Cosmopolitan's*] misstatements, especially one which has been interpreted as a reflection upon a man to whom the science of paleontology owes much." His article then provides a very brief history of the discovery and collection of American Jurassic dinosaurs, based in part on data supplied by Williston, the victim of the *Cosmopolitan* piece. The purpose of Riggs's article appears when his history reaches the summer of 1899: "The valuable deposit [Quarry Six] worked out by the Field Columbian Museum party had not previously been passed over by 'a Kansas University professor' as stated by the author of the *Cosmopolitan* article. On the contrary the quarry had been located and worked for some time by the Kansas University men."[62] If it was Williston himself who took exception to the *Cosmopolitan* slander, a conclusion supported by the fact that he supplied the relevant historical content for the article in *Science*, then Riggs was undoubtedly mortified for his beloved former professor. In any case, he spent his thirty-first birthday writing the correction.

SIX

The Monster of All Ages

After their Wyoming expedition, Riggs and Menke returned to Chicago's Field Columbian Museum with a large lot of dinosaur bones locked in rock. Farrington, curator of geology, recommended Menke, who had already been assisting in the museum's Geology Department in one capacity or another since February, for a permanent position as a fossil preparator. Although the need for skilled assistance was great, Skiff delayed Menke's appointment until the first day of the New Year.[1] Riggs and Menke then spent the winter and spring months of 1900 preparing their finest specimens for exhibit by slowly chipping away the stone with a relentless tapping of hammer on chisel. Riggs, who was frustrated by the inefficiency of the work, put fossils on exhibit as quickly as they could be made ready. Large, heavy, and ungainly, most of the Wyoming specimens required new floor cases for display. By the summer, Riggs and Menke had completed a number of individual limb bones, vertebrae, and pelvic bones of a variety of sauropod dinosaurs. A fifteen-foot length of the tail of *Diplodocus* and the mounted left hind limb of *Morosaurus*, including foot and toe bones and portions of the pelvis, were the most impressive of the new installations.[2] These specimens complemented a modest display of Jurassic invertebrate and plant fossils, a few marine reptiles, and a small number of dinosaur fragments, all acquired from the World's Columbian Exposition. Taken together, the museum's exhibits conveyed, at best, a very incomplete idea of the (often) great size and (always) strange forms of Jurassic dinosaurs.

Farrington was dissatisfied with this rudimentary display. Eager to secure sufficient material to mount a complete sauropod dinosaur, he tried to inspire museum administrators with his excitement. In a letter requesting an appropriation for a renewal of summer fieldwork in 1900, he revived his argument from the previous year, insisting that dinosaur fossils were rapidly being depleted by an increasingly competitive field of museum paleontologists. He also made a novel appeal to institutional vanity and patriotism: "[To reconstruct] one of these fossil monsters 70 feet in length and standing 18 feet high . . . would be an achievement such as no other museum in the world has yet attained. . . . [I]n no other country . . . are remains comparable to these in size to be found." Farrington explained that two "very promising" localities beckoned, one on the western slope of the Rockies, on the outskirts of Grand Junction, Colorado, and one near Buffalo, Wyoming. He drafted a rough budget, estimating that a paltry $700 would cover the cost of sending a field party to reconnoiter both places.[3]

Higinbotham approved the request in part only, slashing the original budget.[4] Skiff relayed this news to Farrington, explaining that "the President having in mind the appropriation made last year for this purpose was disinclined to authorize more than $500 for this expedition, presuming that by exercising economy this amount would be sufficient."[5] Farrington was irked, and wrote a detailed but deferential letter to Skiff justifying his earlier estimates. He argued that the previous summer's expedition was a poor model for judging the probable cost of another field season in Wyoming and Colorado. First, because prospecting new territory for fossils would involve considerable traveling with a rented team and wagon. Collectors had worked a familiar field the previous summer, digging primarily in one spot and thereby keeping travel expenses to a minimum. Second, to work more efficiently, the next expedition would require an extra hired hand for cooking and running the camp. Much valuable time for digging had been lost the previous summer in taking care of camp. Farrington warned that one could not reasonably expect to collect a comparable amount of material again for a similar expenditure. But Higinbotham was not convinced. According to Skiff, "It is Mr. Higinbotham's wish that when these funds are exhausted the expedition will return."[6]

No doubt Riggs was discouraged by the museum's penny-pinching.[7] With so small a budget, plans for summer fieldwork had to be scaled back significantly. The Wyoming locality, a dinosaur prospect that Farrington had scouted back in 1898, was dropped from the agenda altogether, probably because the rancher who first discovered the specimen demanded a $100 finder's fee in exchange for permission to excavate.[8] Summer success thus depended entirely on the circumstances of the western Colorado locality, about which almost nothing was known with certainty.

Grand Valley Dinosaurs

In the nineteenth century, dinosaurs were virtually unknown west of the Rocky Mountains, although numberless explorers, pioneer naturalists, and government surveyors had been canvassing the region and investigating its geology since before midcentury. By the late 1890s, the correlation of strata west of the Rockies with their better-known counterparts on the eastern slope was a pervasive research problem among U.S. Geological Survey geologists working in western Colorado. The discovery of characteristic Jurassic vertebrates in these beds would have helped to resolve this problem. Nevertheless, only one dinosaur, *Dystrophaeus viaemalae*, collected from the western slope had been named and described in print, and its stratigraphic position was problematic.[9] Meanwhile, several of Osborn's collectors had scouted parts of the western slope for dinosaurs. J. L. Wortman explored Utah's Montezuma Canyon in 1893 in the company of Richard Wetherill, an amateur archaeologist. O. A. Peterson examined a dinosaur prospect in northwestern Utah in the same summer. In 1899, James Gidley, a Princeton student—Gidley and Riggs overlapped briefly during the latter's fellowship year—investigated a dinosaur prospect in McElmo Canyon, in southwestern Colorado (see chapter 5). All three paleontologists found fragmentary Jurassic dinosaur remains, but none of them publicized their results. Riggs's expedition was headed into a part of the country that any number of experienced field geologists had traversed, but from which no one had described any dinosaur remains in the scientific literature.

It was the railroad that first attracted Riggs's attention to western Colorado.[10] In the spring of 1899, while he was making plans to collect Eocene fossil mammals in northeastern Utah, Riggs had written letters to the mayors of several railroad towns along the Rio Grande and Western Railroad, a line that linked Denver to Salt Lake City. He inquired about an easy trail north from the railroad into Utah's Uintah Basin. (Traditionally, fossil collectors reached this important locality from southwestern Wyoming by way of difficult trails over the Uintah Mountains.)[11] A

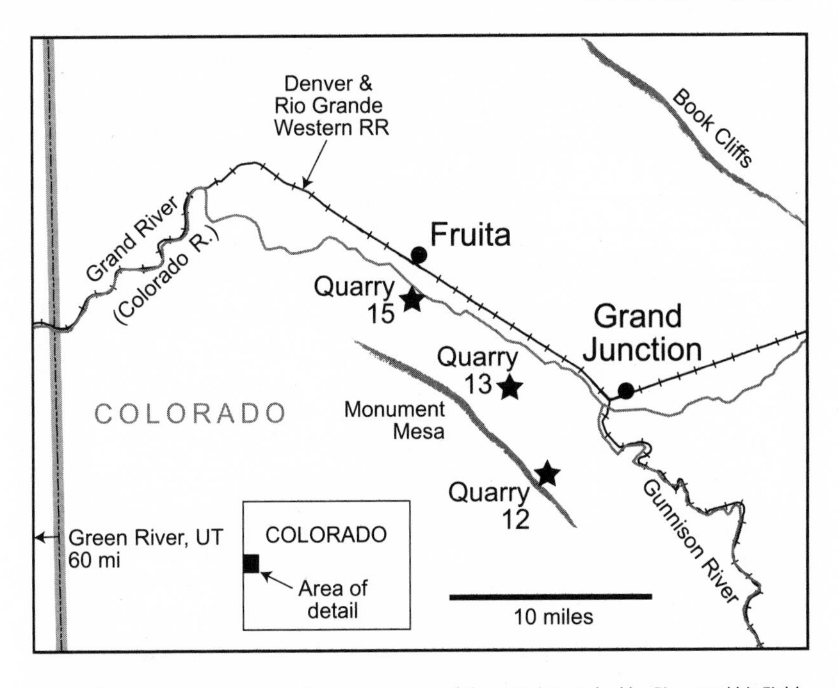

Figure 19. Map of western Colorado showing three of the quarries worked by Riggs and his Field Columbian Museum party in 1900–1901. Map drawn by Cathryn Dowd.

promising reply came from Dr. Stanton Merrill Bradbury, a pioneer dentist and amateur scientist from Grand Junction, Colorado. Located at the confluence of the Gunnison and Grand (later the Colorado) rivers, Grand Junction in 1900 was a small city of thirty-five hundred residents (see figure 19). Bradbury, who played an important role in promoting natural science on the western slope, was one of its most prominent citizens and boosters.

Bradbury's letter was a surprise. Instead of an easy route into Utah, it hinted of a rich and relatively untested collecting ground just west of town and very close to the railroad. Riggs was skeptical at first. As a precaution, he consulted a geological survey report published in 1878 that described the widespread exposures of Mesozoic sediments in this region. Its author declared these beds barren of vertebrate fossils.[12] Bradbury, however, had an assortment of locally collected fossil bones in his possession. Under the circumstances, Riggs was inclined to suspend his skepticism. The classic dinosaur localities of southeastern Wyoming were being culled by a crowd of collectors from rival museums. Aware of the potential significance of finding an entirely new locality, Riggs accepted the judgment of the local authority and decided to scout western Colorado for dinosaurs.[13]

Riggs started late for the field, leaving Chicago by rail on June 16, 1900, and heading for Grand Junction. Menke joined the party again as field assistant and photographer. Riggs hired Victor Hugo Barnett, a younger cousin of his from south-central Indiana, to serve the expedition as cook and camp assistant. Barnett wanted to be a geologist, and he was willing to work for low wages.[14] Arriving on June 20, Riggs and his small party went directly to Bradbury's downtown office, which was decorated with relics belonging to the fledgling Western Colorado Academy of Science. Bradbury showed the collection to his guests, proudly explaining each item's provenance. A small assortment of chocolate-colored dinosaur bones, including a limb bone and a caudal vertebra of *Diplodocus*, was the main attraction. This settled a number of vexing questions regarding the new locality: Riggs could now rest assured that Jurassic dinosaur bones in a good state of preservation had been found in the sediments near Grand Junction. But he still needed to explore the ground himself to determine how much fossil material could be found there. Bradbury, the Academy's president, agreed to guide Riggs to a place where he and his colleagues had collected some of the bones.[15]

The next morning, on rented saddle horses, they followed Bradbury west and south out of town. They crossed the river, which was lined with cottonwoods and flanked by irrigated fields, and passed quickly into the barren, rocky Redlands. Riggs took note of the general features of the landscape while they rode, later recording his observations in a short descriptive paper. He noted a series of benches rising some two thousand feet from the southern bank of the river to the top of the Uncompahgre Plateau. The plateau was composed principally of horizontal layers of red Triassic and pale-colored Jurassic sedimentary rocks, and in many places was surmounted by an erosion-resistant brim of lower Cretaceous Dakota sandstone. The pine-topped heights of Red Mesa to the south, Grand Mesa in the east, and the Little Book Cliffs to the northeast defined the valley's edges. Between the river and the high escarpments there were numerous isolated buttes comprised of Jurassic sandstones, clays, and shales, conspicuous by their brilliant bands tinted green and purple. Occasional thick lenses (a thin bed of limited extent) of cross-bedded sandstone established the terrestrial origin of certain of these beds. A few miles south and west of Grand Junction, in places where the hard Dakota sandstone cap was completely eroded away, the exposed Jurassic beds had weathered into fantastically sculpted badland topography.[16] This was their destination.

They halted near the mouth of No Thoroughfare Canyon, about five miles southwest of town. Bradbury indicated a spot on the saddle between a pair of conjoined hills. Scrambling up the shallow slope, they found weathered fragments of dark fossil bone scattered across the surface. Higher up, dinosaur bones protruded from the undisturbed clay. Ill prepared for collecting, they left this locality empty handed. Riding back to town along the Gunnison River, they found a fossil turtle lodged in a boulder, but it was too heavy to collect. Convinced by the potential of this new locality, Riggs decided to settle in for a longer stay.[17]

The Goat Ranch

The following morning, they bought essential supplies and hired a team and wagon to haul them out to the spot where they had found their first traces of dinosaur bone. They established their first camp at the ruins of an abandoned homestead, a place they called the Goat Ranch. Once settled in, Riggs penned a long letter home to Helen Mosher, a Chicago school teacher he had begun courting the previous spring. In the letter, he described his unusually comfortable field accommodations:

> Friday morning [we] got our effects together and in the afternoon began to pitch camp on a deserted ranch some ten miles southwest of Grand Junction and five miles south of the Grand River. There [is] . . . a fairly good spring which the former resident took pains to wall up and to build a stone spring-house over, with a stone milk trough attached so as to receive the overflow. This formed for us an oasis in the desert of sand covered valleys and bare clay hills about us. The spring and house were soon put in order and a camp planned so cozily that we all claim credit for it.[18]

Riggs continued at length about the campsite, perhaps to make an impression of efficient domesticity on his Chicago sweetheart:

> The stone house 8 x 12 with door in side faces south. Opposite it and some ten feet away we pitched our tent. This left an open passage connecting the two which was soon converted into an enclosed porch by stretching a tarpaulin from roof to roof and down the east end. At the west end near the house we placed the camp stove. The mess chest shelved and mounted upon stakes against the wall between stove and spring-house door serves a[s] cupboard, while its falling door does duty as kitchen table. At the other end and near the eastern wall of this enclosure the loosened door from the spring-house was mounted upon stakes, covered with a black gum cloth and made to do duty as dining table. Inside the spring-house the crates of canned

> goods are stacked against the wall with one side of each opened to afford easy access. The milk trough receives and keeps cool our butter, cooked fruit, etc., while its lower extremity does famous duty for washing photographic plates. An inverted box serves Menke as developing table, while the whole, curtained and used at night[,] forms a first-rate dark room with running water.[19]

No one liked to linger in the tent:

> We really have little use for the tent except for storing personal effects. Everybody sleeps in the open air by preference, and for a sitting room the porch . . . is much pleasanter. Even now Menke is at one end of the improvised dining table writing many letters to his best girls . . . while I with a candle shielded on one side by a box and on the other by a board, am struggling between wind and the flapping canvas at my elbow . . . to write something that may interest a young lady who expects much in the way of letter-writing.[20]

Comfortably situated at the Goat Ranch, Riggs and his party set out to work, but they did not have much luck, at first, locating quality fossils. Each collector worked alone, fanning out from camp in a northerly direction, searching the Jurassic exposures between the river and the sheer face of Red Mesa to the west. They looked carefully for fragments of bone, traced these back to their sources, then dug test holes, hoping to strike a well-preserved specimen buried just below the surface. But five good prospects in succession turned up nothing worth keeping. Meanwhile, the conditions for prospecting were steadily worsening. June rains departed in early July and the daytime temperature soared to over one hundred degrees. Spring flow at the Goat Ranch slowed to a foul-tasting, alkali trickle. Hot wind blowing over bare earth raised noxious clouds of dirt, spawned dust devils, and piled two-foot drifts of sand on their campsite. Frustrated by the fruitless search, and suffocating in the heat and dust, Riggs described his first three weeks in western Colorado as "the most discouraging . . . period that I have known in seven years collecting."[21]

Their luck changed suddenly for the better in July. A sixth prospect, designated Quarry Twelve, yielded two tons of fossil bones from a single sauropod dinosaur, later identified as *Morosaurus*. This skeleton included a fragmentary sacrum, dorsal and cervical vertebrae, a broken scapula and coracoid, a number of broken ribs, and a few isolated teeth, possibly theropod. The weathered fragments of three dorsal vertebrae, found at the surface, led collectors to the specimen.[22] Riggs did not record the location of this quarry. However, Al Look, a local enthusiast for paleontology and an early chronicler of Riggs's Colorado fieldwork, gave the base of the present-day Trail of the Serpent as an approximation.[23]

Their seventh prospect would ultimately prove even better. Menke, exploring independently on the Fourth of July, returned to camp triumphant, boasting that he had found "the biggest thing yet!"[24] Riggs agreed to investigate the prospect the following day. In the morning, Menke guided him three miles north from camp to a low hill capped with sandstone, its steep shale slopes peppered with sandstone boulders. Only a short way up the south slope, he indicated the badly weathered end of a limb bone protruding from the bedrock. Broken fragments of this bone, trailing down the slope like breadcrumbs, had caught Menke's attention and drawn him to the prospect. Riggs made a quick and sober assessment. The specimen was a big one, just as advertised, but it also showed signs of bad preservation. He was interested but unexcited. He wanted to finish up with Quarry Twelve, which was then showing signs of great promise, and return to this specimen later for a second look. Remarkably, Riggs later confessed that his first instinct had been to abandon Menke's prospect as a probable *Brontosaurus*, too commonplace and too poorly preserved to be of any value.[25]

By Thursday, July 26, with Quarry Twelve exhausted and filled in, exploratory work resumed on Menke's prospect, dubbed Quarry Thirteen. A single day's digging, starting at the exposed limb bone, was sufficient to raise Riggs's expectations significantly. The true form of the bone was, at first, a puzzle. The weathered end was a jumble of displaced fragments that fit loosely on the hard, chalcedony stump of bone still planted firmly in the bedrock. Hacking away the hard clay overburden with pickaxes, they dug into the hillside, exposing the bone's crushed and distorted length. Its great size—nearly seven feet long—strongly suggested a sauropod's hind limb. Riggs, suspecting the bone was a femur, snapped a humorous photograph of it lying in situ with Menke, arms folded and wearing a pith helmet, posed alongside for scale (see figure 20). The superlative length of Menke's mystery limb bone had Riggs scratching his head in wonder. Back in Grand Junction that night, grooming himself for a public lecture on Friday, he penned an excited letter to Skiff, announcing that "we have uncovered various parts of a skeleton among which is a thigh-bone (femur) which measures six feet ten inches in length. This is longer by eight inches than any limb-bone, recent or fossil, known to the scientific world. Other parts of the skeleton seem proportionately large."[26]

It was the size of the thing, initially, that fired Riggs's enthusiasm. In letters to Skiff and Farrington, his museum superiors, Riggs reported

Figure 20. Menke poses humerusly with the mystery limb bone at Quarry Thirteen. Courtesy of the Field Museum, Negative #CSGEO3934.

with pride on the size of his specimen.[27] Newspaper articles about the discovery, many based in whole or in part on data supplied by Riggs, all referred explicitly to the dinosaur's great size. "Chicago Has the Largest Land Animal that Ever Lived," read a headline in the *Sunday Times-Herald* (Chicago). "Bones of the Largest Known Animal Found," trumpeted the *Sunday Tribune* (Chicago). The *Boston Journal* dubbed Riggs's dinosaur "The Monster of All Ages."[28] The basis for this claim was a comparison between the length of the new specimen's enormous limb bone and the length of the next largest femur known, which, according to Riggs, was eight inches shorter. Gigantic, nine-foot-long ribs bolstered this claim. Directly or indirectly, Riggs made these figures available to the press, and many reporters incorporated them into their stories about the discovery. The implication was that size does matter in dinosaur paleontology. Most Jurassic dinosaurs were large, but this specimen was especially significant because it was the largest ever discovered. The public read this idea in the newspapers, and the press, apparently, adapted it from statements made by Riggs. The *Sunday Tribune* (Chicago), for example, wrote, "It was when this last thigh bone was laid bare so that it could be measured that the importance of the find was made certain. . . . Professor Riggs' dinosaur is the largest of them all."[29] The *Sunday Times-Herald* (Chicago) expressed the same idea in their fictionalized account of the initial discovery:

> One of the first bones encountered was a mighty rib. They happened to strike it near the middle and dug both ways. Day after day went by without reaching either end, and the increasing enthusiasm and wonder of the eager searchers may be imagined. When the rib was finally uncovered from end to end it was measured with an almost breathless interest. When found to be nine feet five inches long and eight inches wide Professor Riggs realized he had made a great discovery.[30]

Riggs corroborated this dramatic account, in part, by his own admission that the unusually large size of the ribs compelled him to collect this specimen in the first place.[31] What's more, this idea was not contrived strictly for popular consumption. Riggs wrote three papers for his scientific peers describing this dinosaur. Two of them, one published in *Science* and the other in the *American Journal of Science*, emphasized the dinosaur's superlative size.[32]

Returning to camp on Saturday, Riggs and Menke continued to develop the new prospect over the next several weeks. Near the freshly exposed end of the limb bone—later determined to be the proximal end—they found an unmistakable coracoid, a bone from the animal's shoulder girdle. Other bones from the pelvis were fifteen feet away. The broad sacrum, right ilium, and two caudal vertebrae rested upside down, closely associated, and somewhat weathered from partial exposure at the surface. An unbroken series of thoracic vertebrae reached forward from the sacrum, giving false hope that the complete anterior part of the skeleton awaited. More bones came to light as they excavated. They found several long ribs, some braided together, lined up along the length of the spine like driftwood. A well-preserved right femur, almost identical in length to the first limb bone, crossed underneath. But the clay layer in which the skeleton was embedded was gradually thinning forward of the sacrum. At the seventh presacral vertebra the clay pinched out, giving way to a massive bed of sandstone. The coarse-grained, pebbly texture of the sandstone, together with the displacement and uniform orientation of certain bones, suggested a strong current of running water. Riggs concluded that the rest of the skeleton had been carried away and lost forever before the carcass had been entirely covered by sediment. It was a fantastic specimen of heroic size, but it was not nearly complete.[33]

Welcome and Unwelcome Attention

Riggs wrote to his girlfriend with an embellished account of the new dinosaur prospect and she was suitably impressed. She replied, "I am very glad you are having so much success in finding fossils. . . . It is a pity that you were unsuccessful . . . when you first went out, and if I were not sure of an indignant denial I might add a little homesick too, but no doubt good luck has cured all that."[34]

Bradbury was also impressed. He was convinced that dinosaurs would make Grand Junction famous and he did what he could to promote their discovery. To that end, he opened up his office to a select local audience invited to hear Riggs lecture on vertebrate paleontology. The presentation, illustrated by lantern slides projected onto a makeshift canvas screen, took place on a warm Friday evening, July 27, 1900, before a small but attentive crowd. At least one member of the local media attended. Riggs explained the paleobiology of dinosaurs and extinct mammals, even touching on the topic of evolution. He mentioned that Eastern institutions were eager to acquire fossils, the larger the better. He also discussed modern methods of finding specimens, excavating them, and preparing them for exhibition, thus giving his listeners a vivid idea of what he and his Field Columbian Museum colleagues were doing in the Redlands. According to a favorable newspaper review, this was the most interesting part of the lecture. That museums "spared no expense whatever to secure [dinosaurs]" was probably a flourish added by the reviewer, since it most certainly did not apply in Riggs's particular case. Liberal applause, according to the reviewer, followed the talk.[35]

Riggs wanted to cultivate a good relationship with Bradbury, whose position as president of the Western Colorado Academy of Science made him a valuable ally, but Bradbury's agenda with respect to the dinosaur was sometimes in conflict with the interests of the Field Columbian Museum. In his lecture, for example, Riggs was careful to withhold any data that might attract rival collectors to his new locality. Consequently, there was nothing published in the newspaper review pertaining to the enormous specimen then being excavated. Riggs explained his reticence to his girlfriend, writing, "I don't intend to give the shop away until we have cornered it."[36] In a progress report to Director Skiff he was more explicit:

> While I am glad to report our success, I wish to make one reservation. There are half a dozzen [*sic*] parties collecting fossils in the west who would eagerly turn to such a new region if they had wind of it. They must not have opportunity to learn, through the press or otherwise, until the end of the season when we shall have had time to prospect the valley and take the cream of it.[37]

Bradbury, on the other hand, was less restrained. A second article in the *Grand Junction News*, published a week after Riggs's lecture, described the new specimen in tempting terms, celebrating it as one of the largest dinosaurs ever discovered. Worse, the article aimed what was tantamount to an invitation or a challenge at Riggs's competitors: "There is a great deal of vigorous strife between the colleges and museums for [dinosaurs] and beyond a doubt there are vast beds of these and similar petrifactions in the hills south of this city."[38] Riggs clipped this article and enclosed it with a letter to his Chicago sweetheart (she wanted the clippings for her scrapbook), commenting, "I picked up one or two notices from local papers which you may find almost as inaccurate as my letters" (referring to his penchant for exaggeration). Despite Riggs's precautions, news of the new dinosaur locality soon leaked to the national press. A *New York Times* article of August 14, 1900, reported that Riggs had recovered "nearly a perfect skeleton" from the bank of the Gunnison River near Grand Junction.[39]

News of the dinosaur also attracted curious locals to their campsite. Flattered, at first, by the attention, Riggs was an accommodating host. But he was soon overtaxed by his demanding guests. In August, Riggs complained that

> [o]ur solitude has given place to a procession of visitors which has grown more tiresome than the former. My lecture may have been responsible for part of it, but the local newspapers boomed the thing, while hearty, jovial, old Dr. Bradbury, who has a hobby for collecting and who is the patron saint of the local Academy of Science, pushed it along.[40]

On the Sunday after Riggs's lecture, Bradbury brought the school principal and a few other local dignitaries to see the dig. A smattering of

Figure 21. Throngs of Sunday visitors arrive at the Goat Ranch campsite, August 1900. Courtesy of the Field Museum, Negative #CSGEO3920.

teachers and minor government officials dropped in during the following week. Riggs noted that one official, gazing appraisingly at the huge pelvis still lying in situ, remarked, "He's broader across the back than a $200 mule!"[41] On Sunday, August 5, Bradbury organized a picnic and invited local families (see figure 21). Dozens of people made the trip from town on horseback or in wagons or buggies. Some rode their safety bicycles, a wildly popular pastime at the turn of the twentieth century.[42] According to Riggs, "Fifty-odd souls besides ourselves lunched on the sand-stone ledge back of the spring-house. The women were shy of coming in, probably because Victor [Barnett] was baking pies beside the entrance, until a sudden shower came on and then the tent was full of sodden millinery. After lunch all adjourned to the quarry a mile away."[43]

Visitors trickled in throughout the next week, sometimes as many as five or six carriage or wagonloads per day. On Friday, a group of young people came out in the evening and held a moonlight picnic on the rocks, singing college songs late into the night. Riggs was having a good time with his guests, but two weeks of entertaining took a heavy toll. Expecting another large Sunday crowd the following week, Riggs outfitted himself with mattress and pillow and repaired to a favorite nook among the rocks, leaving it to a beleaguered Menke to supervise the visitors. For Riggs, the attention was a mixed blessing:

> I enjoyed having visitors and took pains to explain things to everyone interested, especially people of intelligence, but when the idlers began coming just to have some place to go, and the souvenir fiends began prying off pieces whenever we were not around, it became tiresome. Add to that the fact that they no longer came to camp, but went straight to the quarry which is out of sight from camp . . . and you can see the sort of sentinel duty necessary even on a Sunday. The souvenir hunters mean no harm to be sure, but as one man put it . . . they "would steal the halo off of Christ's crown!" We shall finish the quarry before another Sunday. [Once] the plaster jackets are on they can't harm things much.[44]

Despite the attending inconveniences, Riggs was excited by the popular enthusiasm for dinosaurs. A letter he wrote to museum officials boasted about the spectacle and referred to his western Colorado outpost as a "Museum annex." He reported a peak of forty-seven visitors on the busiest Sunday and more than half that number on some weekdays. Riggs intended these figures to work to the advantage of his upstart program. When he joked about the impossibility of collecting door receipts, for example, he was foreshadowing the crowds of curious Chicagoans who would pay to see the dinosaur exhibited at the museum.[45] Yet if this was the message Riggs wanted to send to Director Skiff and other museum officials, then the administrative difficulties he encountered at work in subsequent years suggests that they never bought into the idea.

By August 17, there was very little left for visitors to see. Once Menke's big dinosaur was fully jacketed in plaster and removed from the ground, operations at Quarry Thirteen were closed for good.[46] The party packed up their gear, replenished their store of oats, and set off on a ten-day prospecting trip on horseback. Although insufficient time remained for any serious excavations, Riggs wanted to determine whether or not a return expedition to this same area would be worth the trouble and expense. Some promising-looking exposures beckoned to the west, so they followed the river, which wrapped around the south and west edges of Grand Junction and headed down toward a small settlement called Fruita. They stayed on the west bank, diligently searching the Jurassic exposures for more fossils. They discovered several new prospects on this quick reconnaissance. Riggs found the incomplete foot of a sauropod dinosaur that he later referred to *Morosaurus*. He also found and excavated a nearly complete forelimb of the same genus. Except for the foot bones, this specimen had been completely weathered out of the soft green shale when it was discovered, and certain parts of it were lost. But the remaining fragments were well enough preserved that the specimen could be restored accurately.[47] Meanwhile, Menke used the opportunity to take a series of dramatic landscape photographs. He also made a small collection of extant reptiles for the museum's Zoology Department, but only what he could fit into a single, formalin-filled jar. Because he made this collection without first obtaining official museum sanction, he had some difficulty recovering the dollar he spent for supplies from the tightfisted administration.[48]

Their late-season reconnaissance yielded some big returns, as well. South of Fruita, on the steep slope of a small hill on the southern bank of the river, one of the collectors, probably Menke, spied a bone protruding from the bedrock. Riggs recognized it as the last cervical vertebra of another large sauropod. The bone was badly weathered, and not much to look at, but some exploratory digging revealed that it articulated with (at least) two well-preserved dorsals. Nearby, a row of ribs preserved in their natural positions, their distal ends broken and exposed in cross-section, promised a partial articulated skeleton lying in wait. With the exception of the exposed cervical and the very tips of the ribs, the bone seemed beautifully preserved. The flint-hard sandstone matrix that protected it from weathering—and that stubbornly resisted the Marsh pick—was the reason. Given the lateness of the season, and the depleted state of the expedition's funding, Riggs had no choice but to abandon the specimen, banking on the opportunity to return and recover it the following summer. Before leaving, Menke wrote his name and the date on a piece of paper, placed it in a bottle, and buried it at the prospect under a pile of stones.[49]

Their prospecting tour at an end, the party returned to Grand Junction on August 27, 1900. Riggs hired a team and wagon to gather up all their specimens and haul them back to town; this took five full days. They rented rooms in town from September 2nd through the 9th and spent that week building wooden crates to freight the fossils safely back to Chicago. Riggs had promised to repeat his lecture at the local high school if he remained in town late enough and he might have made good on his pledge during his weeklong stay in town. On the 10th, with the help of a local laborer, the party loaded their prizes into a boxcar. The aggregate weight of their thirty-eight crates was 12,500 pounds. They left their camp equipment for safekeeping with a local rancher, and then boarded a train for Chicago, arriving September 15. Passengers and cargo all traveled for free, thanks to the enlightened self-interest of the railroads.[50]

A Near Miss

Any regret Riggs felt about leaving a prize specimen and departing the field was tempered by the expectation of returning to western Colorado for a second season. The promise of an as-yet-uncollected dinosaur would make an effective lure for hooking additional funding from the museum, he felt, especially given the newspaper hyperbole inspired by the gigantic dinosaur recovered from Quarry Thirteen. Media attention, however, was a mixed blessing. With a choice specimen still in the ground, Riggs was anxious to avoid attracting any rivals to his new Colorado locality. Yet Bradbury, for his part, was more interested in promoting the discovery than in protecting the Field Columbian Museum's priority; he would have willingly aided any qualified collector who showed an interest in Grand Junction geology. Moreover, Bradbury and his colleagues at the Western Colorado Academy of Science had already shown an active interest in collecting fossils for themselves. As early as 1885, local enthusiasts had taken to fossil hunting in the badlands south and west of town.[51] In fact, in a retrospective account of his Colorado fieldwork, Riggs admitted that a *Morosaurus* forelimb that he had claimed for the Field Columbian Museum had been found and partially excavated by another party.[52] But local amateurs were only a small part of the problem. Riggs was more concerned about the movements of his professional competitors from other museums. In a letter to Director Skiff, respectfully declining the opportunity to deliver a public lecture at the museum on his recent field successes, Riggs explained, "In view of the fact that the Carnegie and American Museums have been pushing their paleontological work most vigorously during the past year . . . and that these museums are constantly on the alert for new fossil-bearing localities, it has seemed desirable to keep silent in regard to the promise which this field still holds for us."[53] Riggs was right to worry. Osborn was then sending his American Museum collectors far afield to scout Jurassic localities whenever and wherever they seemed likely to yield exhibit-worthy dinosaurs. Likewise, the Carnegie Museum was expanding its dinosaur explorations well beyond the confines of southeastern Wyoming.

Yet of all the rival paleontologists collecting Jurassic dinosaurs in the West that summer, only the American Museum's Walter Granger posed any direct threat to Riggs's expedition. With Albert Thomson, Granger left New York on June 7, 1900, heading ultimately for the San Juan Basin of southwestern Colorado, an area that Jacob Wortman (in 1893) and James Gidley (in 1899) had already prospected for Jurassic dinosaurs with indifferent results. Osborn wanted them to take one final look. They stopped first in Denver on June 13, a few days ahead of Riggs. Granger called at the office of Arthur Lakes, Marsh's former collector, intending to make arrangements with him to visit the abandoned Jurassic dinosaur quarries near Cañon City, Colorado. Osborn was then entertaining the possibility of sending a DVP party to this classic locality, first discovered in 1877 and worked profitably by Cope and Marsh for several years. He and his collectors were confident that with their superior methods they could safely extract fossil material that their predecessors had ignored because it was deemed too fragile or too fragmentary. An exploratory visit to the old quarries in the company of a veteran collector such as Lakes was essential for Granger and Osborn to estimate their future potential. But Lakes was absent from the city. Anxious to get into the field, Granger and Thomson left Denver that night for points west. If Granger wanted to visit Cañon City with Lakes, it would have to wait until his return from the San Juan country.[54]

Mancos, Colorado, was the railroad town closest to the fossil beds. They detrained there, but they could not find a suitable outfit for hire at a reasonable price. So they ventured fifteen miles west to Cortez, skirting the famous ruins of Mesa Verde to the south. There they hired saddle horses, three pack animals, camp equipment, and a guide, Sterl P. Thomas, formerly the local sheriff. A light pack outfit would make for more rapid cross-country movement, but it would be impossible to collect Jurassic dinosaur remains without a heavy wagon. They headed west from town, following a dry creek bed into McElmo Canyon. There they stopped to examine the dinosaur prospect scouted by Gidley less than eight months earlier. They found a humerus and two other indeterminate limb bones, and, some distance away, a short string of vertebrae associated with a femur. Probably these represented two individual specimens. The bones were preserved in an "exceedingly hard, coarse, green conglomerate." The ends of the limb bones were badly eroded, and the surface of the shafts spalled off with the matrix when they tried to develop the prospect. A few days of searching in the immediate vicinity uncovered only a few scattered and badly weathered bones. Granger determined that nothing there was worth the trouble of returning with a wagon.[55]

They then headed farther west to Montezuma Canyon to reexamine a prospect Wortman had visited in 1893. But the heat, the dangerous lack of drinking water, and the scarcity of feed for the pack animals drove them off after a single day's exploration. They headed back down the canyon to the San Juan River, then back through McElmo Canyon, stopping wherever practicable to search the Jurassic beds for more signs of dinosaurs. Once again, the pickings were very slim. Granger and Thomson were back in Cortez by June 25. They spent another week prospecting near town without success. Eager to rejoin the comparatively comfortable and productive field party at Bone Cabin Quarry, Granger made tracks for the railroad.[56]

An exciting but untimely letter from Oliver Perry Hay awaited him at Mancos. Hay had begun his itinerant career in vertebrate paleontology at the Field Columbian Museum, but was dismissed in 1896 after a row with Skiff. He joined the staff of the American Museum in the spring of 1900, and he was quick to make use of his extensive network of naturalist colleagues for the benefit of his new institution. Hay had heard a talk at an 1891 meeting of the Indiana Academy of Science entitled "On a Deposit of Fossil Vertebrates in Colorado," but he could not recall all of the relevant details of this interesting fossil prospect. He shared what he knew with Granger, who was leaving for the field, and promised to follow up his tip with some more thorough intelligence. Not long after Granger's departure for Colorado, Hay attended a June meeting of the American Association for the Advancement of Science in New York City. There he exchanged favors with an old Indiana associate, Amos William Butler, the very man who had given the presentation about western Colorado vertebrate fossils.[57]

Butler, it seems, had visited a promising fossil vertebrate locality across the Grand River from Fruita, Colorado, sometime before December 1891. In May of that year, a small number of gentlemen met in Bradbury's office in Grand Junction to establish an amateur scientific society devoted to the study of local geology and natural history. Bradbury, instigator as well as host of the meeting, was elected president. In the fall, a small delegation of the new society's membership visited the fossil beds south of Fruita. There they found two enormous limb bones. They secured a portion of one of these bones for their natural history collection, housed in Bradbury's office. (The delegates returned to the fossil beds the following April and collected an assortment of small bones as well as the remaining limb, apparently intact.[58] These could be the *Diplodocus* specimens that Bradbury showed to Riggs in June 1900.) Butler undoubtedly visited this same locality, perhaps in the company of one or more obliging members of the local academy of science.

When they met in New York, Butler gave what little information he had on the locality to Hay, who passed it on to Granger in Mancos. Hay referred Granger to two residents of Fruita, Frank and Benjamin Kiefer, who had been childhood friends and neighbors of Butler's in Brookville, Indiana, before they emigrated west in the early 1880s. The Kiefers were prominent local businessmen and landowners who may or may not have been involved with the Academy of Science. But they could tell Granger where to find the bones. Moreover, according to Hay's letter, some of these bones were available for close inspection at the "Natural History Society at Grand Junction."[59] If Granger stopped over in Grand Junction or Fruita either on his way out to Mancos or on the way back, he never recorded his impression of the prospect. He had the requisite information with which he might have anticipated one or more of Riggs's best field discoveries. He had the time for exploring and the convenience of railroad passes. He also had an imperative from Osborn to find and lay claim to any promising new Jurassic dinosaur localities on behalf of the American Museum of Natural History. But he apparently decided to move on quickly to more familiar ground in Wyoming, leaving western Colorado entirely to the Field Columbian Museum.

SEVEN

A Monkey and a Parrot of a Time

Determined, confident, hardworking, and often very temperamental, John Bell Hatcher was the most talented and respected field paleontologist of his generation, but he struggled mightily to find his own level in the hierarchical world of American vertebrate paleontology.[1] Born in rural Illinois in 1861, and raised on a farm in Iowa, Hatcher was a frail child who made the most of a few opportunities. He worked in a coal mine to save money for school, entering Grinnell College in 1881. Later he transferred to Yale, where he studied geology and botany. After graduating from Yale's prestigious Sheffield Scientific School in 1884, he offered his services as a novice fossil collector to O. C. Marsh, who gladly accepted. He spent the next nine years toiling almost continuously in the field, mastering his craft and helping to make Marsh's collection of vertebrate fossils one of the largest and finest in the world. The rigors of fieldwork, and the maniacal way in which he drove himself, exacerbated a host of chronic health problems, the worst of which were Hatcher's oft-repeated bouts with crippling inflammatory rheumatism. At the same time, his uncanny success at finding fossil vertebrates and collecting them with great care and ingenuity earned him a wide and well-deserved reputation as a model field-worker. A knowing contemporary proclaimed that Hatcher had a "veritable genius for collecting."[2] In an era when the building of fossil vertebrate collections was paramount in the eyes of museum administrators and patrons, Hatcher's expert services were in very high demand.

Hatcher had an itinerant career. Like many other paleontological assistants at Yale, he often bristled under his employer's autocratic rule. Marsh, he felt, was slow with his checkbook, too miserly in doling out credit for work accomplished, and unreasonably prohibitive in terms of opportunities for qualified assistants to publish their own results. In 1892, when his lucrative U.S. Geological Survey appropriation dried up, Marsh had to rein in expenses by scaling back his paleontology field program. Hatcher jumped ship in January 1893 and enlisted at Princeton as curator of vertebrate paleontology and assistant in geology. At Princeton's geology museum, he assembled another priceless collection of fossil mammals, despite the hardships of an embarrassingly slim budget. He led field parties of undergraduate students through some of the best fossil localities in the American West. Many became his enthusiastic friends and admirers. He also planned and executed three ambitious fossil collecting expeditions to Patagonia in the late 1890s. His new supervisor, William Berryman Scott, a close friend and colleague of Henry Fairfield Osborn, was delighted with Hatcher's productivity.[3]

In 1890, Osborn had tried unsuccessfully to lure Hatcher to the American Museum, but was more or less content to see him collecting fossil vertebrates for Princeton for two reasons. First, Osborn could benefit directly from Hatcher's labors by obtaining duplicate specimens and crucial locality information from his Princeton friend and collaborator, Scott. Second, and far more important, a Princeton position was infinitely more agreeable than having Hatcher working for one of Osborn's deep-pocketed museum rivals in Chicago or Pittsburgh. So, when Hatcher returned from the last of the Princeton Patagonian Expeditions late in 1899, and began to grumble disapprovingly about his lot at Princeton, Scott and Osborn were both chagrined. Hatcher was upset about what he perceived as a lack of adequate appreciation and respect from his superiors at Princeton, although the precise nature of his complaint is not known with certainty. Osborn made an effort to smooth Hatcher's ruffled feathers, but he flew the coop anyway, leaving Princeton for good on March 1, 1900. Osborn commiserated with Scott, lamenting Hatcher's move as "a great loss to us both."[4]

Hatcher alighted at Pittsburgh and assumed the Carnegie Museum post vacated by an outraged Jacob Wortman in late January 1900. Museum benefactor and namesake, Andrew Carnegie, had not been happy when an embarrassing (and erroneous) article appeared in a New York newspaper on the petty squabble that precipitated Wortman's resignation.[5] He expressed his displeasure in a cutting and sarcastic note to museum director William J. Holland, who acted quickly to repair the damage. Holland placed Olof August Peterson in temporary charge of work in paleontology. A recent acquisition from Princeton and a former employee of Osborn's at the American Museum, Peterson was the most experienced and most well-rounded member of the staff. Holland assured the museum's Board of Trustees that the other men of the department assented to the new pecking order cheerfully, but it was a half-truth at best. William Harlow Reed was rankled at being passed over again for another younger man, and he vowed to Holland that he would never work under Peterson's direction in the field, a place where he felt he was fully as competent as any man.[6] Holland then hastened to Princeton early in February to interview Hatcher, Peterson's brother-in-law, as a possible permanent successor to Wortman. The director was profoundly impressed with his candidate. To a colleague he wrote, "Prof. J. B. Hatcher . . . is a wonderfully interesting and exceedingly able man—I think far superior in general culture and knowledge of his subject to even Dr. Wortman."[7] More important, Holland believed that Hatcher was a man "with whom pleasant relations can always be maintained, and who will not in any case transcend the limits which are imposed by considerations of official courtesy."[8] In short, Holland felt that Hatcher would know his place and could be expected to act accordingly. Congeniality was an indispensable qualification for the Carnegie Museum post, even in the eyes of outsiders. When Wilbur C. Knight wrote to congratulate Hatcher on his new curatorial post, for example, he opined, "You have a splendid place *if you can handle Holland*" (emphasis added).[9]

Hatcher, apparently, was undaunted by the prospect of working under Holland. He accepted the Carnegie Museum position and relocated to Pittsburgh within a day or so of leaving Princeton. Holland, naturally, was pleased with his latest acquisition. "[T]hings are going along in the most pleasant manner at the Museum," he soon explained to Carnegie, "and I am greatly delighted with the business-like and very intelligent way in which Prof. Hatcher has taken hold of his duties as a successor to the 'late lamented' Wortman. It has, indeed, been a very fortunate change that we have made."[10]

"Big Things" for Mr. Carnegie

The first order of business on Hatcher's watch as the Carnegie Museum's new curator of paleontology was to hold a "council of war" to formulate plans for the coming field campaign.[11] This done, Holland sent an excited account of the aggressive fossil collecting strategy to Carnegie.

According to Holland, work would continue in earnest at Camp Carnegie, in the Jurassic beds of Wyoming, where the museum purchased land from the government to ensure their rights to a number of dinosaur prospects spotted the previous summer. Hatcher's idea was to work this site with Reed for a time in order to test the latter's fitness to work independently. If and when Reed's work should prove satisfactory, the new curator would leave him to carry on with one or two assistants. Hatcher would then "make a dash" after horned Cretaceous dinosaurs at a locality he had once worked successfully for Marsh on the eastern side of the Laramie Mountains. Peterson, meanwhile, would go after fossil mammals in the Bridger Basin.[12]

Keeping three parties in the field for up to six months was an expensive proposition. But his shoestring budget days at Princeton were over; Hatcher now had access to a well-heeled benefactor. He asked for, and received, almost $2,500 to meet his expenses for the summer. This sum did not include salaries for the three principal collectors, which were paid from a special fund supplied directly by Carnegie; nor did it include the cost of travel and shipping to and from the West, which was provided free of charge by the railroads.[13] At the Carnegie Museum, finding money to do fieldwork virtually ceased to be an issue for Hatcher.

Holland was more cognizant of the competitive nature of museum paleontology than his steelmaking patron, so he appended his letter about plans for fieldwork with a well-meaning word of caution. "The Field Columbian Museum and the American Museum of Natural History are red hot after . . . fossils," he warned Carnegie.

> I do not care to have you . . . say anything to [your scientific friends] about our plans at present. We have the three ablest bone sharps [in Hatcher, Peterson, and Reed], so far as working the field is concerned, in our employment, and the other museums would be very apt, when they found out we were moving in a certain direction, to follow in our footsteps. We do not care to be trailed, and we are keeping our plans quiet.[14]

Hatcher slipped quietly out of Pittsburgh, arriving in Wyoming on April 10, 1900, in the teeth of a raging spring blizzard. Accumulated snow kept him pent up indoors for twelve tedious days in Laramie, where he stayed with Reed's family and was a perfect guest. At the University of Wyoming's geology museum, he looked over some of Reed's previous work very carefully and found much of it wanting. When the weather finally cleared, he left for Medicine Bow driving a new wagon and team. On the 22nd, he pulled out for Camp Carnegie on Sheep Creek with two assistants, William Patten and Reed's son, Willie. The elder Reed,

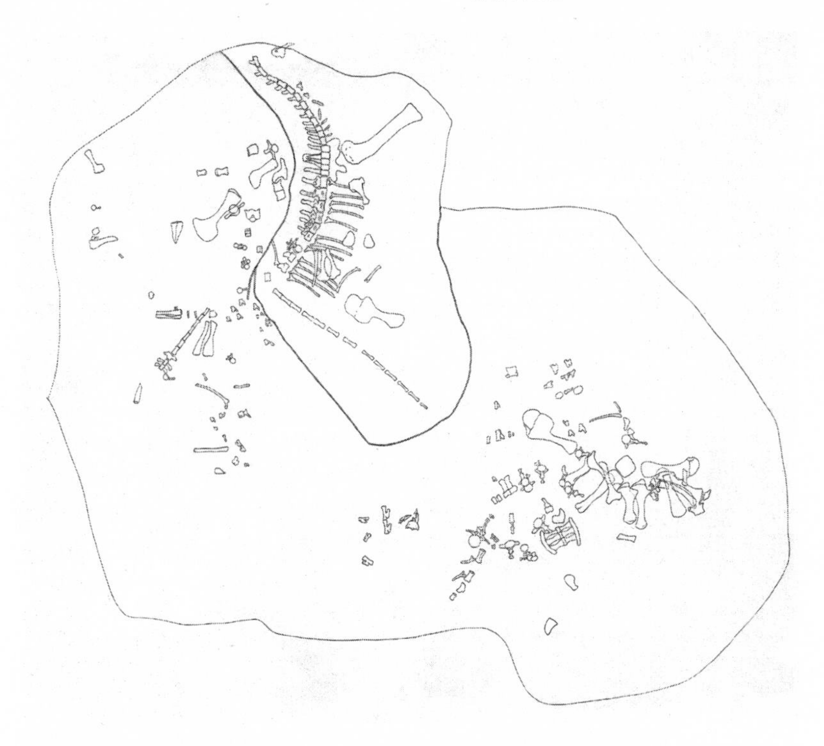

Figure 22. The Carnegie Museum's *Diplodocus* quarry map. From Hatcher, 1901.

however, would not be joining them. Instead, the veteran collector had suddenly remembered a cache of five fossil mammal skulls he had stowed away for safekeeping in 1885 at a place he called Picket Lake, somewhere northwest of Rawlins, Wyoming. With Hatcher's permission, he headed west to join Peterson to guide him to the locality. Meanwhile, with the idea of recovering anything that might remain of the *Diplodocus* skeleton collected by Wortman's party in 1899, Hatcher and his assistants spent a busy week plowing and scraping a three-thousand-square-foot space adjacent to the old excavation (see figure 22). They reached the bone-bearing layer in short order, uncovering seven fine caudal, dorsal, and cervical vertebrae mingled together indiscriminately. A broken and water-worn humerus and a few other unidentified bones also turned up. Hatcher was pleased with the productivity, but he feared that the bones did not belong to the *Diplodocus* collected previously. He also examined a few other prospects in the vicinity, and though they looked promising, he did nothing to develop them. A procession of rain, snow, and sunshine marked his days. After more than a week of this work, he was anxious to get away to his Cretaceous dinosaur locality, but there was as yet no sign of Reed. Yet he was loath to leave the excavation in Reed's care without spending a few days with him first.[15]

On May 3, with Reed still at large, Hatcher decided to make his break. He had an urgent errand to run. So, in order to buy himself a little extra time, he hired a man to drive his outfit northeast over the Laramie Mountains to the town of Lusk, Wyoming. Then, leaving Patten and the younger Reed working at the Sheep Creek quarry on their own, he boarded a southbound train for Cañon City, Colorado. There he negotiated a deal with Marshall P. Felch, a local rancher who, with his brother Henry, had collected an abundance of spectacular Jurassic dinosaurs for Marsh near Garden Park in the 1870s and '80s. Hatcher leased the exclusive digging rights at Felch's locality for three years at a cost of $25 per month, provided that choice dinosaur material yet remained in the abandoned quarries. Felch, naturally, assured the Carnegie curator that such was the case. It only remained for Hatcher to arrange for a competent field-worker to take charge of the excavation.

After a sweeping inspection of the local deposits, Hatcher was struck by the peculiar advantages of collecting at Garden Park, Colorado. Here, the collecting season could be extended well into the winter months in relative comfort and safety. More important, the locality was historically a very productive one for obtaining Jurassic dinosaurs from a number of different horizons. Best of all, the long, uninterrupted succession of rocks at this locality meant that the stratigraphic positions of the dinosaurs collected here could be determined easily and accurately. This was a decided advantage to Hatcher, a paleontologist with a greater geological bent than most of his contemporaries. It also would be invaluable for determining the possible phylogenetic relationships among the dinosaurs collected here.[16]

Hatcher sped back to Wyoming, arriving in Lusk on May 8, 1900, the same day as his outfit. He was eager to continue on into the Hat Creek Basin, fifty miles north of town, to begin the search for Cretaceous dinosaurs, but a pair of disturbing letters from Reed and Peterson, forwarded to Lusk from Medicine Bow, demanded his immediate attention. Reed reported a "total falure [*sic*] of finding the bones in the picket lake locality." He nevertheless remained hopeful of finding an extremely prolific fossil mammal deposit nearby.[17] Peterson likewise acknowledged their failure to locate Reed's cache of fossils, but he was less prosaic than his partner about the expedition as a whole. "I am much disgusted with this trip so far," he wrote. ". . . [T]he whole locality is a most unlikely looking place to find fossils."[18] Hatcher was already predisposed against Reed after inspecting the latter's amateurish work at the University of Wyoming. He quickly concluded that the veteran collector had fabricated his entire account of the cache of skulls and the elusive fossil mammal locality. "I fear the entire story is fiction & that Peterson is loosing much valuable time," Hatcher wrote contemptuously to Holland.

> From the tone of his letter [Peterson] was evidently much disgusted from being hauled about over the country for nothing. . . . Mr. Reed's veracity is not to be depended upon & for my part I do not like to be needlessly fooled or tricked by one of our own men who should have our own interest at heart. Nevertheless I am still willing to give him a fair chance as a collector, but from the nature of his work for Prof. Marsh & at the University of Wyoming I fear it is bad judgement to leave him in charge of so important work as that at Sheep Creek. . . . I trust you will not think me prejudiced in the matter. . . . I do not honestly believe Mr. Reed sufficiently methodical in his work to be in charge.[19]

A follow-up letter from Peterson regarding Reed's alleged fossil mammal locality reinforced Hatcher's conclusion. "I have now chased arond [*sic*] with Mr. Reed from locality to locality," Peterson wrote, "without he being able to show me a place of the least interest or a place with any evidence of fossils at all. . . . My opinion is that the whole thing from beginning to end is a romance. But that it would be carried to such an extent is indeed surprising."[20] He sent substantially the same letter to Holland.

The news provoked a crisis of confidence with Holland, who would have to answer directly to Carnegie for any fieldwork failures. "I confess that your letters and Peterson's have had a very disturbing effect upon my mind," he wrote back to Hatcher. In advising his curator on what to do with respect to Reed, he explained Carnegie's wishes as explicitly as possible, and then shifted the greater part of the burden for pleasing their patron directly onto Hatcher:

> It is, as you know, of the utmost importance that our Museum should succeed in obtaining a fine display of showy things at the outset. Mr. Carnegie has his heart set on dinosaurs—'big things'—as he puts it. In conversation with him in New York the other day he wanted to know particularly how we were getting along in securing material for the restoration of a [Jurassic] dinosaur. This is his fancy, and we must please him. . . . I would be better satisfied if I knew that you and not Reed were in charge of that work on Sheep Creek. If not you, I would be better satisfied to know that Peterson were there. . . . The matter is largely trusted to your judgment. . . . Of course we would like to get a collection of Bridger material and we want material from the Laramie, but having put our hands to the plow in the Jurassic, we ought by all means to work the latter field

carefully and well. . . . [T]his summer's work in the Jurassic ought to show a wonderful advance if possible over that of last year.[21]

"A Good Soldier Obeys His Chief"

After reading Holland's letter and reflecting on the situation, Hatcher reluctantly decided to abandon the Bridger for the present and send Peterson to work instead in the Jurassic with Reed, while he continued to pursue dinosaurs in the Wyoming Cretaceous. A telegram from the curator requested Peterson to return to Medicine Bow with haste. Hatcher felt that it was "a great waste of money" to keep two well-paid collectors at the same locality. "For such wages we should get thoroughly reliable & experienced collectors & preparators, who we would be able to send to any fossil field whatsoever with full confidence," he stressed in a letter to Holland.[22] It seems clear that Hatcher wanted Reed gone.

Whatever Hatcher's true feelings were in this matter, his letter of instructions to Reed was certain to provoke the latter's wrath. "I have directed Mr. Peterson to proceed to Medicine Bow and take charge of the work in the Jurassic," he wrote to an unsuspecting Reed.

> You will work under his direction during the season, & will no doubt be able to do good work with him. . . . I trust you will not take this in an offended spirit, as it is done as we think in the best interest of yourself & the museum & with the approval of Director Holland. I think a years work in the field with Mr. Peterson will be of great value to you & I sincerely trust that your relations will be pleasant.[23]

But rather than send this letter directly to Reed, Hatcher elected to enclose it with a registered letter to Peterson, instructing the latter to give the letter to Reed in person.

Arriving in Medicine Bow on May 25, 1900, Peterson opened Hatcher's letter and discovered his fate. If he was satisfied with his place in the pecking order, he was less pleased with the way in which Hatcher arranged to make it happen. Peterson wrote him to express his displeasure:

> I wish to again remind you of the fact that Mr. Reed expressed his unwillingness of working under my instruction shortly after Dr. Wortman's resignation. . . . Since my last experience with Mr. Reed . . . I can not have the confidence and respect I formerly had for him and since he undoubtedly know[s] my feeling toward him I leave to your own judgement what the result may be. I will however take your enclosed letter and go to his camp as soon as I can get the use of a horse.[24]

Peterson needed two full days to steel himself for a possible confrontation, find a suitable horse, and then ride out to Camp Carnegie. On Sunday morning, May 27, he handed Reed the letter from Hatcher. Author Tom Rea describes the scene vividly: "Reed boiled over. The country around the camp and the *Diplodocus* quarry is wide and dry, flanked by hills on the horizon to the north and east. Wide enough, and dry enough, that even in May it would quickly have sucked up and muted any words Reed said, there being no cliffs for echoes."[25] Reed cried foul, and refused to work under the direction of his younger colleague. He also refused to budge. Peterson would have no words with him and managed to avoid a fight, although he fully expected one. Perhaps the presence of Reed's wife, who was in camp to do the cooking, and his son, who was also working the quarry, helped to defuse the tense situation. Nevertheless, relations in camp were very bad, and they did not appear likely to improve as long as both men remained. Peterson found himself in a very awkward position, and he appealed to Hatcher to make amends.[26]

Reed, on the other hand, appealed directly to Director Holland for justice. "You will remember you told me you were my friend," he wrote, "and if ever I had any reason to think I was unjustly treated, to come to you for advice, and you would deal justly with me. Well, now the injustice has come, and like a thunderbolt." He described the circumstances of Peterson's arrival in camp, and the contents of Hatcher's letter. "I cannot fully obey [Hatcher] and retain my self-respect," he continued.

> . . . Last year I allowed myself to be crowded back and did it without protest, as I desired to learn the improved methods of taking out and packing the fossils, but now I have served my apprenticeship and am confident I can do exactly as good work as Mr. Peterson. . . . And this is my reward! . . . *I will never strike a pick in the ground* under Mr. Peterson,—not that he is not a good man or a good collector. Until I hear from you I will prospect around the country and try and open up some new things. But I shall not trespass on Museum grounds.[27]

Holland was alarmed at the open rebellion expressed in Reed's letter of protest, but he declined to do anything to mollify the veteran fieldworker. He wrote a carefully worded letter to Reed in which he washed his hands of all responsibility for his employee's predicament. As long as Hatcher was directly responsible for the results obtained in the field, Holland explained, then he would also enjoy full authority to arrange the details of fieldwork. Holland would not interfere. Instead, he encouraged Reed to reconsider his decision and return to work under Hatcher's authority. "A good soldier obeys his chief," he noted.[28] To Hatcher he sent a letter detailing how he would have dealt with the situation differently, in order to avoid a conflict. "However, here we are, and you must work out the problem," he insisted.

I have no intention whatever of interfering with your authority. . . . I believe that there is work enough at those Jurassic exposures to justify putting both Reed and Peterson into the field at that point. . . . If you can modify your order in some way so as to bring about harmony, do it. If not, you will have to stand upon the issue as drawn, even if it leads to a final severance of relations with Mr. Reed.[29]

This is almost certainly what Hatcher had in mind when he returned to Medicine Bow on June 5 for a final showdown. He thought little of Reed as a fossil collector. Moreover, he was angry that Reed had gone over his head by taking his grievances directly to Holland. In a quiet room in town, the two men met for nearly an hour, trying to tie up loose ends. When Reed reiterated his refusal to work under Peterson and denied that anyone save Holland had any authority over him, Hatcher considered the matter closed. Reed, as far as he was concerned, had severed his connection with the Carnegie Museum by abandoning his post and refusing to return to work.[30] He wrote to Holland with the news, and the director was relieved with the outcome. "Reed is afflicted, as unfortunately so many partially educated and 'self made' men are apt to be, with an exaggerated idea of their importance and the value of their attainments," Holland reflected in his reply to Hatcher. "I feel greatly relieved to know that he is no longer on our force. . . . Now . . . we are free to . . . utilize the money which we were spending upon his salary to far better advantage." Hatcher had made this same point previously and had suggested the names of two younger men who had experience collecting vertebrate fossils and had shown some ability. Both would be cheaper and more agreeable than Reed, and could be easily trained to fill the dual role of collector and preparator at the Carnegie Museum. Holland agreed, but in spelling out exactly the kind of subordinates he wanted in his employ, he shackled his curator with a thinly veiled warning about museum order:

Let me entreat of you, my dear Mr. Hatcher, to be very careful in the selection of men. Docility, willingness, the disposition of the soldier, who obeys orders, are needed in an institution of this character as much as in the army or the navy. Men who imagine that the holding of a minor position in an institution of this sort entitles them to assume the airs and to talk in the tone of men who have attained to scientific distinction . . . are to be avoided. We have had unfortunate experiences here in the case of one or two men who we have had with us in the past, who, having published a page or two of their exceedingly insignificant observation, have suddenly blossomed out in their own estimation as full-fledged scientists. . . . We wish . . . intelligent, capable, willing men possessed of good common sense. I know such men are somewhat scarce, but I look to you to . . . find them. We wish no more Reeds, no more Wortmans.[31]

Following Holland's plan, Peterson took charge of the Jurassic dinosaur work at Sheep Creek Quarry once Reed and his family cleared out. In order to be rid of him altogether, Peterson paid what he considered an exorbitant price for the title to Reed's wooden shanty. Patten, whom Reed had originally employed, stayed on the Carnegie payroll to help Peterson with the heavy labor. Some excellent material was turning up in the new excavation, opened first by Hatcher, and then worked briefly by Reed. Before his untimely departure, Reed had reported twenty-four vertebrae, some ribs, two sternal plates, and the right scapula and coracoid. The bones, although in a good state of preservation, were found disarticulated. Hatcher suspected that they had a general quarry consisting of the scattered and mixed remains of a number of different sauropod dinosaurs. Finding more material to complete the *Diplodocus* skeleton remained the highest priority, so he directed Peterson to abandon the new excavation for the present and to work the old quarry to the east and west.[32]

Unfortunately, Wortman's party had left a great pile of tailings on the edge of the original *Diplodocus* quarry in 1899. Peterson had to remove this first. Next, he extended the quarry along its northwestern side. His new excavation stretched approximately thirty feet wide by sixty feet long. Soon enough, not far from the spot where the first skeleton was taken out, Peterson located a scapula, a coracoid, a single dorsal vertebra, a pelvic bone, and numerous rib fragments. He was sure the bones belonged to the *Diplodocus* skeleton collected the previous summer. He also found the ilium, scapula, coracoid, two vertebrae, and one dorsal plate of *Stegosaurus*. The mid-June weather was ideal for collecting, but it was not to last. With so much success coming so early in the season, Peterson buzzed with satisfaction.[33]

"Bones . . . for the Millions"

The wagon route from Camp Carnegie to Medicine Bow skirted some of the DVP's most productive dinosaur localities. So when Pete Kaisen and Paul Miller—now both working for the American Museum—arrived in mid June to reopen Bone Cabin Quarry, it did not go unnoticed. The neighborly thing to do, when traveling to town for resupply, was to stop for a visit. Peterson and his assistants all made a habit of this practice. Members of the DVP field party visited Camp Carnegie on at least two occasions, also, including a holiday soiree on the Fourth of July, 1900. Both parties used these opportunities to spy on their rivals. Peterson shared his observations of the DVP's activities with Hatcher, but never with Holland. The tone of his commentary was usually bland, objective, and largely disinterested. "The Am. Muse. party have run on to some good Stegosaurus bones in their old quarry," was a typical remark of Peterson's.[34] "[Walter] Granger . . . has at last struck it rich in the old Como bluff," he wrote on another occasion. "[J]ust came from the old camp [at Bone Cabin Quarry] where they are packing up to leave for good as they are taking down the shanty to make boxes. They have 15 cervicals and dorsals [of *Diplodocus*] going in the bank at Aurora. They live in the old section house and I expect they will stay with it untill [*sic*] they have him out."[35]

Hatcher was undoubtedly interested in news of the DVP's successes. But he never reacted to such news in the same competitive spirit that seemed to motivate Osborn and Holland, or even some of Osborn's subordinates. In fact, Hatcher's attitude about rival collectors was surprisingly—even uniquely—generous and liberal-minded. In the summer of 1891, for example, he had written a reproachful letter to Marsh, arguing,

> I think the idea of keeping a corner on fossils of any kind should be given up. . . . Of course it is the proper thing to go on collecting[,] for only by such work are the new things brought out. But if Osborn or Cope or anyone else see fit to send collectors into this rich field (which they have a right to do) there are bones here for the millions and it would be the utmost folly for me to attempt to keep them from getting some of them. I should not have written as I have only for the repeated mentions you have made in your letters about poachers.[36]

The following summer, in 1892, when he happened upon Wortman and Peterson, who were then hunting for fossils on Osborn's behalf in one of Marsh's hotly contested Cretaceous mammal localities, Hatcher was as good as his word. To honor the occasion, the collectors brewed some punch together and drank to the health of their rival bosses, Marsh and Osborn. According to Wortman,

> We found [Hatcher] an exceedingly pleasant and affable fellow and he gave us much valuable information in regard to localities. He said that he recognized that we have just as much right here as he has and that instead of being opposed to it he would be

pleased to see us get a good collection. He told me exactly where he had been and what ground had not been looked over.[37]

Hatcher was as eager for fossils as any field paleontologist—maybe even more so. But he did not seem to resent the successes of other collectors.

Holland, meanwhile, was also keeping tabs on the DVP. In New York periodically on business or to meet with Carnegie, he often made a point of visiting the American Museum to see what Osborn and company were up to. These visits sometimes inspired a knee-jerk competitive reaction from the Carnegie Museum director. Late in April, for example, he had a long discussion with William Diller Matthew, who was in temporary charge of the DVP during Osborn's absence. During the course of their conversation, Holland mentioned that the *Diplodocus* skeleton that Wortman had collected in 1899, which was then being prepared for study and exhibition at the Carnegie Museum, and which boasted a remarkable series of vertebrae stretching from midtail to the nape of the neck, included only ten dorsals. Matthew confirmed that the DVP had come to the same conclusion with respect to the dorsal formula for *Morosaurus*. When he returned to Pittsburgh, Holland wrote a brief article on the dorsal formula of *Diplodocus*, and submitted it for publication in *Science*. "I think our Museum ought to have the credit for making the first contribution toward the settling of this disputed question," he explained in a letter to Hatcher, "and I could not wait for you to return as I am morally certain that our friends in New York will not delay rushing into print with their discovery."[38] Hatcher had no problem with Holland's article per se, but he was concerned about its implication. He wrote back to Holland to explain his position in no uncertain terms:

Under the circumstances I have no objection to your publishing [the article in *Science*], provided of course that you bear in mind that one of the fundamental conditions under which I came to Pittsburgh was that all matters relating to the publication of papers bearing on vertebrate paleontology should be in my charge, in other words I shall expect to be the actual & not the nominal head of the Department, always consulting yourself as The Director in all important matters, & I do not fear but that we will get along harmoniously, for there will be more than work enough for us all & plenty of room for good, wholesome discussion on all such questions, free from petty jealousies.[39]

If Holland felt the sting of this mild rebuke, he would be positively mortified when Hatcher revisited the issue later in the pages of *Science*.

Meanwhile, back at Sheep Creek, ten more days of exploring in his new excavation sufficed to rattle Peterson's confidence. He was no longer certain that the bones he was finding pertained to the *Diplodocus* collected the previous summer. In fact, *Brontosaurus* and *Stegosaurus* remains appeared in great numbers, hopelessly intermingled with the bones of *Diplodocus*. Peterson's new excavation, like the one opened a month earlier by Hatcher, seemed to be another general quarry. Peterson hauled two loads of boxes to Medicine Bow on June 26, 1900, stopping briefly en route to visit with the DVP party at Bone Cabin Quarry. He wrote to Hatcher with the latest interpretation of the contents of the quarry and also with some interesting news from Medicine Bow. It seems that someone had blown the safe open at the Cosgriff Bros. General Store, escaping in the ensuing chaos with a small amount of booty. Peterson kept an account at that store and left some of his expedition money there for safekeeping. Fortunately, Peterson's cash was untouched.[40]

Near the end of June 1900, Hatcher made a trip to Laramie to meet with Charles Whitney Gilmore, who had just completed his junior year of studies in mine engineering at the University of Wyoming. Knight, his professor and mentor, recommended Gilmore highly as a skilled paleontological assistant. So did Charles Schuchert, of the U.S. National Museum, who spent some time with Gilmore in the field in the summer of 1899. Hatcher found Gilmore to be a "modest, capable & willing young gentleman," with a natural talent for paleontological fieldwork. He hired him to work with Peterson and Patten for a period of four months. Gilmore showed up for work at the Sheep Creek locality at noon on June 27th, accompanied by his mother. Peterson dismissed his cook summarily, paid him off, and pressed Mother Gilmore into service. Grateful for help at the quarry, where there was no end of work, Peterson wrote to Holland that he found Gilmore to be a "practical and willing young man" who was careful in taking out bones. Holland, of course, had no idea that Gilmore (along with Knight) was the very man who, according to Wortman, had allegedly "gouged out" his coveted sauropod prospect "in a most primitive and unskillful manner" the previous summer.[41]

Work in the quarry kept the Carnegie crew busy for most of the summer. But by early August they had taken out every bone in sight. Peterson still hoped to find another rich pocket of bones, but the Sheep Creek *Diplodocus* quarry was already showing clear signs of giving out completely. Although a miscellany of *Brontosaurus* and *Camarasaurus* bones turned up in the quarry, their takings consisted mostly of *Diplodocus*, including a complete pelvis, a complete hind limb with foot, a scapula, and some fifty-odd vertebrae. Best of all, the party found the rear portion of

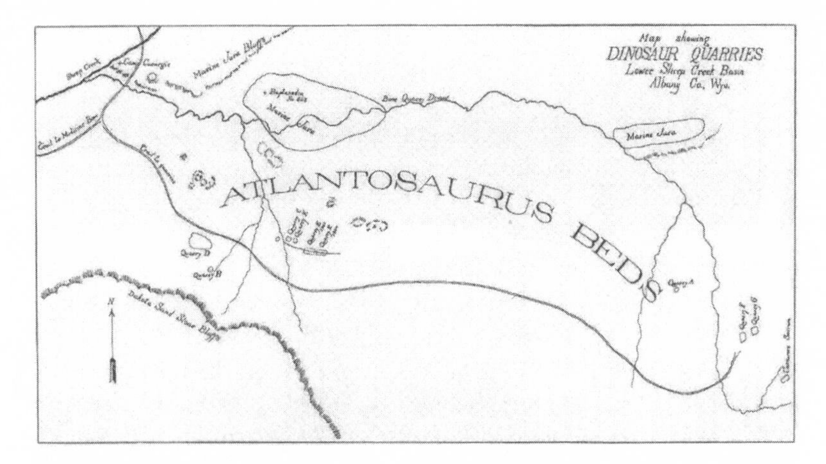

Figure 23. Map of the Sheep Creek quarries, Carnegie Museum, ca. 1900–1903. From McIntosh, "Catalogue."

a skull, with a dentary and a single maxilla, all exceedingly rare discoveries. Peterson was somewhat remiss in claiming that "if we now could find humerus[,] radius[,] ulna[,] and fore foot I think we could begin to talk restoration of Diplodocus."[42] Visionary talk about a fully mounted restoration of *Diplodocus* had been blowing at the Carnegie Museum for quite some time already.

Fortunately, there were several other prospects nearby (see figure 23). They spent two days at a place that Peterson called Big Bone Quarry, for example, but collected only the head of an enormous femur there. Across the draw from this site they opened up another prospect—a probable *Brontosaurus*—finding a femur, tibia, and a few vertebrae. Peterson put Gilmore to work at this spot, while he went in search of new Jurassic dinosaur prospects in the area around Troublesome Creek. But Peterson found nothing on his prospecting trip, thanks to a soaking rain that drove him off after only one day of searching. By August 9, the *Diplodocus* quarry, or Quarry D, as it had recently come to be called, was abandoned. So was Gilmore's *Brontosaurus* prospect, which yielded a few promising bones, and then nothing but useless fragments. They began extensive plowing and scraping, and opened up two new quarries. One quarry yielded a great number of poorly preserved bones of *Brontosaurus* and *Diplodocus*. Hoping to secure a useful forelimb of *Diplodocus*, Peterson decided to have Gilmore work this quarry anyway.[43]

In the other quarry, probably the site designated Quarry E, they found a scapula and a string of ten to fourteen caudals of *Brontosaurus*. All were poorly preserved. But with nothing better in sight, Peterson also decided to take this specimen out. He followed the caudals into the bank and in a week he recovered fifteen or so more, plus a partial pelvis and a complete hind limb. The foot, he hoped, would follow. Also, a large block of pebbly matrix appeared to contain one or more posterior cervical vertebrae. Notwithstanding the poor condition of the bones, Peterson was very enthusiastic about this specimen. Gilmore, meanwhile, continued to work at his prospect, Quarry C, finding a general mixture of large sauropod bones. By September 10, with Gilmore's quarry still producing, Peterson began to doubt that the work at Sheep Creek could be finished by the close of the season. Indeed, he and Patten opened up another new prospect, possibly Quarry F, consisting of various, poorly preserved carnivore and stegosaur bones. These gave out quickly, but they led the collectors to a partial pelvis and six caudal vertebrae of another *Brontosaurus*. Late in September, Gilmore's Quarry C produced fourteen presacral vertebrae, many ribs, a scapula, a femur, and thirty-eight caudals of a single individual dinosaur, possibly a *Brontosaurus*, their fifth of the season. Peterson was now hopeful of acquiring sufficient material to make a complete restoration of this animal. Three days of stormy weather slowed the work toward the end of September. Peterson, nevertheless, was confident of shipping two full railroad carloads of Jurassic dinosaurs for the season.[44]

Hatcher arrived in Medicine Bow on October 4, 1900, for a short visit with Peterson at the Sheep Creek locality. Together they discussed plans for the following year and made arrangements for shipping the present season's takings back to Pittsburgh. Bad weather then prevailed, so they closed up the quarries for the season, despite the fact that "a vast quantity of bones" remained in the ground. Hatcher had not been well, complaining to Holland of a terribly sore throat—he left Peterson to attend to the loading. With only a single boxcar at his disposal, Peterson had to stack up his precious crates, ninety-six in number, like so much bailed hay. He feared the load was dangerously top heavy, but a railroad freight agent reassured him. When a runaway coal car collided with Peterson's boxcar, the same agent advised him that the damage to the car's exterior was of no consequence. Despite the setbacks, Peterson's car was ready to go by October 17. It arrived in Pittsburgh some weeks later without further incident. Peterson left Medicine Bow on the same day, stopping briefly in Laramie to file claims on the property worked by the Carnegie Museum party. After a brief vacation, he returned to Pittsburgh late in October 1900.[45]

Holland, who had spent part of his summer in Europe conferring with Carnegie, was extremely gratified by the good news respecting Jurassic dinosaurs. He was particularly excited by the prospect of a *Diplodocus* skull. Over the summer, in the museum's fossil preparation lab, Arthur Coggeshall had been working on a block of matrix containing several anterior cervical vertebrae collected from the *Diplodocus* quarry in the summer of 1899. In the process, he discovered an excellently preserved axis—the second vertebra of the neck. Holland mistook this bone as the first vertebra of the neck, and expected it would articulate nicely with the skull his field party found in Quarry D. He lauded Peterson's accomplishments in a letter, and reflected at length on the museum's recent successes:

> With what we possess . . . we ought to be in a position to effect a restoration of Diplodocus which will be the envy of palaeontologists throughout the world. Indeed our work has been very successful, and is very grateful in review. I am charmed as I look back to see how in a few brief months we have, through the energetic labors of those in the field, been able to acquire already a comparatively strong position among institutions devoting themselves to palaeontological inquiry. If the work goes on as it has begun it will be . . . a short time until we have assembled a collection of which any institution might be very proud.[46]

Getting Started at Cañon City

Hatcher hired William H. Utterback, an old acquaintance from his collecting days at Yale, to assist him in the field through October, and then assigned him to take hold of the work of reopening the historic Marsh quarry at Cañon City, Colorado. He was exactly the kind of field-worker Hatcher wanted working under his charge at the Carnegie Museum. "He is a good workman," Hatcher claimed, "industrious, quiet & unassuming. . . . He is a man that can be trusted anywhere, [and] he is already a good collector."[47] He arrived at Cañon City on the evening of November 1, 1900, and then went out to survey the bone beds the following day. He followed a road that led east from town and then turned north through a canyon watered by Four Mile Creek. Nine or ten miles distant from Cañon City, the road emerged onto Garden Park. The quarry was situated just short of this point, on a narrow sandstone shelf high above the canyon bottom, but below the top of the bluff. Felch, the local rancher and former bone digger, who was now in poor health, had agreed to show Hatcher's collector the lay of the land. Utterback was impressed by the vast amount of heavy pick and shovel work required at this locality merely to get down to the bone-bearing layer. He was less impressed with the accommodations. Felch's place, he thought, was too dirty and miasmic to board. Instead, he determined to raise a tent over a wood-burning stove and do his own cooking. He wrote to Hatcher requesting a per diem arrangement for expenses in order to avoid the scrutiny of museum authorities. "[S]ometimes one wants something that might not be approved of," he explained.[48] He reluctantly lodged with Felch while waiting for Hatcher's reply. Although the daytime weather was ideal for outdoor work, nighttime temperatures dropped so low that he began to rethink his plan of living in a tent. When Felch offered him the use of a run-down shanty near the quarry, he gladly accepted, even investing $10 of Carnegie's money in some basic repairs and weatherproofing.[49]

Digging commenced on November 4. The narrow shelf of hard sandstone that served as Utterback's foothold on the canyon wall was also the fossil-bearing layer. In order to quarry this layer for bones, he first had to strip off the hard rock overburden. Since the bone layer was relatively flat-lying and cut straight back into the slope, it stands to reason that the further he extended his dig into the bluff, the more overburden he would have to remove in order to reach pay dirt. At the extreme edge of his dig, for example, he estimated the thickness of the above-lying rock at something over thirty feet. Given the steepness of the slope, and the hardness of the rock, plowing and scraping would be impossible. Almost every inch of overburden would have to be blasted, picked, pried up, and carted off in a wheelbarrow. It was tedious, backbreaking, time-consuming work. The magnitude of the job was a surprise to everybody. In early December, Utterback reported, "Am making good headway by myself but I tell you J. B. she does gain going into that hill. . . . There is one Museum employe whose hand will not tremble when he signs for Nov wages. Even worked Thanksgiving and had bacon for dinner."[50]

Utterback quarried through the winter, despite some fearfully cold and snowy weather, which slowed or halted the work from time to time. The rock was extremely hard on steel, so he bought a forge, anvil, and hammers to sharpen his own tools on site. In February, he hired another man to help with the heavy labor. Together they cleared away an enormous amount of overburden, exposing about fifteen hundred square feet of the fossil-bearing sandstone. There they found numerous dinosaur bones in splendid condition, ready to be taken up. Word of Utterback's success reached Holland in March 1901, just in time to influence his annual report. Holland was "exceedingly happy" about the news. "It is very probable that we shall obtain as the result of this undertaking a large amount of fine material," he predicted. He had recently hosted a visit by Osborn, his American Museum rival, who assured him that "such progress, as has been made . . . if continued for a few more years, will bring the Carnegie Museum into the front rank as a center of paleontological inquiry and research."[51] Of course, Osborn was unwilling to see that happen if he could help it.

Hatcher, meanwhile, settled in Pittsburgh for the winter of 1900–1901. He needed time to recuperate from the wear and tear of another long field season, but he was sick again in January.[52] Moreover, there was a vast amount of work to be done at the museum: arranging work spaces, managing the affairs of the department, and getting down to the business of research. Now that they were coming into such close contact, he and Holland crossed swords. Wortman, who was friendly with Peterson, and still bitter about his own experience under Holland, gleefully described the situation at the Carnegie Museum as a "'Monkey and Parrot' of a time" to a smug Osborn, who replied that he was "not at all surprised at the condition" of affairs in Pittsburgh.[53]

Hatcher and Holland, it would seem, were not well suited for one another. Hatcher was impetuous and strong-willed. When he disagreed with the director he was very free with his opinions. Holland did not take well to censure of any kind, particularly from one who was supposed to be his subordinate. But Holland recognized Hatcher's genius for paleontology and he needed his expertise if he was ever to please Carnegie, so he tried to accommodate himself to his curator's temperamental nature whenever possible. He sent gushing letters expressing his "entire satisfaction" with Hatcher's work and bestowing his "highest commendation."[54] But accommodation sometimes proved impossible. Holland complained to Hatcher in March, for example, "that no amount of kindness or willingness on my part to aid you and assist you is capable of satisfying you."[55] They bickered repeatedly over the wording of employee rights and privileges in early drafts of the museum's by-laws.[56] They fought over Hatcher's tentative plans to absent himself from the museum for up to two years to collect fossils in Antarctica; Holland indulged his curator's scheme to a point, but he did not want to see him go south. Hatcher could be extremely single-minded about his scientific ambitions and he resented what he considered Holland's half measures to assist him.[57]

For Holland's part, the major sticking point in their troubled relationship was undoubtedly an article Hatcher published in *Science*, in November 1900, on the vertebral formula of *Diplodocus*. Only six months earlier, Holland had published an article in *Science* on the same subject that was based on the very same fossil material. The director, Hatcher felt, had overstepped his authority and trampled on Hatcher's rights as curator by authoring a paper on vertebrate fossils. And now Hatcher let him have it in print. He corrected several minor errors in Holland's article. He had to split hairs to arrive at the conclusion that *Diplodocus* boasted *at least* eleven (instead of ten, as Holland had claimed) dorsal vertebrae and he later recanted the substance of his argument,[58] but he had made his point: Hatcher, and not Holland, would have the final word on vertebrate paleontology at the Carnegie Museum.

Yet there was usually an undertone of mutual respect and tolerance even in some of their most heated exchanges. Hatcher, for example, penned this sarcastic complaint to Holland and affixed it to the end of his monthly departmental progress report:

> I have been deeply grieved to find that my conduct of the Department . . . has been so unsatisfactory as to call forth the personal abuse visited upon me, by yourself on Nov. 7th. I am also much affected by the further abuse you saw fit to administer on Nov. 28th when you called me a jack-ass & a d—d fool. Such language, it seems to me, cannot but tend to destroy the harmony, enthusiasm & interest so essential to the welfare of the institution & should therefore be discouraged. I shall earnestly strive in the future as I have in the past to so conduct myself & my department as not to merit such unqualified disapproval.[59]

Holland ignored the complaint, and then three weeks later he sent his curator a Christmas "trifle" intended to meet the cost of "a good, fat turkey."[60] Work at the museum continued apace, with Hatcher and Holland sniping at one another like siblings. That neither man ever exceeded the other's narrow limits goes to show how much they needed each other.

EIGHT

Fossil Wonders of the West

Henry Fairfield Osborn resented every success the Carnegie Museum enjoyed in the Wyoming Jurassic in 1899, a locality entered previously by his own institution. He particularly envied them their acquisition of a beautiful, exhibit-quality skeleton of *Diplodocus*, a feat his own DVP had yet to accomplish, despite their two-year head start. The fact that it was Jacob Wortman, Osborn's embittered and recently departed former field foreman, who had engineered the Carnegie coup added insult to injury. Yet Wortman's resignation from the Carnegie Museum in January 1900 only served to make matters worse, as William J. Holland, director of the Carnegie Museum, filled his shoes (first) with Olof A. Peterson and (later) with John Bell Hatcher, both recently returned from the last of the Princeton Patagonian Expeditions. Osborn must have been mortified by the news. Peterson was a skilled collector who had once worked profitably for the DVP, and Osborn wanted him back. And Hatcher had been the apple of Osborn's eye since the latter first opened negotiations for his own position as curator at the American Museum in 1890. Now, in 1900, Peterson and Hatcher were both bound for Pittsburgh.

Still, talented collectors were only a means to an end; what Osborn was really after was a comprehensive collection of vertebrate fossils. His chief interest, for the moment, was in gigantic Jurassic dinosaurs, and losing Wortman, Peterson, and Hatcher to a rival museum did not necessarily put these objects out of reach. Osborn's DVP boasted a stable of young, talented, and increasingly experienced collectors who were eager to please their employer. Information about potential new fossil localities flowed freely to Osborn through a vast, far-flung network of scientific colleagues, amateur collectors, and even complete strangers, who were equally eager to please the famous and influential scientist. The resources, financial and otherwise, to follow up this information were also available to Osborn through his social network of wealthy New York businessmen. Osborn's DVP controlled an extremely productive Jurassic locality at Bone Cabin Quarry in southeastern Wyoming. Thus far, it had failed to produce the complete sauropod dinosaur that Osborn was after, but continuous digging over the next several seasons was more than likely to fulfill this goal. Another key to ultimate success was to prospect aggressively for new Jurassic dinosaur localities wherever they were likely to turn up. This was especially important as long as Osborn's competitors in Chicago and Pittsburgh were attempting to do the same. Years later, Osborn wrote to one of his collectors, Barnum Brown, with advice on how best to pursue fieldwork:

> I think you ought to spend the larger part of your time prospecting, taking in as large a radius of the country as you can, rather than quarrying and taking out specimens. . . . [I]t is poor policy for us to do any more digging than is absolutely necessary in order to see that specimens are well taken out; so I say once more, prospect, prospect, prospect. [Walter] Granger was digging instead of prospecting when he let Wortman slip in and find that big *Diplodocus*.[1]

Not only does this accurately encapsulate Osborn's feelings about fieldwork throughout the second Jurassic dinosaur rush, but it also lays the blame for failure directly on his collectors.

Return to Bone Cabin Quarry

Work resumed at Bone Cabin Quarry on June 10, 1900. Granger and Albert Thomson were then headed for their brief and fruitless reconnaissance of the San Juan Basin of southwestern Colorado, in search of a new Jurassic dinosaur locality. Brown, too, was absent in Patagonia. So Peter Kaisen, in only his third season with the DVP, assumed temporary charge of the field party in Wyoming. Paul Miller, who had worked for the Carnegie Museum the previous summer, served the expedition as cook and laborer. A tall, muscular Dane of twenty-three years, with an anchor tattooed on one wrist and a frigate on the other, Miller was recovering from a bad bout of wanderlust when he happened upon the Carnegie Museum field party at a Wyoming depot in the summer of 1899. Wortman then needed unskilled help at the Sheep Creek quarry, so he hired Miller to dig, mow the grass for packing material, and make boxes. He was "ingenious," according to Wortman, "and a good careful worker." Fired with enthusiasm for vertebrate paleontology by Wortman's impromptu campfire lectures, Miller hoped to dig for dinosaurs again the following summer. But when Wortman resigned his position at the Carnegie Museum, Miller's window of opportunity seemed to close prematurely. Fortunately, Kaisen, a fellow Dane, tapped him to work for the American Museum party. Together they began the heavy stripping and cutting at Bone Cabin Quarry. Once this was accomplished, Kaisen put Miller to work excavating bones for the first time. Both developed ultimately into magnificent field-workers.[2]

Frederick Brewster Loomis assisted the expedition as a volunteer digger. He was exceptionally well qualified for such a position, having recently completed a PhD (1899) in paleontology at Munich, where he once had the pleasure of meeting Osborn, who was visiting with Loomis's mentor, the German paleontologist Karl Alfred von Zittel. Returning to the United States, Loomis landed a position as a biology instructor teaching comparative vertebrate anatomy to eager undergraduates at Amherst College, his alma mater. A young man of twenty-six years with boundless energy and enthusiasm, Loomis wanted to acquaint himself with the great fossil beds of the American West. He also wanted to learn the art of fossil collecting. Richard Swann Lull, a DVP volunteer in 1899 and an Amherst neighbor, encouraged Loomis to write to Osborn and offer his services. Naturally, Osborn accepted, and sent him west to join the DVP party at Bone Cabin Quarry. He traveled via Denver, where he stopped to call on Arthur Lakes, who, only one week after Granger's attempted visit, was still out of town. In Laramie, he paid a visit to Wilbur C. Knight, of the University of Wyoming, and in Medicine Bow he met William H. Reed, recently of the Carnegie Museum, but now working independently; both were headed for the Freezeout Mountains to collect fossil vertebrates. He joined Kaisen and Miller late in the second week of the field season, arriving in camp on June 23, 1900. He was a welcome addition. Granger described Loomis as "an excellent camp companion, a bad-weather man, who could be depended upon to maintain his wonderful spirit no matter what happened."[3]

They started working in the northwestern corner of the quarry, at a point where work had been abandoned the previous summer, in 1899. Immediately they struck a rich pocket of bones. Kaisen began developing a very good *Stegosaurus* prospect with numerous vertebrae, limb, and foot bones, and at least one characteristic dorsal plate, while Miller worked a confusing jumble of *Brontosaurus* bones. In mid-July, near the northern limit of the season's excavation, someone in the party found the incomplete hind foot of a small, birdlike, carnivorous dinosaur. Later, a large part of the skeleton of this same animal turned up, including a complete skull and one lower jaw. In size and structure it resembled Marsh's genus *Coelurus*. Nothing like it had previously been recovered from Bone Cabin Quarry.[4]

Loomis was excited about the work, and he wasted no time getting started at the quarry. He had incredible beginner's luck, finding the rear half of a small dinosaur skull within two hours of his arrival in camp. Working slowly and deliberately, he had it ready to be plastered and taken out in three days. Osborn, of course, was delighted by the news, and he wrote to congratulate the volunteer collector on his good fortune. "I suppose that Peter [Kaisen] has told you that we have found very little skull material. . . . Skulls are very much needed in our collection, in fact they constitute at present our greatest need, so that every worker in the quarry should keep this constantly in mind." Loomis then had so little experience as a bone digger that Osborn felt compelled to send him some needlessly elementary instructions. "[A] skull is very sure to be a bad looking prospect, that is fragile and incomplete," Osborn advised. "For this reason picks . . . should not be driven recklessly at any point unless one is quite sure that there is no bone showing."[5]

In this part of the quarry, seventy-some feet north and west of where it was first opened, the soft blue clay matrix gave way to a hard sandstone layer, which was much more difficult to work. The quality of the bones in this layer, on the other hand, was considerably better. Sometime over the course of the summer, the party developed a new improvement on the plaster field jacket. The plaster sometimes adhered to the bone so tenaciously that when it was removed in the lab, bits of the bone's surface would sometimes come away with it. To eliminate this problem, the field party introduced a separating layer of tissue paper or flour paste bandages between the plaster jacket and the exposed surface of the bone. By this technique, and with the painstaking delicacy of novice collectors, the field party recovered a respectable quantity of well-preserved specimens.[6]

Granger and Thomson soon returned from their Jurassic dinosaur reconnaissance in southwestern Colorado, arriving at Bone Cabin Quarry on July 5, 1900. They were one day too late to participate in a holiday soiree at Camp Carnegie, on Sheep Creek, where the DVP party joined their Pittsburgh rivals, including Olof A. Peterson, Charles Gilmore, and William Patten, for a friendly celebration. The visit was also an opportunity to spy, and Osborn encouraged Loomis to make a report on Peterson's progress. Granger also submitted a report on his recent prospecting failures in Colorado. Osborn was "disappointed but not very much surprised" that Granger's trip had been a bust—no one in the DVP had as yet heard any word on Riggs's recent success near Grand Junction. More than anything else, he wanted his collectors to search new Jurassic exposures, apart from Bone Cabin Quarry, in order to find and collect a complete sauropod dinosaur. He was determined not to let the Carnegie Museum collectors outcompete his own field parties this season, especially in light of Wortman's "ten-strike" the previous summer, and particularly because Hatcher and Peterson, each of whom he wanted for his own staff, had both chosen to work in Pittsburgh instead. For Osborn, a man accustomed to having his own way, this must have been excruciating. In any event, he detailed Granger to seek for new localities. "You had better go right to work prospecting with Thomson," he insisted. "Strike off into some new country as I think there is very little use prospecting over the old ground."[7]

Granger disagreed. He thought it best to scout thoroughly the area around Bone Cabin Quarry before moving on to less familiar territory. Osborn's letter (of July 6, 1900) was addressed to Colorado and long delayed in being forwarded to Medicine Bow, so with no explicit instructions to the contrary, Granger felt that this was the wisest course of action. "Don't you think this is our best plan?" he asked in a letter to the curator.[8] Osborn hastened to respond to Granger's question with a letter outlining some very specific expectations. "I do not think it worth while to spend very much time on the old ground," he wrote. "It is better to strike off into new exposures, if you can find them. It seems to me that the Jurassic should turn up again in a locality like that which Wortman found last year [at Sheep Creek]. This is what we should look for. Go ahead steadily and I am sure you will be rewarded with success." He also "very much wanted" Granger to find some examples of sauropod fore feet with the bones preserved in their natural positions (in order to test a speculative reconstruction that he had already made).[9] This was a repetition of a request made previously in letters to Kaisen and Loomis, where he also stressed the imperative of finding more skull material. Given that these were results over which Granger and the other collectors had no control, it was unrealistic for Osborn to expect them to be fulfilled.

The major responsibility for meeting Osborn's expectations fell squarely on Granger, who felt anxious to produce satisfactory results for his patron. The tone of most of Granger's letters to Osborn during the 1900 field season was strongly apologetic, which suggests that he feared his work was not up to snuff. Regarding the Colorado expedition, for example, Granger wrote, "I made a mistake in not waiting until [fall]. . . . I was sorry to have to write that discouraging letter from Cortez and hope I shall not be obliged to do it again this season." On the eve of his prospecting trip with Thomson, Granger felt keenly the pressure to perform. He confessed to Osborn that he was "anxious to report some success . . . you may depend that I am doing whatever I can here."[10]

Eager to make a worthy discovery, Granger and Thomson left camp at noon on July 17 with a light outfit suitable for prospecting. They stopped first in Medicine Bow for mail and supplies, and then began sweeping through the local Jurassic exposures, including those in the Freezeout Mountains and along Sheep Creek and the Little Medicine Bow River. Other collectors were already at some of these same localities collecting dinosaurs for rival institutions. After his showdown with Hatcher, and his resignation from the Carnegie Museum early in the 1900 field season, Reed had gone out collecting again on his own. He opened up a new prospect near Rock Creek, less than twenty miles from Bone Cabin Quarry. Granger and Thomson stopped in to see him and to examine his excavation. There were many specimens in sight, including parts of *Stegosaurus* and an unidentified carnivore, but Granger—for now—was not terribly impressed. The bones were disarticulated, and many appeared to be poorly preserved. Reed was asking $1,000 for the rights to his quarry, but Granger turned him down cold, preferring instead to trust to providence.

Unfortunately, Granger had no luck at all locating new prospects. That is, until Thomson found, at long last, a very promising *Diplodocus* specimen at Como Bluff, on August 10. They established camp in the old abandoned section house in Aurora, where H. W. Menke spent a lonely winter in 1898–99, and began at once to develop their new prospect. It seemed, at first, to be exactly what Osborn was hoping for. According to Granger, "It was as fine a looking *prospect* as I would ever care to have, fifteen cervicals and dorsals, closely connected, with all of the neck ribs and the first two or three back ribs in position."[11]

Granger was so encouraged that he sent instructions to the men at Bone Cabin Quarry to close down their dig and join him at Como Bluff. Yet interesting and valuable material was still coming to light there. Indeed, the discovery of one of the best specimens of the entire season, another rare sauropod skull fragment, occurred at the upper limit of the last cut. There was an unexplored area of approximately five hundred square feet adjoining this cut that Granger suspected could produce more skull material or more specimens of the small carnivorous dinosaur found earlier in the summer. At the same time, however, there were some disturbing signs that the quarry was nearing exhaustion.[12] There also was evidence that Granger's confidence in the quarry was on the decline. Peterson, who passed by regularly when he traveled between Camp Carnegie and Medicine Bow, assumed that the DVP party was abandoning Bone Cabin Quarry for good when he observed them tearing down their wooden shanty to make boxes.[13] In any case, Osborn had set a premium on obtaining a complete sauropod dinosaur skeleton, so Granger decided to bring Kaisen, Miller, Loomis, and the entire DVP outfit to bear on the new prospect at Como Bluff. Some unfortunate circumstances conspired to make the new dig a difficult one, and Granger and Thomson needed a few extra hands.

Work on the new *Diplodocus* prospect was "somewhat arduous," due to the difficulty of access to the quarry site, and because of the steepness of the overhanging bluff. To enlarge the prospect, Granger decided to make a vertical cut more than twenty feet down from the top of the bluff through the above-lying sediments. All this work had to be done by hand. But after five full days of hard labor and a thorough search of the new cut, all the party uncovered was one additional dorsal vertebra and a few disarticulated dorsal ribs. Discouraged by his long run of bad luck, Granger wrote to Osborn with the sorry news, lamenting that the "*Diplodocus* . . . has not come up to expectations." Their work was not entirely in vain, however. "I feel that the specimen will be of some value to us," Granger reasoned, "especially as it does not in any way duplicate Brown's specimen."[14]

Osborn was "very much disappointed that the specimen played out." But, like Granger, he was hopeful that it could be combined with the Brown specimen to make a composite mount. He was also sensitive to Granger's frustration and sent him a candid assessment of the summer's results. "I have no doubt you have done your very best in the matter of prospecting," he wrote, "and I hope you will be rewarded by some very good luck before the season closes." So the pressure was still on. "On the whole it seems that you have had a fair season and should not feel discouraged," he added.[15] But it is difficult to imagine that Granger found much encouragement in Osborn's mixed message.

By September 20, the new *Diplodocus* specimen was collected, boxed, and ready for shipment. The Bone Cabin Quarry dinosaurs were also ready to go. Most of the specimens were already stored in Medicine Bow, awaiting the arrival of a freight car. Osborn, it seems, was having some difficulty securing free passage for the fossils. He urged Granger to use the delay to do some more prospecting. So the entire party—minus Loomis, who left for Amherst a week earlier to resume his teaching duties—continued to search for dinosaurs all over Como Bluff. Granger resented the lost time, and the repetition of effort in the fossil beds, but he carried out his instructions faithfully. "We microscoped Como Bluff while we were waiting for our car," he reported to Osborn. Raw, blustery weather made the work unpleasant—even dangerous. Thomson was actually blown off the bluff one day, but escaped unscathed. Still, they found two or three mediocre prospects, although they did not attempt to take them out. On October 1, Thomson left for a new Jurassic locality in the Black Hills with a team, wagon, and extra saddle horse. Miller was dismissed the same day. On the 10th, a tardy freight car turned up on a siding at Medicine Bow, a reluctant gift from Osborn's uncle, J. Pierpont Morgan. Granger and Kaisen loaded up the season's takings, packed in forty-seven boxes and weighing ten-and-a-half tons. Granger departed that same day by rail for Sturgis, South Dakota. Kaisen packed up the camp equipment, boarded the horses in winter quarters, and left for New York two days later.[16]

The Sturgis Rally

In mid-September, George Reber Wieland, a newly minted PhD (Yale, 1900) with a research affiliation at Yale's Peabody Museum, stopped in New York to confer with Osborn about collecting Jurassic dinosaurs in the Black Hills. Wieland had collected specimens of the gigantic but poorly known sauropod dinosaur *Barosaurus* for Marsh in 1898. Now he was offering to do the same for Osborn. Lively, small in stature, but with a booming voice, Wieland wielded an infectious enthusiasm for fossil fieldwork and Osborn was completely taken with him. He was experienced and well educated. More important, he brought a tempting dinosaur prospect—in all probability, another *Barosaurus*—to the negotiating table. Fed on a steady diet of disappointing news from his collectors in Como Bluff country, and still smarting from the loss of Wortman, Peterson, and Hatcher to the Carnegie Museum, Osborn was inclined to try something new. He wrote to Granger with a tentative plan and requested his immediate reply; with summer almost over, there was no time to lose. The plan was to send Granger, with Kaisen or Thomson and one complete camp outfit, to meet Wieland in the field and collaborate with him on a two-month reconnaissance of the Jurassic beds flanking the Black Hills. If, on the other hand, the DVP party had lately discovered a worthy prospect in Wyoming, then all bets were off. Osborn thought it would be wise for his collectors to gain some valuable experience in a new locality, given that the prospects for returning to Bone Cabin Quarry were then looking so bleak. He may also have been eyeing Wieland as a possible addition to the DVP staff. But he would not close the deal with the Yale collector without hearing first from Granger.[17]

Granger received the news of Wieland's impending appointment with good grace and generosity. "I am much interested in the *Barosaurus* question and it seems to me that we should not loose [*sic*] the chance to get Wieland's services," he wrote to Osborn. He added that, with two teams, two wagons, and plenty of camp equipment, there was ample material to support an expedition to the Black Hills without hindering the party at Como Bluff, now or in the next season.[18] A less modest man in Granger's position might have felt threatened by an interloper of Wieland's stature. But Granger was, apparently, his usual accommodating self. Impressed by his confidence, as well as by his pedigree, Osborn placed Wieland in charge of the Black Hills expedition. Osborn explained to Granger that "[Wieland] is very confident of success. I expect to hold him responsible for this part of the trip and will . . . place him in charge. . . . You will therefore refer matters to him." He softened the blow by insisting that they would work together pleasantly. Wieland, Osborn wrote, "seems to be a thoroughly nice fellow and I do not anticipate any friction."[19] If Granger harbored any bitterness about this decision, he did not share these feelings with the curator. Possibly he greeted the news with a sense of relief at no longer being the party most directly responsible for meeting Osborn's great expectations. Now the burden would be Wieland's. Osborn was taking a risk in promoting the Yale collector above his own time-tested field-worker. Among other things, it shows that he had a great deal of confidence in Granger's good nature.

Wieland had a healthy appreciation for the value of his own skills and experience. When Osborn tendered his first offer of relatively modest wages, supplemented by the singular honor of working this locality for the DVP, Wieland deflected it easily. He preferred to be compensated at a rate equal to that of the best collectors. "I regret to feel how all important the financial side is to me," he wrote with feigned apology. "I wish you could feel disposed to name some better sum as wages, for I expect direct results." Osborn was anxious to have Wieland's services, and raised his offer to $225, plus expenses, for two months' work. If this sum did not match Wieland's idea of what is "right and fair," Osborn was willing to consider the question further, but he warned that anything more might delay the expedition until the following spring. But Wieland called his bluff, asking for $250 and urgently recommending the advantage of entering the field right away. Osborn relented, but withheld $50 of Wieland's wages pending a satisfactory outcome to the expedition.[20]

With railroad passes secured, Wieland left New York on September 28, 1900, headed for Sturgis, South Dakota, a small town on the northeastern rim of the Black Hills. A dome-shaped uplift with ancient slate and schist peaks at the center, the Black Hills are surrounded by concentric rings of progressively younger sedimentary strata that envelop them like a fringing reef. One of these strata, now known as the Spearfish Formation, is a conspicuous bed of soft brick-red clay, easily eroded, which forms a valley that loops around the hills. Overlying the red clay is a series of erosion-resistant beds of sandstone, shale, and limestone, which forms a distinctive ridge around the hills from three to five hundred feet in height. The lowermost of these beds are Jurassic, some of them bearing vertebrate fossils.[21] Wieland's errand was to prospect these beds to find their most productive exposures. He stopped en route at Buffalo Gap, South Dakota, a railroad junction southeast of the hills where the Jurassic ridge takes a sharp turn to the north. There he found dinosaur bones in several locations. All the bones were small and isolated, and none merited collecting, so he pressed on to Sturgis, arriving October 3. The weather was "splendid."[22]

Until Granger and Thomson arrived with the horses and camp outfit, Wieland's mobility would be somewhat limited. Two years earlier, while dinosaur hunting for Marsh, Wieland found bones weathering out of a site three or four miles south of Sturgis, on the Fort Meade Military Reservation. He revisited this site on foot, and found another one nearby where there were "portions of very large bones well preserved." He also learned—how, he did not record—that a student from the South Dakota School of Mines (in Rapid City) had been nosing around the site in the summer and had taken a number of bones belonging to several different animals. According to Wieland, "[the student] very likely destroyed some valuable traces." He also found a large bone on the farm of one Henry F. Wells, another former collector for Marsh who now lived in the Black Hills. Bones proved difficult to find in place because the steep-sloped Jurassic exposures were so often covered with a thick layer of talus. In many places, the talus slope had failed, and in sliding down it dislodged and scattered bones, obscuring their points of origin. He found at least two old landslides containing traces of dinosaur bones. Eager to explore the Jurassic beds on the north and west sides of the hills, Wieland was getting impatient for some company, and more especially for a horse.[23]

Thomson arrived in Sturgis with a team and wagon and an extra saddle horse on the 10th of October. Granger arrived by rail two days later. They set up camp a few miles south of town, near Wieland's dinosaur leads. Together they did some preliminary exploring. Someone found a new prospect with one large femur, in place, and several other limb bones apparently associated. Also, they located the quarry worked previously by the S. D. School of Mines student, which had hitherto escaped Wieland's notice. It turned out to be a small site of some promise, covering an area of approximately 150 square feet, and containing an abundance of disarticulated bones of various large dinosaurs. The bone-producing layer was within two or three feet of the surface and could be worked with ease, or so Wieland reported. Granger, on the other hand, thought the bones were unremarkable. Even so, given the scarcity of workable localities on the eastern slope of the hills, Wieland thought it might be wise to work this quarry for the DVP. But he would not break ground there without explicit instructions to do so from Osborn.[24] Perhaps he was reluctant, after all, to poach on another man's prospect. With a few of their own more or less promising specimens in sight, Granger and Thomson started digging, while Wieland left to explore.

If Granger thought the burden to produce now rested entirely with Wieland, he was mistaken. A letter from Osborn turned the pressure back on him. "I hope you and Thompson [*sic*] will do your very strongest prospecting with Mr. Wieland, so that we can establish a good base for our Dinosaur work next season," Osborn wrote. "Report from Pittsburg is that Hatcher has had a wonderful season's work, sending back three car loads of fossils, if this is true, he has probably done better than we have."[25]

By October 14, Wieland was headed north and west on horseback, hoping to round up some better specimens. To the north, between Whitewood and Spearfish, the dinosaur-bearing beds thinned out and he found nothing. He continued west and north, dropping down into the valley of the Belle Fourche River, crossing into Wyoming, and then climbing over the Bear Lodge Mountains. He ventured as far northwest as the valley of the Little Missouri River, where the Jurassic beds dip below the surface. He found extensive exposures of Jurassic sediments along the course of the Belle Fourche and especially in the many side canyons and creeks that branched away from the river. On Hulett Creek, for instance, he located three or four prospects of beautifully preserved bones. He also heard reliable accounts of bones at other localities to the south.[26]

After almost a week of exploring, a cold, driving rainstorm chased Wieland indoors, forcing him to give up the search for fossils a few days earlier than planned. Even so, he felt that he had seen enough during his hasty survey to justify a return trip to this same area during the succeeding field season. He wrote to Osborn from a warm and dry lodging in Aladdin, Wyoming, with an enthusiastic prognosis of the area's potential. "The country is very picturesque and will be pleasant to work," he predicted. The dark red Spearfish Formation, the same stratum that forms a valley around the Black Hills, was here exposed in vertical cuts that lined the banks of the Belle Fourche River like a brick wall. Dark green pines crowned the heights above, while thick grass covered the valley bottom. Above all was Devil's Tower, a thousand foot column of rock standing like a sentinel in plain view of the fossil localities on Hulett Creek. Although the area had never been worked before for fossil vertebrates, Wieland was certain about future success in this locality. "I shall not be surprised if this area should to some extent rival the old Como region," he claimed. But he also hinted at the possibility of competition. While mailing letters in Hulett, Wieland made inquiries after gigantic bones of the local postmaster, who was just then sending off some rock samples to Wilbur C. Knight, state geologist of Wyoming. Perhaps the obliging postmaster would make mention of Wieland's interest in local dinosaurs.[27]

Following his discoveries along the Belle Fourche, Wieland returned to base camp on October 23, 1900, with a new, gloomier outlook on the relative value of working the eastern slope of the Black Hills for dinosaurs. He spent a few days with Granger and Thomson working on their several prospects, all of which were turning out poorly. He was none too pleased with the materials obtained, so far, and wrote to Osborn with the disappointing news. "We shall take what we [can] get from two of the [prospects], but a really first class Dinosaur we have not thus far found," he lamented.[28] Granger, too, was frustrated by the lack of good specimens in this locality. He sent a progress report on the 31st, which may have been meant to trick Osborn into treating the field party to a welcome change of scenery: "According to [Wieland] there is great promise in the new country northwest of the Hills, and it will be a great satisfaction to get into a new section. Judging from Wieland's description and from samples of bone which he brought back, it is, as he says, another Como. I hope so, anyhow." Meanwhile, the weather was turning colder, bringing a slight flurry of snow.[29]

Osborn's reaction to Wieland's discouraging news was surprisingly coddling, especially in light of the former's inflated expectations. "I am entirely satisfied with what you are doing," he wrote to Wieland. "It is only by this careful prospecting and working up of the country, which is always more or less disappointing and discouraging [that] the final discoveries are made." With respect to the prospecting, Osborn provided some instructions, which must have come as a disappointment for Granger, once Wieland handed them down the chain of command: "I think you will do well to continue your work in the South [near Sturgis], reserving the more serious exploration of the North for another season. Even if you do not make any remarkable discovery, you will have at least reached a satisfactory conclusion . . . that these beds are not promising." He closed his letter to Wieland with an unprecedented endorsement: "Remember that I have entire confidence in your energy, good judgment and ability to find anything, if it is to be found."[30] In all of Granger's years at Bone Cabin Quarry, he had never received anything quite so extraordinary.

Wieland left again in the first week of November for a site to the south approximately three miles north of Piedmont. He found a new sandstone horizon there, only a few feet thick, with isolated bones of several different kinds of dinosaurs, both large and small. On an uncomfortably cold and windy day, he opened up a trial excavation and secured a good femur, one foot long, which he identified as a possible new species related to the duck-billed dinosaur *Camptosaurus*. He also reported the presence of a string of small vertebrae and a few other bones, possibly of birds, although he did not bother to collect them. In the talus, on the slope below his excavation, he also found traces of much larger bones. Piedmont was the scene of Wieland's previous work for Marsh, which netted a fine specimen of *Barosaurus*. He was so confident of renewed success in this locality that he left his trunk, together with his bedding, in camp near Sturgis, assuming that he would soon be calling on Granger or Thomson to come join him with the DVP outfit. And although Granger's assistance was never required, still Wieland thought it might be useful for him to see the exposures around Piedmont. Wieland wrote to Granger with this suggestion, imploring, "If you should come down here . . . could you possibly bring my trunk?"[31]

On November 16, 1900, it started to snow heavily and it did not stop for days. Nighttime temperatures dropped as low as zero. Conditions were so bad that Wieland abandoned his prospecting and returned to Sturgis to help with the dig. Meanwhile, Granger and Thomson were waiting for a break in the storm to complete their work and pack up the campsite (see figure 24). They waited a week, then ventured out in the

Figure 24. Field-workers at their wintry campsite near Piedmont, South Dakota, late in the field season of 1900. Image #17976. American Museum of Natural History Library.

cold to rebury Granger's unfinished *Brontosaurus* quarry with a thin layer of dirt—they could return to collect the remaining bones in the spring. Granger reasoned that the bones had been within a few inches of the surface for untold years and he doubted that another six months would make much of a difference. Several valuable fragments of this specimen, including various vertebrae and limb bones, had already been taken out and packed for shipping. Thomson had been working on a *Morosaurus* specimen, which consisted of ribs, chevrons, caudal and dorsal vertebrae, and a good scapula. He had extracted this entire skeleton before the bad weather moved in. Wieland, however, thought there was a small chance of obtaining more bones from this quarry, as the dig extended laterally a scant four or five feet beyond the last bone recovered.

Their spoils filled five boxes weighing approximately fifteen hundred pounds. Wieland sent the shipment to Edgemont, South Dakota, where Barnum Brown, just returned from his stint in Patagonia, received it and forwarded it to New York together with his own shipment of Cretaceous fossils. Granger and Thomson parted ways with Wieland at Minnekhata, an important locality for fossil cycads, on the morning of November 29th.

They expected to be back at the museum during the first few days of December.[32]

Making the Best of a Bad Situation

Plans for DVP fieldwork in the summer of 1901 were more ambitious than ever, and Osborn worried about the exceedingly slim margin he was running on his budget for field expenses. He nevertheless renewed Wieland's engagement as leader of a second Black Hills expedition for Jurassic dinosaurs at the reduced—but still relatively extravagant—rate of $150 per month. (Granger and Thomson would accompany Wieland in the field for $80 and $60 per month, respectively.) Osborn would only commit to Wieland for two months of guaranteed employment, in May and June, with a possible extension into July, depending on results. He urged him to exercise a careful supervision over any expenses and to use his best judgment in the choice of specimens to be taken out and shipped to New York. "In case the locality does not turn out as rich as we expect it may be necessary to alter the plans of the party," he warned.[33]

Accordingly, Granger, Thomson, and Kaisen left New York on the evening of May 4, 1901, and headed west. In Nebraska, on the 8th, they parted ways. Kaisen proceeded west to Medicine Bow, while Granger and Thomson caught a train north to Rapid City, South Dakota, where they retrieved their outfit and purchased provisions. Granger tried to scare up a cook, but failed to find one to his taste. Once the wagon was loaded, they drove it in a northwestern arc along the rim of the Black Hills, over the Bear Lodge Mountains, and down into Hulett, Wyoming, where they made camp on May 15. Wieland arrived by stage the next day and together they moved the campsite closer to his best prospects of the previous fall. At Wieland's instigation, Thomson was pressed into service as the expedition cook. This was merely a stopgap measure while Wieland tried to recruit someone more permanent. In a letter to Osborn, Wieland made reference to some unspecified difficulty with respect to victuals during the previous field season and seemed inordinately concerned about the present fragile state of the field party's morale. Clearly, something needed to be done about the food. All were grateful when Wieland found a cook to start on May 20, although Thomson was still baking bread as late as the 21st.[34]

For the rest of the month they prospected with indifferent results. They searched south along the Belle Fourche River more than twenty miles and west beyond Devil's Tower. One Sunday in camp, Thomson put his boot heel to a rattlesnake and then saved its skin—it was the only specimen collected during this entire unlucky period. Near the end of May they made their best camp of the season on the west side of Devil's Tower, where they awaited two special visitors: Osborn and his traveling companion, German paleontologist Eberhard Fraas.[35]

Osborn was making the rounds of the DVP field sites while guiding his German guest to some of the most renowned fossil vertebrate localities in the American West. Wieland's camp was his most important stop. "I am greatly interested in this camp," he explained in a letter to his wife, "for upon its success depends a large measure of our record for the season. I am not over sanguine." Nevertheless, he endured unusual privations. A luxurious Pullman Palace Car delivered Osborn and Fraas as far as Deadwood, South Dakota. There they boarded a frontier train, "as primitive . . . as you ever saw," which brought them to Aladdin, Wyoming, and the railroad's terminus. The train ran very late, so Osborn and his companion suffered a hurried dinner, then a cold buckboard stagecoach ride to Hulett. Granger fetched them to camp on the morning of June 1st, while Wieland and Thomson were off prospecting. After lunch, Wieland took his guests west to examine some Jurassic sections, which were overgrown with grass and trees and looked generally unpromising. Osborn had a serious discussion with Wieland about the very disappointing results of his survey to date. Wieland was so discouraged that he offered to leave the party at once, but Osborn refused him. The next day—a Sunday—all hands except Thomson went out four miles to the southwest to survey two prospects, including a *Brontosaurus*, and a bone bed tucked inconveniently below a hard ledge of sandstone. Wieland, no doubt, made the most of what little the DVP party had in sight. They returned to camp late in the afternoon, ate another hasty dinner, and then Granger drove Osborn and Fraas back to Hulett in the wagon, through a cold drizzle. At Hulett they turned in. Osborn slept very badly, and then rose at 3:30 a.m. for a torturous five-hour stagecoach ride. Tormented by wind and rain, they slogged their way back to the railroad at Aladdin. They had originally planned to spend two or three days exploring the White River Badlands with Hatcher and a party of U.S. Geological Survey scientists. "I ought to spend more [time in the Badlands] but I am so anxious to get home to you that I will cut it short," he explained sweetly to his wife. In fact, on account of the cold, he decided to cancel the visit altogether. He left Fraas with Hatcher on the railroad platform at Hermosa, South Dakota, and then sped east for home.[36]

Despite his romantic rhetoric about the pleasures of camp life, Osborn disliked fieldwork and did it only very rarely. In letters to his field assistants, he liked to wax poetic on the joys of fieldwork, especially the camaraderie, the fresh air, and the rewards of sound sleep and good health. "Your account of camp makes me long to come out and join you," he wrote in a typical letter to Brown.[37] In a letter to Oliver Perry Hay, who was then toiling on Osborn's behalf in the Bridger Basin of Wyoming, Osborn wrote, "I congratulate you on the stiff joints and sore muscles, the sound sleep and the good appetite of camp life. I wish I were enjoying some of these luxuries."[38] But, in reality, he did not adapt well to the rigors of the field. Despite innumerable special accommodations, including luxury travel, a light workload, a flexible schedule, and a staff of hired help at his beck and call, fieldwork more often than not left Osborn feeling uncomfortable, sleep deprived, and profoundly homesick.

Osborn also resented the responsibility of planning and managing the conduct of his field parties, although this was largely a product of his own lack of confidence in his field-workers. He missed Wortman, and he longed for a worthy replacement to fill the role of DVP field foreman. He set his heart on luring Hatcher, long his ideal of a field-worker, away from the Carnegie Museum. "I . . . look forward with great relief to next year when Hatcher can do the worrying and the planning for these expeditions," he confided to his wife. "It will be a grand thing getting him in the [American] Museum—he is anxious to come and I think we will get on smoothly."[39] But Osborn was mistaken. He had inferred something from a conversation he had had with Hatcher during the course of a luxurious rail excursion through Colorado and Utah earlier in the field season that led him to believe that the Carnegie curator "practically" agreed to join the DVP,[40] but it never happened. Hatcher was perhaps the one prominent paleontologist that Osborn most admired and most coveted for his DVP staff, but could never quite ensnare in his patronage web.

Osborn cannot have taken much pleasure in his visit with Wieland, although a polite letter to Granger claims that he did. "I enjoyed my visit to the camp very much," he wrote, "and was glad to find you all working in such good spirits." At the same time, he found it necessary to slash the amount of money originally appropriated for the Black Hills expedition, and was clearly anticipating a reorganization of field parties after July 1, barring some unexpected reversal of fortune. Granger would be compelled to spend the next month helping Thomson root around uselessly among their few unpromising prospects, while Wieland continued to scout for something better. "I am somewhat disappointed at the results thus far," Osborn wrote with measured understatement, "but the prospects I saw . . . convinced me that the Dinosaurs are [there] to be found—if we search hard—and enjoy better luck."[41]

Unfortunately, Osborn's departure was a harbinger of more bad things to come. Prolonged stormy weather delayed Granger's return to camp and cost the party two full days of digging. Then both of their prospects failed in short order. The first, Thomson's *Brontosaurus*, consisted of a washed-out (removed from its original place) and badly weathered hind limb and pelvis partially buried in talus. Thomson traced the source back to a ledge of very hard sandstone, where he found a series of small caudals leading distally into the bedrock. With nothing but the tip of the tail remaining, the prospect was abandoned. On the bone bed, as the term implies, rested a great many specimens, but nothing sufficiently well preserved to bother digging up. This, too, was abandoned. Some unspecified traces of carnivorous dinosaurs also turned up in this locality.[42]

Loomis joined the unlucky party on Sunday, June 9, 1901. His task was to study the local Jurassic sections, collect fossil invertebrates, and help to excavate dinosaurs, when, and if, necessary. With Loomis in tow, the party pulled up stakes and headed south along the Belle Fourche River, stopping when it began to rain and thunder violently. The river flooded, marooning the party for a week while the rain continued to fall. They did a modicum of fruitless prospecting whenever they could, but mostly they were idle and bored. Some of the men took to hunting prairie dogs for sport, or stooped to stoning an unfortunate red squirrel. When the weather finally dried up, they moved farther south, finding extensive Jurassic exposures, but no worthy specimens.[43]

With the end of June approaching, Osborn finally admitted defeat. "I fear the Black Hills is a failure. . . . We have had poor luck and must make the best of a bad situation," he wrote to Granger. His solution was to dissolve the party. He decided to send Loomis and Thomson to join Matthew, just returning from Utah, the scene of his own brief (and unsuccessful) Jurassic dinosaur reconnaissance, in the hunt for fossil mammals in eastern Colorado. He wanted Granger to join Kaisen at Bone Cabin Quarry.[44] Some weeks previously, with the outlook of the Black Hills expedition bleak, and getting bleaker, Wieland sheepishly offered to continue collecting for the DVP for "purely nominal wages, which I would leave absolutely and uncomplainingly . . . at your discretion."[45] Nothing ever came of this suggestion, however, so Wieland quit the party in defeat and returned east on June 30, 1901, with five days remaining on his contract.[46] His failed experiment with the DVP was likely the impetus behind Wieland's future concentration on paleobotany.

A few loose ends remained on the eastern slope of the Black Hills, however, which Osborn had evidently forgotten. So Granger moved the outfit back to Rapid City, arriving July 7. They wasted two days in town waiting for Osborn's blessing. Then they headed north to Piedmont on the 10th. There they spent two days looking over Wieland's sandstone locality—where small birdlike bones had been spotted—but found absolutely nothing. Granger was quite certain that he found the very spot where Wieland unearthed the small *Camptosaurus*-like femur. But all three collectors examined this site very carefully and found no additional traces of bone. They next searched for a mile or more in either direction, again finding nothing. So they moved to Sturgis and reopened Granger's *Brontosaurus* quarry. They found two more dorsal vertebrae, which were so soft and weatherworn that Granger left them behind. The mid-July heat was staggering, over one hundred degrees in the afternoon. They prospected one more day, and then drove back down to Rapid City on the 17th, through a punishing hailstorm. Loomis and Thomson continued south by wagon to rendezvous with Matthew. Granger caught an evening train in Rapid City for Wyoming.[47]

NINE

Watch the Dinosaur Shrink!

Plans for a second season in western Colorado in 1901 for the Field Columbian Museum party profited from their experience of the previous summer. For instance, E. S. Riggs wanted to add a fourth man to the party, in order to better handle the hard labor and the heavy lifting associated with collecting gigantic sauropod dinosaurs. He also favored a much earlier start date (in mid-April rather than mid-June), expecting to take advantage of cooler working weather in the spring and to "forestall prospective rivals."[1] O. C. Farrington, curator of geology, seconded Riggs's recommendations, emphasizing the competitive advantage of an early spring departure. The Field Columbian Museum party had located a very promising-looking Jurassic dinosaur late in the previous field season and all concerned were eager to lay claim to the specimen. In a letter to Director F. J. V. Skiff, he urged, "There is every reason to believe that [the new prospect] will turn out to be even more valuable than [the dinosaur collected] last summer. The desirability of the Museum securing [it] in advance of prospective rivals can hardly be questioned."[2] Riggs and his party had taken out a partial, associated skeleton of enormous size the previous summer, but this specimen would not be suitable for mounting because so much of it was missing. The new prospect found late in the field season and left in the ground showed promise of being far more complete. Riggs probably could have used his previous successes to leverage a more liberal appropriation for fieldwork. Instead, he submitted a carefully considered yet very modest budget request of $700—the increase reflecting the cost of one additional collector.[3] His request was granted in full, but so small a sum would leave little margin for error.

Riggs purchased supplies and equipment in Chicago that were either unavailable or prohibitively expensive in the West. He bought camera supplies for H. W. Menke, and wedges, a sledgehammer, a whiskbroom, and gum arabic for taking out the dinosaur. Three new camp beds and two stools would make field accommodations more comfortable for the collectors. Riggs had an ambitious plan for shortening his supply lines in Colorado, and for this he purchased some hardware, including a screwdriver, wrench, carpenter's square, and hand saw. With his own money, he bought a rifle and cartridges, presumably for hunting.[4]

They left Chicago by train in the latter part of April 1901. The railroads again provided free passage for the men of the party, including Riggs, Menke, and V. H. Barnett, along with their cargo, from Chicago to Grand Junction and return.[5] They narrowly escaped being in a disastrous train wreck near Leadville, Colorado. Their train had been divided in half in order to make the steep mountain grade. They chanced to be in the second section when the first was struck hard by a freight train. They were held over on a siding below the wreckage for half a day, and then steamed into Grand Junction on the morning of April 25.[6]

In the morning, Riggs and his party paid their respects to friends in Grand Junction, including the cooperative dentist Stanton M. Bradbury, and then spent the rest of the day in town recovering from the harrowing journey. The following day, they bought two horses, rented saddles, and hired a team and wagon to haul one month's supply of plaster (400 lbs.), burlap, oats, hay, and groceries to their field site, just across the Grand River from Fruita, Colorado. They pulled first for the ranch where their camp equipment had been stored for the winter. A lack of adequate roads in the Redlands made the passage painstakingly slow for the horse-drawn wagon. The next day, the 27th, they "overhauled" all their baggage and brought one full load to their new field site near Fruita. They made permanent camp on the south bank of the Grand River, at a spot selected the previous summer both for its convenience to the prospect and for its charms. Riggs designated the site Camp Cottonwood and described it in detail in a letter to Helen Mosher, his Chicago girlfriend:

> Our camp is becoming quite inviting as camps go. Imagine a deep and rocky gulch with a dry creek bed leading down from the hills to the river. On its [west] bank [there is] a little projecting point of level ground bearing an isolated clump of four cottonwood trees between a steep bluff and the river bank. A few clumps of willows line the

bank of the river up and down where there has been sufficient soil lodged to afford a footing. This jutting point of level ground scarcely 40 by 100 feet and eight feet above the water's edge is our homestead. Following the tortuous and bowlder [*sic*]-strewn bed of the arroyo three or four hundred yards we come to the scene of our labors. This is at a point 75 or 80 feet above the creek bed on a slope facing eastward and commanding most of the valley. A fair enough prospect now that the weather is fine and the snow lying on Grand Mesa to the eastward and a faint line of white showing on the high Uintah range away to the northward make one forget of what June and July have in store for us on this naked and unprotected slope.[7]

The late April weather was still raw and wet. Riggs caught a severe cold his first night camping out at the ranch and was miserable for his first few days at Camp Cottonwood. When the teamster arrived in camp with his trunk, Riggs was finally able to self-medicate with bromo-quinine—he would be back on his feet in a matter of days. Meanwhile, the rain continued. With their leader stretched out on his camp cot, convalescing, there was little for the rest of the party to do. They spent several days making themselves comfortable and prospecting half-heartedly between cloudbursts in the hills near camp. Most important, they verified that their dinosaur prospect was still available for the taking. The Grand River, mud brown and swollen by snowmelt and steady rain, surged four feet. It continued to rain through the first week of May, but the water, which threatened briefly to drown their campsite, crested and then receded harmlessly.[8]

For the sake of greater efficiency, Riggs wanted to shorten his supply line. It was twenty miles in the saddle along a slippery cow path to get from Camp Cottonwood to the bridge at Grand Junction—even longer with a heavy wagon. Meanwhile, the general store, post office, telegraph, and railroad depot in Fruita were forbidden to the party by the impassable river. Riggs embarked on a plan to build his own boat. There was an unused ferry cable strung across the river at a convenient point less than one mile from Camp Cottonwood. The ferry crossing had been abandoned some years previously after a number of fatal accidents. Locals thought Riggs was foolhardy for trying to operate a ferry on the Grand River during spring flooding. "Everyone warns me of the danger in attempting a ferry, but I still think I can make it a go," he reasoned in a letter to Mosher.[9]

His plan was to build a large scow, twenty-four feet in length and nine feet wide, with a flat bottom and broad, square ends. Back in Grand Junction, at a local sawmill, he placed an order for two-inch lumber, specifying that the edges should be dressed smooth to make close-fitting joints.

He received delivery of his order at a spot where the river lapped calmly at the edge of town, and where a gently sloping shore provided space to work and easy access to the water. He hammered a frame together, and then nailed the bottom, the sidewalls, and the flooring to the frame. Next, he installed a homemade windlass to control the vessel's movement through the water. Once finished, the boat was levered up and set on rollers and then carefully pushed down the slope and into the water.[10] It rode the waves like a dinner plate.

The following day, May 14, 1901, Riggs ordered some bulk supplies, including canned goods and more plaster and hay, and had them delivered to the boat. With the heavy quarrying work about to begin in earnest, Riggs asked around town about the local labor market and was referred to a penniless German traveler named Alfred Jeager, who was then wandering the streets of Grand Junction looking for work. Riggs hired him, and together they piloted the boat, christened *Mary Ann*, down the river toward Fruita. Riggs and Jeager, who was handy with ropes, fixed their craft to the abandoned ferry cable and set off for camp.[11]

Once installed and fully operational, the ferry was a real boon for productivity. By crossing the river, Riggs was able to reach Fruita by a relatively easy three-mile journey, instead of the arduous twenty miles required to get to Grand Junction. Essential supplies, especially those needed in bulk, like plaster and feed for the horses, could be conveyed far more conveniently by water than by wagon. Better still, transporting the specimen at the end of the field season, which was expected to consist of several tons of mineralized bone and sandstone matrix wrapped in plaster jackets, would be far easier and faster by using the depot in Fruita. The diggers also realized a welcome improvement in their diet, including fresh meat, eggs, milk, cookies, and fresh fruit, as well as Menke's Bull Durham pipe tobacco, thanks to the convenience of shopping for groceries in Fruita.[12]

While Riggs was away building his boat, Menke and Barnett began working at the quarry. With Riggs and Jeager added to the force, work progressed more rapidly. First they removed the overburden, a heavy layer of hard sandstone reposing above the specimen. As they picked and shoveled some ten, then twelve, finally fifteen feet back into the slope, more dorsal vertebrae and ribs were exposed, only slightly displaced from their natural positions. Dynamite helped with the removal of excess rock, especially as the facing of the quarry climbed eighteen feet up the slope. To protect the exposed vertebrae from falling rocks, they were bandaged with a thick protective layer of burlap strips soaked in plaster and allowed to set up firm like a cast. A thin layer of paper prevented the plaster from sticking to any exposed bone. The bones were later removed in large blocks, fully encased in rigid plaster jackets. They were then carried away from the quarry to a safe spot to await packing and shipping.[13]

The dinosaur was preserved lying on its right side and was remarkably complete and undistorted. The left hind limb and most of the ribs from the left side were missing entirely, while the left ilium was preserved as fragments. Riggs believed the entire skeleton had been present when it was first buried in sediment and fossilized, but that the neck, head, and forelimbs had been subsequently eroded away. Unfortunately, they found very few surface fragments— perhaps some of the more substantial pieces had been carried off by other fossil hunters. Most of the elements of the vertebral column, beginning with the badly weathered last cervical and continuing through to the thirteenth caudal, were preserved in their natural positions. The next ten caudals recovered were found more and more displaced distally. Bones of the pelvis were found more or less in place. The right femur was just slightly removed from its articulation with the pelvis.

By the time the dorsal vertebrae, pelvis, and femur had been uncovered, the quarry stretched more than twenty-five feet from end to end. The caudal series, unfortunately, curved backward at the pelvis and tailed directly into the hillside. To chase it down, the party resorted to tunneling. Accordingly, Riggs sent a man north to the coal mines at the Little Book Cliffs to borrow a rotary drill (see figure 25). The drill was mounted on a tripod and fixed in place by turning a crank that forced a threaded, sharp-ended vertical shaft into the roof of the tunnel. The bit, secured into a gear shaft with stout cranks at either end, was drilled into the rock by two men turning the cranks in tandem. A stick of dynamite, with a long fuse, was then inserted in the hole, the hole tamped to the surface, and the charge detonated. By this means, the party blasted out a tunnel that reached twenty feet deep into the hillside, just above the level of the bones. They rigged a sheet of canvas over the mouth of the tunnel to provide extra shade, and then resumed quarrying the bone layer on the floor of the tunnel with excellent results. The displacement of the caudal vertebrae increased distally, while their size and complexity steadily decreased. Once Riggs decided that the point had been reached where the effort to recover the distal-most caudals and their associated chevrons was greater than the reward of obtaining them, he called a halt to the tunneling.[14]

Because Riggs abandoned the tunneling before the end of the tail was reached, there was a chance of recovering more bones from this speci-

Figure 25. Two men set up the drill preparatory to tunneling for the tail. At the bottom left, several dorsal centra are exposed in situ, covered with a protective layer of burlap and plaster. Courtesy of the Field Museum, Negative #CSGEO4005.

men on a follow-up expedition. In 1992, paleontologists from the Museum of Western Colorado in Grand Junction returned to his site to try to recover more bones from the tail. With the aid of a backhoe and air-driven tools, they found five additional pieces, including four chevrons from the tail and an eight-inch segment of rib. Relics of Riggs's field party turned up, as well. The first day of digging revealed rounded stick matches, nails, bits of plaster, burlap, and newspaper shreds. Further digging uncovered Riggs's original mining timbers, a peanut shell, a sawed-off and well-worn shovel, and a whiskbroom.[15]

Shipwreck!

A mishap with the ferry coincided with the start of tunneling. On the afternoon of May 29, 1901, Riggs and another member of the party—probably Barnett—rode down to the south landing, crossed the river, and walked into Fruita for supplies. There, Riggs purchased a boatload of plaster in ten 100-pound sacks. He also bought dynamite and food, both of which were in critically short supply. A courier delivered these items at the north landing and helped load them onto the ferry. Riggs and his companion then launched confidently into the river. Nearing the south landing, a strong current veered into the shore. Riggs was on the bow heaving at the windlass that hauled the boat slowly toward the bank. His mate was sitting idly at the stern with the heavier part of the cargo. From behind, Riggs heard a cry, "Look there!" He turned and saw a plume of water jetting through a seam that had opened up in a sidewall. (Riggs, an experienced carpenter, blamed the hot summer sun for warping the boards.) In response to the crisis, Riggs ordered bags of plaster shifted forward to the bow. Regrettably, this only served to raise the stern sufficiently to let the bow, now facing upstream, nod into the waves and take on more water. Riggs stayed at his post until one of the guy lines snapped, casting him into ice-cold water up to his chin. The river swept over the boat, carrying away most of the plaster. When the second line snapped, the boat righted itself and began drifting freely downstream. Riggs stayed with it, while his mate struggled to the south shore.[16]

Riggs climbed back into the half-swamped craft, grasping for something to bail it out with. He tore up a loose floorboard and laid it across his knee to flip the water out, but this was practically useless. Watching for an opportunity, he drifted more than three quarters of a mile. When the current swept him close to the south shore, he seized the oars and rowed vigorously for land. Then, with rope in hand, he leapt ashore, soaked to the skin but unhurt. The boat, it seems, was also in reasonably good shape. The cargo, on the other hand, was a total loss. He tied up the boat and resumed bailing until his wet companion rode up, leading an extra saddle horse. They tied up the boat, and then rode back to camp, changed, and walked to the quarry with dramatic news of their shipwreck.[17]

The next day, all hands rode down the river to rescue the boat, which required more than half a day's effort to tow back to their landing. Fortunately, Jeager, who had once been a sailor, was proficient with knots and rope and fearless on the ratline. He offered to scurry out over the river to reattach the boat to the cable. They then lost another full day at the quarry in putting the *Mary Ann* back in working order. This done, the whole hungry party repaired to Fruita for rest and resupply. In town, they read a wildly exaggerated and inaccurate report of the accident that had been published in the *Grand Junction News*. It read, "A party of government geologists attempted to cross the Grand River, Wednesday, and the cable broke, precipitating the party into the stream. Luckily they were all saved, narrowly escaping a watery grave."[18] Riggs thought it wise to send a reassuring telegram to his nervous museum superiors. Evidently, he learned a valuable lesson about the load capacity of his boat, returning to work at the quarry with only half the amount of plaster lost in the accident.[19]

By early July 1901 the digging was finished. Riggs hired a workhorse to haul the specimen from the quarry down to the boat landing, then began ferrying the blocks across the river, a few at a time. Once the specimen had been entirely removed, he shifted camp to the north landing, where the party began boxing the specimen for shipment to Chicago. In Fruita he bought straw for packing and wood and nails for building crates. Later, he hired an extra man to help haul his spoils to the depot in Fruita. Altogether, there were fifty-one crates of fossils, plus six parcels of camp equipment and supplies, totaling 18,400 pounds. By July 17th, they were ready to return to Chicago. Riggs paid Jeager a measly $73 for two months and three days of labor, they parted ways amicably, and then he, Menke, and Barnett boarded an eastbound train. With the money remaining on his appropriation, Riggs bought two upgrades for Pullman berths for the long trip from Denver to Chicago. Somebody, maybe Menke, probably Barnett, made the long trip home on a seat in coach.[20]

Riggs was saving his money with a purpose over the field season, returning to Chicago with $300 to his name. In August, he took a quick and frivolous excursion to Buffalo, New York to see the Pan American Exposition and narrowly missed history on September 6, 1901, when an anarchist fired two rounds at President William McKinley. But Riggs had history of his own to make. He hurried back to Chicago where he married Helen Mosher in a small, private ceremony on the 11th, in a house on Chicago's South Side. The house was decorated in palms and smilax and the bride carried a bouquet of lilies of the valley. When the happy couple left on their Lake Michigan honeymoon, the president seemed to be recovering from his gunshot wounds, but he took a turn for the worse. McKinley died from gangrene early in the morning on the 14th, and Vice President Theodore Roosevelt, hero of the Spanish-American War and a boyhood friend of Osborn's, became the twenty-sixth president of the United States.

Riggs's Closest Competitors

Riggs and the Field Columbian Museum were pioneers in Jurassic dinosaur fieldwork west of the continental divide, but the competition was creeping ever closer. In April 1901, Osborn received an unsolicited tip from Wilbur C. Knight, of the University of Wyoming, who had been surveying for oil in west-central Utah when he made a rich strike in Jurassic dinosaurs instead. He was confident that this discovery represented one of the most extensive exposures of Jurassic dinosaur-bearing strata ever found in the Rocky Mountain region. Unable to capitalize on this surprise discovery for his own institution, he passed on the relevant information to Osborn, including a sketch map and a geological section, and urged him to send out a collecting party as soon as possible. "I was impressed with the great thickness of the beds and what appeared to be the great vertical range of the fossil bearing strata . . ." he wrote. "I do hope that this region may turn out something of great interest. Had I had the means I would have worked it this season myself."[21]

Osborn then unknowingly strayed very close to Riggs's locality in late May 1901, when he and John Bell Hatcher, then working for the Carnegie Museum, boarded a special excursion train in Cañon City bound for Green River, Utah, where they both made a preliminary examination of the Jurassic exposures described in Knight's letter. Although they found a number of dinosaur prospects at Knight's locality, neither paleontologist recorded any general impressions about its potential productivity. Perhaps, after all, dinosaurs were not the chief objective of this particular trip, which differed markedly from the typical turn-of-the-century field expedition. Edward T. Jeffery, president of the Denver & Rio Grande Railroad and a close friend of Osborn's, provided the use of his luxurious private railcar for the occasion, complete with a personal attendant, chef, and gourmet provisions. Twice they steamed through Fruita, Colorado, within a mile or so of Riggs's quarry, puffing their cigars, drinking wine, and enjoying the scenery. No one noticed the Field Columbian Museum party scratching at their hillside just across the Grand River. Osborn excelled at this privileged style of fieldwork, but to a hardscrabble geologist like Hatcher, chasing fossils from a Pullman car must have seemed somewhat distasteful. Although this excursion appears to be a cooperative venture involving prominent paleontologists from rival museums, it was, in fact, strictly predatory: Osborn had invited Hatcher to participate because he was then attempting to recruit his services for the American Museum. Hatcher must have been tempted by Osborn's advances, but he remained loyal to the Carnegie Museum.[22]

To get an adequate assessment of Knight's locality, Osborn sent out another DVP representative, William Diller Matthew, to make a more thorough investigation. Matthew, who went to Utah for a week of exploring in early July 1901, searched "with some care" the same Jurassic exposures that Osborn visited near the confluence of the Grand and Green rivers. There he found "a good many bones," but he did not think the locality was very promising. The difficulty was that most of the prospects seemed to consist of individual bones only, rather than partial skeletons. Worse still, nearly all the specimens he located were found in a prohibitively hard sandstone or conglomerate matrix. Discouraged by the field season's misfortunes to date, he left Utah to link up with his DVP colleagues on the eastern slope of the Rockies. A relative newcomer to fossil vertebrate fieldwork, he worried about playing the bad luck role of a Jonah. If Matthew followed Knight's instructions to the letter, then he would have made his rail connection for Green River in Grand Junction, Colorado. Bradbury's office was only a few blocks from the depot. Gathering information there about Riggs's Jurassic localities in western Colorado could have been easily done, but there is no record that Matthew made the effort. He did take the trouble to note from the window of his train that Jurassic exposures extended eastward into Colorado along the Grand River. If permission could be obtained to use a handcar along the Denver, Rio Grande, and Western tracks, it might pay to prospect these beds further.[23] What Matthew did not know was that Riggs was just then working those very beds.

That same summer, Osborn received an unsolicited tip from Willis T. Lee, a high school principal in Trinidad, Colorado, and yet another University of Kansas graduate, about some Jurassic exposures extending several miles on the eastern slope of the Rockies near the town of La Junta and along the Las Animas River. Lee knew Osborn personally. They had met in the Freezeout Hills of Wyoming in the summer of 1899, when Lee had been a participant with S. W. Williston's KU expedition. Osborn sent Barnum Brown to follow up on Lee's lead. Over a period of five days, Brown found four scattered dinosaur prospects, but nothing significant enough to justify an extended stay or an additional hire. Lee, it seems, had offered to work these beds during his summer vacation. And though he had some bone-digging experience under Williston, Brown reminded Osborn that he had not approved of some of Lee's work (how Brown could have known this is hard to imagine, however). Moreover, working conditions at Lee's locality were quite bad. The Jurassic exposures that Brown examined were on very steep slopes and would be difficult or impossible to work with horse and wagon. Worse, the slopes were covered with a dense growth of grass and cedars, making the fossils difficult to find. He advised against working in this region while there yet remained more favorable localities in Wyoming and elsewhere, including western Colorado. Brown had learned of Riggs's work through the press and envied his former classmate's success. In a letter to Osborn, he confessed: "I don't know Rigg's [*sic*] exact locality. It must be rich from the newspaper stories. I wish we might have a show at it."[24] Brown wanted to venture farther west in search of dinosaurs, possibly joining Matthew in Utah, but a discouraging letter from the latter urged him to head for Wyoming instead.

"Watch the Dinosaur Shrink!"

In mid-June 1901, while Riggs was away in Colorado collecting his dinosaur, officers at the University of Chicago were celebrating their tenth convocation with great fanfare. The occasion sparked much talk and excitement about the rapid growth and future expansion of the university. William Rainey Harper, the president of the university, envisioned an immense endowment and had ambitions to build the largest and finest institution of higher learning in the country.[25] The Field Columbian Museum, which was then located in the old Fine Arts Palace in Jackson Park, had always had a loose affiliation with the University of Chicago, only a few blocks away. For the sake of greater scientific effectiveness, increased economy, and lesser duplication of effort and material, Harper harbored plans for a closer, more formal association with the museum. In September, he and the museum's director, F. J. V. Skiff, negotiating in conformity with instructions with Marshall Field himself, worked out a tentative agreement for confederation and submitted it to their respective trustees for formal approval.[26] Although Skiff was in favor of the merger, some of the trustees were not, and it never passed. Meanwhile, rumors of Harper's expansionist policies and word of Riggs's new dinosaur reached the *Chicago Daily News* at roughly the same time. On June 21, 1901, both were lampooned in a front-page editorial cartoon (see figure 26) drawn by Luther Daniels Bradley entitled: "Watch the Dinosaur Shrink!" In it, Harper reacts to Riggs's discovery, saying: "Very neat Riggs, very neat! I may let your little place run as a side show to mine."[27] Director Skiff, whose sense of humor was squelched by his strict sense of order and duty, was probably not amused. Nor, likely, did he appreciate the suggestion that the Field Columbian Museum was Riggs's "little place." Still, to be mocked so prominently, and to be placed in the lofty company

Figure 26. A *Chicago Daily News* cartoon satirizing University of Chicago President W. R. Harper's expansionist ambitions. Nothing ever came of the proposed merger of the Field Columbian Museum with the university. From *Chicago Daily News*, "Watch."

of President Harper—albeit, in a cartoon—were sure signs of the rising prominence of Riggs and his paleontology program.

Despite the lack of official sanction, Skiff and Harper continued to work toward a closer relationship between museum and university. In the winter of 1902, for example, when Harper was contemplating a belated replacement for vertebrate paleontologist George Baur, deceased since 1898, he hatched the idea of hiring Riggs's KU mentor Samuel Wendell Williston as a new faculty member with a joint appointment as curator at the Field Columbian Museum. At the university, paleontology was a separate and distinct department, with tenuous ties to geology and zoology.[28] Williston was already a prominent scientist who would be brought to the university as a full professor. To take him on at the museum, also, it would seem necessary to make him a full curator. Therefore, Harper proposed that the museum create a new Department of Paleontology, separate from Geology, to make ample room for a scientist of Williston's stature. Skiff was sympathetic to this idea, writing to Higinbotham: "I am very much in favor of dignifying Paleontology to the rank of a department. Prof. Williston . . . is as capable a man for the place [of curator] as is available and stands very high in his science."[29] Higinbotham, however, was reluctant to approve such a sweeping change, particularly as it involved the creation of a new department, without the agreement of the other members of the executive committee, then absent from the city on personal business. That is, until it was made perfectly clear to him that the university would share the cost of bringing Williston to Chicago. The finer administrative details of the appointment were left undecided until a quorum of trustees could be convened, but Higinbotham approved the expense. Beginning in October 1902, the Field Columbian Museum would have the part-time benefit of Williston's services for the bargain basement rate of $1,500 per annum.[30]

Unfortunately, problems with this arrangement arose even before Williston's arrival in Chicago. For example, a May 1902 letter from Director Skiff explained to Williston that his paleontology research program would be administered as a division of the museum's Geology Department, at least for the present. Under this arrangement, Williston would be answerable directly to Farrington, curator of geology. In fact, Skiff had asked Farrington to advise him on the character and scope of Williston's work, and the director had enclosed a copy of this report in his May letter.[31] In time, Williston would develop a strongly negative opinion of Farrington's shortcomings with respect to the biological aspects of paleontology. It seems likely that these feelings originated with Farrington's report, which does not survive. More bad news arrived in late July when, at a quarterly meeting of the board of trustees, the motion to create a new Department of Paleontology at the Field Columbian Museum was defeated. Williston's position as associate curator of geology, under Farrington, had been fixed permanently in place.[32]

Williston's subordinate position in the museum hierarchy was a source of great frustration for him, and he developed a deep resentment of Farrington's oversight. He was particularly bothered by his lack of direct access to Director Skiff and worried that Farrington might be misrepresenting the needs of the paleontology division to higher administrators. The result, as Williston saw it, was a plague of problems, including a lack of adequate museum funding for fossil vertebrate fieldwork and a misguided preoccupation with developing costly exhibits. To his friend J. B. Hatcher of the Carnegie Museum he confided:

> I may say (privately) that I suppose my relations with the museum will terminate soon. It was hoped that some effective combination could be made between the Museum and the University for mutual advantage, but I have no expectation that anything will result in the immediate future. . . . And I am rather glad of it. Until vertebrate paleontology is put on an independent footing in the museum it will be vain to expect much advancement. . . . I think that I can get as much money from the University as I could from the Museum for collecting, and I shall be largely free from the worry and trouble of expensive installations.[33]

Another Follow-up Expedition to Colorado?

Riggs had no opportunity for Jurassic fieldwork in the summer of 1902 because he was busy attending to the disposition of the several dinosaur specimens recently unearthed in Wyoming and Colorado.[34] Itching to return to the West, he spent the winter months of 1902–03 scheming about a renewal of fieldwork. He was curious about the dinosaur exhibition plans at rival institutions and wrote to Wilbur Clinton Knight, who was always a candid correspondent, inquiring after plans for his new museum building at the University of Wyoming. Was he thinking of attempting to mount a complete sauropod dinosaur? Knight responded that he intended to mount in parts only, at least for the time being. Riggs's other object was to learn a more precise locality for a dinosaur recently found on the eastern slope of the Big Horn Mountains of Wyoming by a party of Carnegie Museum explorers. Knight provided the information requested.[35]

Riggs finished a summer fieldwork proposal and submitted the same to Farrington on March 11, 1903. His plan called for two field parties with separate objectives: one risky venture close to his research interests piggybacked onto another Jurassic expedition devised to appeal to museum administrators. To win Farrington's unqualified endorsement, Riggs parroted his superior's pet exhibit project, writing, "In view of the value of large specimens for museum exhibition I would recommend that the collecting of [Jurassic] Dinosaurs [in western Colorado] be continued with the purpose of ultimately mounting a complete skeleton of one of the larger forms." Deposits similar to the dinosaur-bearing rocks across the river from Grand Junction proceeded far west along the Grand River and south along the valley of the Little Dolores River. Interested locals, including the Grand Junction dentist Stanton M. Bradbury and his Academy of Science colleagues, had given Riggs information about fossils found in several places in this region. Riggs proposed to take Menke and, if possible, Jeager west from Grand Junction with team, wagon, and provisions starting around May 1, 1903. They would prospect the Jurassic exposures along the banks of the Grand River until they reached Moab, Utah, and then they would follow the Little Dolores River back into Colorado. Expecting that a promising specimen could be located in one month's time, and confident that Menke could readily handle its exhumation, Riggs planned to leave Colorado to join a second party in Casper, Wyoming, on or about the first of June. The second party, composed of Riggs, James B. Abbott, a new fossil preparator at the museum, and William J. Baumgartner, one of Williston's new graduate students in paleontology at the University of Chicago, would work north along the eastern slope of the Big Horn Mountains. Numerous Jurassic dinosaurs had been found in this area, including the Carnegie Museum specimen pinpointed in Knight's letter and the skeleton scouted by Farrington back in 1898. Best of all, if a prize dinosaur could be bagged early in Colorado, Riggs could divert the second party to the Big Horn Basin to collect Eocene fossil mammals while still assuring a rich haul of exhibit-worthy Jurassic fossils for the museum. Riggs estimated the probable cost of the two-pronged expedition at a hefty $1,347.[36]

Farrington approved the proposal and passed it up the chain of command to Skiff, but not with the enthusiasm that Riggs might have wished or expected. He attached a brief and muted cover letter recommending the full appropriation, but his letter failed to provide any compelling reinforcement for Riggs's proposal. Instead, he cried foul about the lack of funding for fieldwork in vertebrate paleontology the previous summer and then recited this uninspired refrain: "The promising localities are being rapidly culled and hence collections should be made in them ere it is too late."[37] Skiff, then absent from the museum, declined to recommend the expedition to President Higinbotham without even having seen Riggs's proposal; probably the unprecedented expense—more than double the cost of the expedition of 1901—gave him pause.[38] Unfortunately, news of Skiff's reaction was never passed along to Riggs, who stewed hopefully about his expedition for more than a month. In mid-April, with only two weeks remaining before his proposed departure, he lost his patience. He wrote an ill-advised letter directly to Skiff to "beg to ask for information as to whether or not I shall be enabled to carry out my plans for field work." He gave good reasons for his exigency, writing, "If it is not practical for the party to reach Grand Junction before May 15th, it will become necessary to abandon all plans for work there this year. With that will go the immediate prospect of completing the skeleton of the large dinosaur."[39]

The letter violated Director Skiff's strict sense of museum order, and he responded (tellingly, in a letter to one of his own assistants, rather than to Riggs) more to the perceived impertinence than to Riggs's urgent request:

> With reference to the fossil expedition to Colorado [and Wyoming], the first thing that occurs to me is that I do not care to have Mr. Riggs make recommendations to me direct. They should be through the Curator of the Department. I am still in doubt as to the necessity for the expedition, but I am willing to take the matter up, if it is considered advisable, when I [return to] the Museum.[40]

Skiff's assistant presumably related the director's feelings to both Farrington and Riggs, giving some offense, but also offering a slight hope for affirmation. But two more weeks of restless waiting only netted a final refusal on April 28th.[41]

Skiff's rejection spelled the end for Jurassic dinosaur fieldwork at the Field Columbian Museum, at least for the time being. Riggs had enjoyed his field successes in Wyoming and Colorado and he harbored ambitious plans for sustaining—even expanding—his fossil vertebrate collecting program in order to keep pace with the competition in New York and Pittsburgh. Skiff wrecked those plans in an instant. Moreover, Skiff's contemptuous tone put Riggs and vertebrate paleontology back in their place near the very bottom of the museum hierarchy. From that place it would prove difficult to rise. Indeed, Riggs's Jurassic dinosaur venture at the turn of the twentieth century was an early high-water mark for vertebrate paleontology fieldwork that would not be equaled again at the Field Columbian Museum for more than two decades.

Williston's official connection with the Field Columbian Museum expired in October 1903, and he made no proactive effort to get it extended, despite the significant loss of salary and concomitant hardships.[42] He sent this gloomy prognosis to Hatcher, his Carnegie Museum confidant:

I am no longer connected with the Field Columbian Museum. I have no faith nor hope that this museum will accomplish very much in vertebrate paleontology—it certainly will not so long as the department is subordinated to geology and in charge of [Farrington,] who knows no more of the subject than he does of the biology of the moon.[43]

Harper, however, was eager to see this arrangement made permanent for the mutual benefit of museum and university. He encouraged Williston to plead his case directly with the museum president.[44] This Williston did in a lengthy and impassioned letter to Higinbotham dated December 28, 1903. In it, Williston urged "the immediate accumulation [of vertebrate fossils by] every possible energy of the museum." Certain materials, he argued, were rapidly becoming scarce and difficult to obtain due to the "extraordinary activity . . . among the chief rivals of the Field Columbian Museum." He likewise recommended that the exhibition of most of this material, a costly and time-consuming enterprise, be postponed until the museum could be relocated to a larger, permanent building. He cited a number of new fossil localities that were ripe for exploitation, and urgently recommended that "the collections from the Jurassic fields of Wyoming and Colorado be intermitted for a few years, because of their great cost, bulky size, but more especially because these fields are the least likely of all to be soon exhausted." He suggested that the museum purchase the fossil vertebrate collections of the University of Chicago, if possible. He advocated a larger budget for fieldwork, $1,400 the first summer and $2,000 for subsequent seasons, and substantial raises for the permanent assistants, including Riggs and Abbott (Menke was then about to transfer to the museum's Zoology Department). Most important of all, Williston explained,

[s]hould you desire me to take charge of the work, I must insist . . . upon entire freedom of all control save that of the Director and Board of Trustees. . . . [W]ith the privilege of unrestricted consultation with the Director . . . I can and will make the Field Columbian Museum one of the five or six most famous museums in the world in paleontology within the next six or eight years.[45]

Williston's recommendations were, for the most part, disregarded. At about the same time, Higinbotham asked Farrington to provide his own plan for the future of the Geology Department. Not surprisingly, the curator was unwilling to see paleontology excised from his immediate control, and he recommended against it: "I believe the present plan of organization the best possible for the growth and efficiency of the Department. Administration can be conducted on the present plan more economically and more systematically than would be possible if the Department were divided into two or more."[46] And so no change was made to the administrative status of paleontology. Williston was let go, and he devoted himself to building a vertebrate paleontology research program at the University of Chicago. He would continue to make use of the Field Columbian Museum collections from time to time, and he continued to mentor Riggs, but his official connection to the museum was severed for good. Riggs, in the capacity of assistant curator, returned to his default position as the nominal head of vertebrate paleontology at the museum, but always under the relatively disinterested supervision of Farrington, curator of geology.

Riggs, too, submitted an outline to Farrington for future work in vertebrate paleontology. His plan for fieldwork showed the strong influence of Williston's ideas. He omitted any reference to Jurassic dinosaurs, for example, despite the fact that this field showed so much promise and had been so good for his career. He also made some resolute general remarks:

> I would say that the work of collecting has been carried on with such limited funds that the parties have been handicapped. The opportunity to spend more time in prospecting would enable the collectors to choose the best specimens and so an appreciable saving of time in the laboratory would result. Moreover, the collectors have not dared risk expeditions for the rarer fossils, lest they should not secure a sufficient showing to make the expedition an evident success.[47]

Riggs was right. Appropriations for vertebrate paleontology field work at the Field Columbian Museum were significantly smaller than the equivalent budgets at his New York and Pittsburgh rivals. At the same time, the goal of hunting fossils for display purposes was holding up the search for rarer, less showy specimens. Although there was much to be said for Riggs's cutting remarks, Farrington chose not to share them with President Higinbotham in his own more optimistic report.

Following his final confrontation with Higinbotham, Williston believed the outlook for vertebrate paleontology at the Field Columbian Museum was very bleak indeed. In a prophetic letter to Harper, he lamented that "the authorities of the Columbian museum do not seem to appreciate the fact, for fact it is, that paleontology is becoming the chief interest in nearly all the great museums of the world—its absence [there] will one day be greatly regretted."[48] In the years that followed, vertebrate paleontology sputtered on at the Field Columbian Museum and continued to lose ground to its better-funded rivals at the Carnegie and American museums.

There is one final point to be made regarding sauropod research at the Field Columbian Museum during the second Jurassic dinosaur rush. When Osborn succeeded Marsh as vertebrate paleontologist for the U.S. Geological Survey in 1900, he inherited the Yale professor's unfinished, government-funded research projects. A few of these projects he assigned to other paleontologists as a way of currying favor. But the choicest plum of the lot was probably the sauropod monograph, which he kept for himself. That Osborn and the DVP were maybe not the best candidates for this assignment was borne out later, when the finest original contributions on this interesting group of animals appeared in publications issued by the Carnegie and Field Columbian museums. (More on sauropod research appears in chapter 12.) It seems that Osborn was so busy with other projects, most of them involving fossil mammals, that he seldom had time to work on the Sauropoda. He could have relinquished the project to a worthy alternate at a rival museum, but competition stayed his hand. The Carnegie Museum, with its fantastic collection of sauropod dinosaurs, its commitment to vertebrate paleontology, and its abundant resources, was probably the ideal institutional home for such a project. Hatcher, however, had been trusted with a U.S. Geological Survey monograph on the horned dinosaurs, and was then working himself to death on myriad other projects, as well. Aside from Hatcher, there was no obviously worthy paleontologist at the Carnegie Museum to take over the work. Peterson was not a very capable researcher and writer, Osborn felt. And Osborn never would have surrendered such a prize project to Holland.

With only a small fraction of the funding and institutional support of his chief rivals, Riggs nevertheless made a number of important contributions to sauropod research. He was certainly a capable paleontologist, and his personal and professional circumstances made him a particularly compelling candidate to take on the sauropod monograph. He had interrupted his PhD work at Princeton in 1898 in order to take a position at the Field Columbian Museum. With Williston's support and encouragement, he began working toward completing his degree at the University of Chicago in 1902. He made special arrangements to use a series of three sauropod dinosaur publications as his doctoral thesis.[49] He never finished, although by the summer of 1904 he lacked only one final publication to qualify. One reason he never wrote the final paper is that funding for Jurassic dinosaur fieldwork dried up at the Field Columbian Museum after 1901. With a mandate for research from Osborn and the U.S. Geological

Survey, coupled with even a small amount of federal financial support, Riggs might have had more favor from the Field Columbian Museum authorities who were so reluctant to support his fledgling paleontology program. More institutional support would have meant more fieldwork, more specimens, and—very likely—more opportunities to publish.

TEN

Hatcher Heads West

Early in 1901, work on a descriptive monograph on *Diplodocus* kept J. B. Hatcher pinned to his Pittsburgh desk, even as the weather warmed out West. When he finished on May 10th, he hastened to Cañon City, Colorado, to survey W. H. Utterback's progress on the Garden Park quarry. He was pleased with the outlook and wrote to Director Holland with an optimistic report:

> Mr. Utterback has a large area of the bone bearing strata uncovered & there are many bones in sight. How good or important they may be it is at present impossible to say. The area stripped ready for taking up the bones is about equal to that worked by Marsh's men during the seven years they worked this quarry. Bones are outcropping everywhere . . . in abundance but as yet they have not been sufficiently uncovered to tell whether they articulate with each other & form skeletons or not. I hope & believe there will be many fine things developed from the quarry as the bones are taken up.[1]

Utterback had exposed a crescent-shaped swath of fossil-bearing bedrock approximately sixty feet wide by twenty-five feet deep. At the western end of the quarry, bones appeared in prolific numbers, some apparently associated together, most in a jumbled heap. Bones appeared at the eastern extreme in far fewer numbers. For the present, it was difficult to tell just what the quarry might ultimately produce.

Hatcher wanted Holland to send him a copy of Marsh's *Dinosaurs of North America* to help him make sense of the tangle of specimens. He started working on what appeared, at first, to be the pelvis of *Allosaurus*, including both ilia and sacrum. Developing these bones, however, proved much more difficult than expected. The sandstone matrix encasing the bones, Hatcher now learned firsthand, was exceptionally hard. He planned to split it using shims and wedges, but it was riddled with fractures and refused to break where it was intended to. Instead, each bone had to be developed in place, using picks, hammers, and chisels, and then removed from the quarry almost completely prepared. It would be a long, difficult task to clean this quarry out. Yet within a week they had a partial jaw, tibia, fibula, and several foot bones of the rare theropod dinosaur *Ceratosaurus* out and boxed. They also had a partial articulated skeleton, possibly *Morosaurus*, in sight, including a pelvis, a femur, and a long series of nineteen caudal vertebrae. The weather was good for quarrying, but uncomfortably hot in the afternoons. Hatcher complained to Holland that he was "very much annoyed by visitors."[2]

On the morning of May 21, 1901, H. F. Osborn turned up at the quarry for a visit, together with a guest, the German paleontologist Eberhard Fraas, a "delightful" man, according to Hatcher, who, a few years later, would be making his own collection of late Jurassic dinosaurs in East Africa. Both were favorably impressed by Hatcher's new dinosaur prospects. Osborn, in fact, expressed regret at not having sent his own American Museum collectors to work in this important and historic locality. He particularly wanted to examine the quarry where Cope's collectors had taken out the type specimen of *Camarasaurus*, now in the American Museum's collections, so they all made a long climb to the top of the bluff for a look-see. Hatcher was harboring plans to reopen this quarry later in the season, and no doubt Osborn's expression of interest sealed the deal. Not one to waste too much time in small talk, Hatcher showed his visitors a copy of his somewhat conjectural restoration of the skeleton of *Diplodocus*, a plate from his recently completed monograph, then in press (see figure 27). He had a few minor misgivings about the position he gave to the feet, and about his justification for establishing the new species, *Diplodocus carnegii*. According to Hatcher's account, both paleontologists admired the work and reinforced his conclusions on every doubtful point. He was immensely pleased by the positive feedback and he let Director Holland know all about it. Later that same afternoon, Hatcher rode back with Osborn and Fraas to Cañon City, and then joined them for a luxurious rail excursion to Green River, Utah, to investigate some purported Jurassic deposits for dinosaurs. They passed en route through Fruita, Colorado, completely unaware that Elmer S. Riggs and a small party from the Field Columbian Museum were then at work on a very

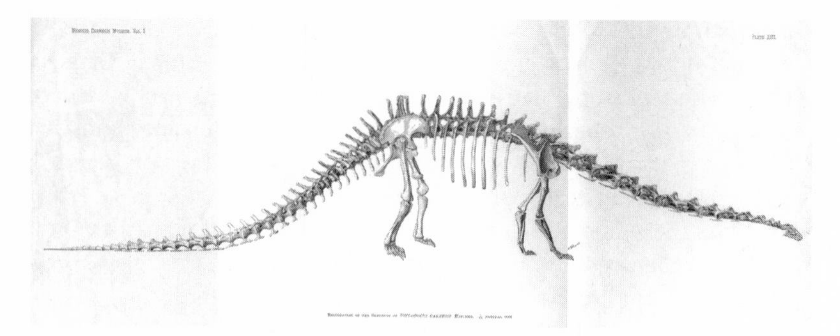

Figure 27. Hatcher's reconstruction of *Diplodocus* made an excellent impression on Henry Fairfield Osborn and Eberhard Fraas. From Hatcher, 1901.

promising dinosaur prospect just south of town. Sometime during the course of their travels, Osborn dropped a bombshell. He offered Hatcher an appointment as associate curator, in charge of fieldwork, at the DVP. According to Osborn, their long conversation was "very satisfactory and friendly." Hatcher, he believed, had practically agreed to come to New York the following January.[3]

Hatcher neglected to mention Osborn's surprise proposition in his next letter to Holland, although he wrote a detailed account of the visit, the talk about *Diplodocus*, and the reconnaissance trip to Utah. But if he really was contemplating a possible move to the American Museum to work under Osborn, Holland's reply must have given him much food for thought:

> I am pleased to know that Prof. Osborn agrees with you in all the points [respecting the restoration of *Diplodocus*] which you name. However, I am satisfied that your judgment in such matters ought to be accepted in preference to that of our esteemed friend, Professor Osborn, who, while he holds a high position and has done much good work, is nevertheless in my judgment, from what little I have observed, not as thorough a student of these matters as either yourself or [William Berryman] Scott. . . . Osborn is not to be spoken of lightly. He is a great man in his way, but I believe that your knowledge exceeds his in many matters.[4]

This must have been very welcome praise indeed. At the Carnegie Museum, Hatcher was a full curator in charge of his own department, answerable solely to Director Holland. He had risen to a position where, in theory at least, he would not have to defer to anyone else on matters pertaining strictly to science. He locked horns with Holland on occasion, but the director's letter went a long way toward establishing Hatcher as the ultimate museum authority on vertebrate paleontology. A move to the American Museum would be a step down to the kind of subordinate situation that he found so intolerable under Marsh at Yale, or under Scott at Princeton. Worse still, he did not have a particularly warm relationship with Osborn. In Pittsburgh, Hatcher had finally come into his own, and he was bound to stay put.

The "Sum and Substance" of Work at Garden Park

The last days of May 1901 brought a weeklong downpour that put a damper on work at the Garden Park quarry. Hatcher departed just as the bad weather arrived, leaving Utterback and his assistant to sit by with the prospects all huddled under a tarp and waiting for sunshine. The prolonged cold and wet conditions were miserable and discouraging. But when the weather cleared up, work resumed in earnest. By mid-June, skeleton A, the tentatively identified *Morosaurus* specimen, was free from the quarry and crated for shipping (see figure 28). Ten feet away, at the extreme western limit of the quarry, a second partial articulated skeleton turned up. This specimen, skeleton B, consisted of a series of vertebrae, including the last two cervicals and ten dorsals, plus numerous ribs, mostly in position. Interestingly, there was no duplication of parts between these two skeletons. Furthermore, a mess of some fifty-odd bones, mostly disarticulated, was spread in a jumbled arc behind the two partial skeletons, just at the north wall of the quarry. Utterback thought this material might comprise a reasonably complete skeleton of *Stegosaurus*. Some of the bones, unfortunately, ran under the hill, and it would be necessary to do more arduous stripping to get them out. A few other minor prospects, mostly limbs, vertebrae, and ribs, some loosely associated, cropped out to the east. Overall, Utterback's outlook was surprisingly gloomy. In July he wrote to Hatcher with a sour report. "Regarding our success in the bone business am at a loss to know exactly what to tell you," he wrote. "While we are doing a great deal of hard labor and getting some good bones the result is hardly satisfactory considering the expense which we are under." Referring to the eastern two-thirds of the quarry he regretted to report, "Considerable of the ground worked over has been completely barren."[5]

C. W. Gilmore graduated from the University of Wyoming with a bachelor's in science degree in 1901 and Hatcher promptly appointed him to a position as collector and preparator at the Carnegie Museum.

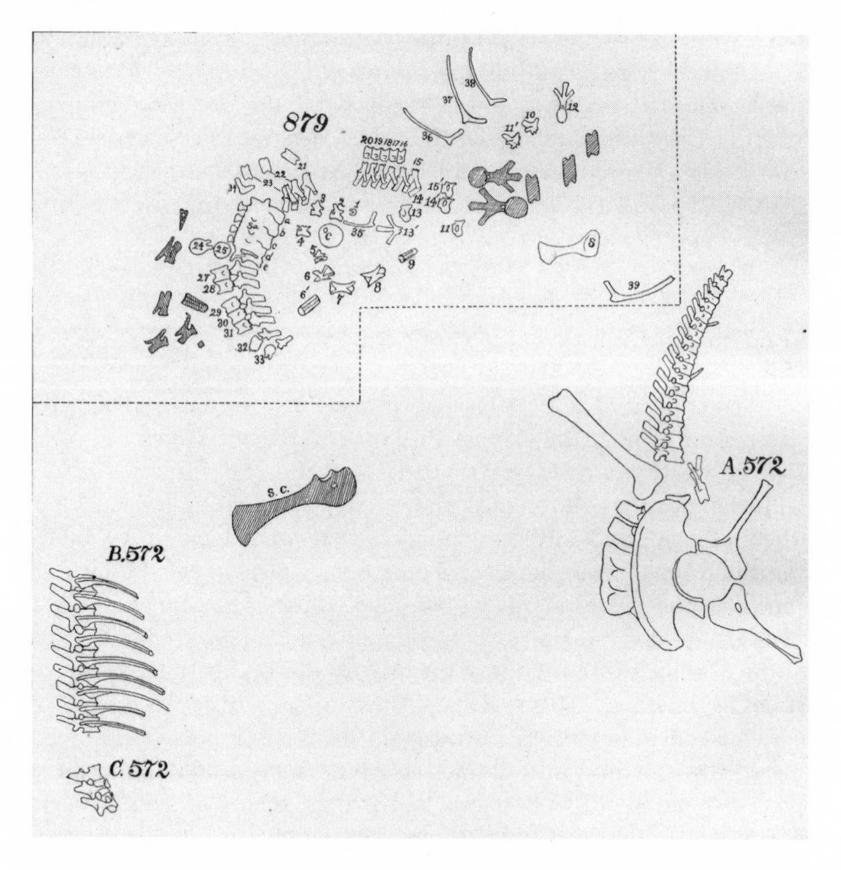

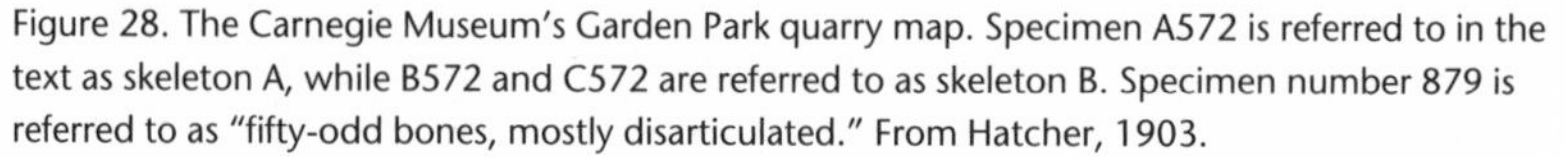
Figure 28. The Carnegie Museum's Garden Park quarry map. Specimen A572 is referred to in the text as skeleton A, while B572 and C572 are referred to as skeleton B. Specimen number 879 is referred to as "fifty-odd bones, mostly disarticulated." From Hatcher, 1903.

Hatcher assigned him to work at the Sheep Creek locality in Wyoming. Regrettably, he penned so few letters during his tenure at the Carnegie Museum that precious little is known about the exact circumstances of his labors in the field.[6] With a friend, G. F. Axtell, as an assistant, he spent the early part of the season reworking several of the quarries opened up the previous summer, including Quarries C and E. Quarry C was a disappointment, giving up only a few curious vertebrae. At Quarry E, on the other hand, they struck the mother lode. This quarry yielded several fragments of *Stegosaurus* and a *Camarasaurus* vertebra. Best of all, Gilmore and Axtell found a large, isolated skeleton of *Brontosaurus*, including nine cervicals, nine dorsals, eighteen caudals, ribs, a well-preserved sacrum and other parts of the pelvis, one right hind limb, and one right forelimb with foot bones preserved in their life positions. Gilmore expected that this nearly complete forelimb—a rare find in sauropods—would be very useful for reconstructing the posture of these enigmatic animals. Somewhat later in the season they opened up an "upper horizon" in Quarry E that yielded a second partial skeleton of a smaller sauropod dinosaur, including cervicals, dorsals, right scapula (without the coracoid), and parts of the pelvis and limbs. Gilmore suspected that it might represent something new to science and was fairly confident that there was sufficient material preserved with which to establish a new genus.[7]

Meanwhile, the rigors of fieldwork were exacting a terrible toll on Hatcher's health. After leaving Cañon City, he joined a U.S. Geological Survey party for a reconnaissance of the White River Badlands in South Dakota. Osborn was slated to join them also, but, homesick and weary from an arduous stagecoach ride around the Black Hills, he bailed out at the last minute. Osborn was not a man to take unnecessary chances in the field. Hatcher, on the other hand, endured two uncomfortable weeks of tramping through the badlands. The weather was unseasonably cool and wet, and he spent several sleepless nights curled up in his damp bedding. The Cheyenne River, which cut across the group's path, was then running high, and Hatcher's outfit got hopelessly stuck midstream. He then spent half a day wading in the cold, muddy torrent, working to free his wagon. He suffered in silence for almost two weeks, and then complained to Holland in early July. He had been planning an ambitious prospecting trip to Montana, but decided to postpone it, in order to spend a week recuperating at a hotel in Hot Springs, South Dakota, instead. Yet Hatcher had no intention of remaining entirely idle—a productive Triassic fish locality was only a short walk from his hotel. Holland, whose son was then gravely ill, was genuinely alarmed at Hatcher's condition, and he urged him to abandon the search for fossils and seek proper medical attention. But, after a week of mineral bath therapy, feeling somewhat better but still not entirely well, Hatcher resumed his hectic field schedule. He went on long rides to chase down dinosaur prospects in Wyoming and Montana, and to visit with Peterson, who was collecting fossil mammals north of Harrison, Nebraska. He felt unwell throughout, but stubbornly persisted with the work, despite extreme discomfort while riding horseback. He decided to cut his field season short in mid-July, but not before making a prolonged supervisory visit to Cañon City.[8]

Hatcher's assessment of the Cañon City quarry was a stunning reversal of Utterback's gloomy prognosis. He determined to his own satisfaction that the bone-bearing horizon there was "decidedly lower," and therefore older, than the Jurassic dinosaur beds at other classic localities, including

Como Bluff and Sheep Creek, Wyoming, and Morrison, Colorado. Moreover, a cursory reexamination of the partial skeletons unearthed since his last visit in May suggested that all their preliminary diagnoses were clearly mistaken. Hatcher now determined that Utterback's dinosaurs represented something wholly unknown to science. He wrote to Director Holland with an excited explanation of his new views and with the basic outline of a grand, new Darwinian research program on dinosaur systematics. "I feel sure we have here the key to the development of the Sauropoda & I propose to corral everything in sight," he wrote. He explained that more than five hundred vertical feet of terrestrial Jurassic sediments were exposed at this locality, and that dinosaur remains had been found at several disparate horizons. "Now I want to pound away here for several years & secure good series from every horizon possible," he continued, "determining accurately the relative positions of the various horizons, which can be done with ease, & then work out the phylogeny of the different genera of Dinosaurs. . . . It is a great undertaking but we are in a position to do it & must not let the opportunity escape us."[9] He spent a few more days in Colorado, tending to business. Among other things, he had to file several land claims, with Holland as his partner, in order to circumvent Felch, the local landowner, who was apparently trying to play the museum false. He left on August 16, 1901, and finally reached home on the 20th.[10]

Following instructions from Hatcher, Axtell left Sheep Creek, Wyoming, where he had been working with Gilmore since June, to reinforce Utterback at Cañon City. Hatcher was clearly eager for results from this important locality. Axtell began work on August 22, prospecting for new material in some of Cope's old quarries near the top of the slope. Hatcher spotted a single vertebra in one of them when he canvassed the area for new prospects. Axtell worked this spot, and a few others nearby, but turned up nothing but broken fragments of ribs, vertebrae, a radius, and a scapula—nothing worth the trouble of taking up. A week's worth of heavy rain rendered the work slow and discouraging. A nice cervical and a rib finally turned up in mid-September, giving Axtell brief cause for optimism, but he found nothing else. It was an inauspicious start.[11]

Cope's quarries were located at the top of the bluff, up-section from Marsh's quarry, in a thick layer of chocolate-colored shale best exposed on the slopes of a small, rounded, sandstone-tipped prominence known locally as The Nipple. At the base of this small hill there was an acre of bare earth littered with fragments of bone. Hatcher wanted to strip this area down to find the bone-bearing horizon.[12] Ideally, he hoped his collectors would recover sauropod dinosaur remains from this bed that would show some kind of intermediate development between the specimens found just below the bluff in Marsh's quarry and the specimens coming from places such as Como Bluff and Sheep Creek.

After Axtell was called away, Gilmore found an excellent replacement in Paul Miller, who assumed his duties with characteristic vigor. Miller was familiar with the nature of the work and with the Sheep Creek locality, having been discovered and trained as a bone digger in 1899 by Wortman, then with the Carnegie Museum. He had also worked in the summer of 1900 for the DVP at Bone Cabin Quarry. In the summer of 1901, he was back working for Carnegie again. On September 8, DVP veteran collectors Walter Granger and Pete Kaisen, along with novice bone digger George Olsen, came calling for a friendly visit and a Sunday meal. President McKinley had been shot two days earlier—perhaps the paleontologists talked about his fate. Work then resumed in Quarry E on the following day. Less friendly (at least as far as Hatcher and Holland were concerned) was a letter to Gilmore from Riggs, offering the relatively new Carnegie collector a job at the Field Columbian Museum with better pay and, allegedly, better prospects. Gilmore wrote to Peterson confidentially with news of the unexpected offer, asking him for a frank opinion about the relative merits of work at the Chicago museum. Peterson referred the matter to Hatcher, his supervisor and brother-in-law, reasoning that it was in Gilmore's best interest to do so. Hatcher's reaction was not recorded. Gilmore, meanwhile, cleaned out Quarry E and opened a new prospect at Quarry J, which netted only a few limb fragments of *Stegosaurus* before giving out. A diligent search failed to turn up any new leads, so he closed up the work and shipped his fifty boxes of dinosaurs on October 23, 1901. He was back in Pittsburgh on November 8, working in the preparation laboratory. He confidently recommended a return expedition to the Sheep Creek locality for the following field season. Gilmore stayed with the Carnegie Museum, at least for the present, and enjoyed a raise to $65 per month (beginning in 1902) for his loyalty.[13]

During the first part of October 1901, while Utterback crated the last of his specimens from the Marsh quarry, Axtell spent his time blazing a steep trail from their campsite to The Nipple (Axtell called it the anthill), at the top of the bluff. They joined forces by the 15th. Apparently, Utterback had taken it for granted that Axtell was already at work plowing and stripping a new quarry at Cope's old locality, as per Hatcher's instructions, and he was much aggravated by the younger man's failure to make any meaningful progress. He wrote to the Carnegie curator, now back in Pittsburgh recuperating, with an unkind evaluation of his colleague. Axtell's worst failing, according to Utterback, was a "lack of ambition and bone enthusiasm." Consequently, he was practically inanimate when left to his own devices. A few days of close supervision showed that Axtell could follow perfectly well, so long as there was someone else around to set a worklike pace. "If . . . you wish us to still work to-gether I will try and do the best I can even if I have to do most of the work," Utterback generously offered. He complained also about his surly team of horses. But he assured Hatcher that he would try to get along with man and horses, as best he could.[14]

Together they began a new stripping more than twenty feet wide and one hundred feet long on the south side of The Nipple, which they finished twelve hard days later. It extended from the east side of The Nipple along the entire length of the south side and connected at a right angle to Cope's old quarry on the west. They searched the ground thoroughly and found only useless fragments—nothing more than a few inches in length. By the end of the month, they were busy making a new stripping of the same dimensions on the north side. Two men who once worked this locality for Cope happened by and explained to Utterback that they had found only a few quality vertebrae at this particular spot, despite the abundance of bone fragments on the surface. Utterback's expectations plunged with these surprising revelations. He wrote to Hatcher for permission to open up a new quarry in another locality to the south, near a place he called Cottage Rock. The beds there were significantly older and lower in the section, being only about twenty feet above the distinctive red colored Triassic beds. "Now while not wishing to create any false hopes will say . . . that the indications [there] are far more favorable. . . . There are a large number of nice looking bones exposed in the ledge for a distance of perhaps 100 ft," he explained. The rock would be hard to work, but there was a far lesser volume of stripping to be done at this spot than there had been at Marsh's old quarry. He also asked permission to discharge Axtell and hire someone else who was not afraid of hard work.[15]

They finished the new stripping on the north side of The Nipple on November 7, 1901, and Axtell was fired the same afternoon. "So long as I had plowing and scraping to do I could get a half way move on him, but at pick and shovel work or prospecting he is no good . . .," Utterback explained to Hatcher. "He takes his discharge good naturedly, and [I] feel confident he realizes he has not done his duty as an employee of the Museum." No bones had made an appearance in the new stripping, thus far, but there was still some ground to be gone over. Meanwhile, Utterback had done some preliminary prospecting west of camp, and he discovered a few small, nondinosaurian bones near the base of the Jurassic. These he sent to Hatcher to sweeten the bad news about Cope's old quarry.[16]

By November 25, the game was up; the reexamination of Cope's quarries near Cañon City had ended in "dismal failure." Several strippings made around The Nipple yielded nothing worth keeping. The Cottage Rock locality, which Utterback opened next, produced nothing but waterworn fragments. Utterback worked a few other prospects with pick and shovel, but failed to find very much of anything. Only one worthy specimen turned up: a single well-preserved femur of *Allosaurus*, which was found about two miles west of camp, near Wilson Creek. Utterback was despondent. "[N]o man ever worked harder than I have the past year to make this thing a success," he protested to Hatcher. "Have given every moment of my time to the work also have parted with quite a few hard earned dollars never charged up. The sum and substance of the entire business is this, the small amount of Jurassic exposure in this locality has been prospected to a finish."[17] He continued to prospect without success while waiting for instructions from Pittsburgh. Hatcher, who must have been deeply disappointed in the results, finally ordered him home. Utterback shipped somewhat more than sixteen thousand pounds of specimens on December 23, 1901, and arrived at the museum on the 30th.[18]

Strange Bone

Like Osborn, his American Museum counterpart, Hatcher wanted to assemble in Pittsburgh a representative collection of fossil vertebrates from as many different times and places as possible. With such an abundance of Jurassic dinosaurs already at hand, he elected to send the bulk of his field forces after other materials in the summer of 1902. He sent two expeditions in search of fossil mammals in Nebraska and Montana. Fossils from these places would be far smaller, and, to some tastes, at least, less showy than Jurassic dinosaurs. To appease Carnegie's fascination for large, spectacular forms, he sent Utterback into the Laramie beds of east-central Wyoming to renew the abandoned search for Cretaceous dinosaurs. But with a mandate from Holland to mount a huge dinosaur, Hatcher could ill afford to neglect the Jurassic entirely. Consequently, he directed Gilmore to return to the Sheep Creek locality to make another sweep for more sauropod dinosaurs.[19] Hatcher would float from place to place, prospect for new localities, and play a general supervisory role over all of his collectors.

After a long overland trip north from Medicine Bow, Utterback established his first camp on Crazy Woman Creek, twenty miles south of Buffalo, Wyoming, on June 5, 1902. There, on the eastern flank of the Big Horn Mountains, the Laramie Cretaceous beds lay exposed in long, lazy (irregular) stripes. The underlying Jurassic sandstones, Utterback felt, were reminiscent of the beds at Cañon City. He found one or two prospects in the Jurassic straightaway, but wanted to hear from Hatcher whether or not he should confine his collecting strictly to the Cretaceous. Meanwhile, he continued to prospect extensively in the Laramie beds, spending the better part of June riding north and south from camp, searching lucklessly for Cretaceous dinosaurs. On the 25th, he made another exciting find in the Jurassic and quit the Cretaceous for good. He struck camp and moved his outfit south to a ranch along the Red Fork of the Powder River, about twelve miles from Kaycee, a lonely Wyoming outpost. The outlook for good Jurassic dinosaurs from this locality seemed very bright. One skeleton he found almost completely eroded out, but believed he could get a few well-preserved bones from it, regardless. Nearby, another skeleton showing a fine, long row of articulated caudal vertebrae and a few limb bones disappeared into the bedrock. He began working the site right away by gathering up the loose fragments on the surface, but a drenching summer squall drove him off. He wrote to Hatcher with an excited report and a caveat. "[Another] good man is needed very badly," he explained, ". . . as it is almost impossible for one man to work alone." For the short term, there were ample small bones in sight to keep Utterback busy, but experienced help would be needed for the plowing and scraping, and for handling the big bones he expected to find deposited in the bank. He despaired of finding good help among the sheepherders and ranch hands who populated the high plains and wondered idly if Gilmore, then at work at the Sheep Creek locality to the south, might be available to join him.[20]

The weather continued bad for another week, but Utterback made good progress, anyhow. He had sixteen caudals out and ready to box by July 3rd. The sacrum, several limb bones, and other vertebrae lay exposed in the quarry, too large and cumbersome to be moved safely by one man. The bones were in excellent shape and, at first, were easy to work out. Utterback was confident of getting the greater part of a skeleton. When word arrived from Hatcher that Gilmore would not be free to join him, Utterback put two local men to work at his quarry.

Meanwhile, Hatcher's movements in the Western states that summer are difficult to reconstruct. He probably visited Utterback in the field, although there is no record that he did. Whether or not he had the opportunity to see it for himself, he was confident enough in the prospect to claim in print that it "gives promise of being the most perfect skeleton of any member of the Sauropoda yet discovered."[21] A letter he wrote with news of Utterback's latest discovery provoked a pang of jealousy with Osborn.[22]

Hatcher's letter to Director Holland, on the other hand, was greeted with delight. Holland's August reply to Hatcher, who was sick and convalescing once again at Hot Springs, South Dakota, indicated that the skeleton had been tentatively identified as another *Diplodocus*. This was welcome news indeed. Holland hoped that Utterback would find those elements of the skeleton that were badly needed to complete a mount, especially the skull and the elusive forelimbs, which were missing from the *Diplodocus* collected by Wortman's party in 1899. But there was bad news at the museum. Sidney Prentice, Hatcher's talented scientific illustrator, had contracted a case of typhoid fever. He came to the office one morning "looking like a living skeleton," and Holland sent him home with orders not to return without first consulting a physician. "Typhoid is too treacherous a disease to permit anyone to trifle with it," he wrote to Hatcher.[23]

Despite the additional help, Utterback's progress slowed to a crawl in the weeks and months that followed his initial discovery. The stripping proved to be extremely difficult work. Three feet of overburden above the bone layer caused the most trouble, every pound of which had to be removed slowly and tediously with hammer and chisel. Bones continued to turn up in prodigious numbers, however, including a hind limb with parts of the foot, more bones of the pelvis, both scapulae, and an abundance of vertebrae. By mid-September 1902, the weather was already turning very cold and disagreeable. Both of his men quit for more lucrative work on the 23rd. By early October, Utterback also was ready to quit. He covered the bones still showing in his quarry with a sheet of canvas and a heavy layer of dirt. He hired an outfit to haul his ten boxes of bones to the depot at Clearmont, Wyoming, loaded on the 24th, and returned to Pittsburgh with half a dinosaur in his baggage.[24]

Gilmore entered the field sometime in the spring and by May 26, 1902, had sent a discouraging report about the slim pickings at Sheep Creek. Hatcher wrote to urge his young collector to be patient—he expected another rich harvest from this locality.[25] Gilmore opened up Quarry K, where he recovered a partial pelvis, femur, and a single dorsal vertebra of *Camarasaurus*. He may also have done some mop-up work in Quarry E, where two partial skeletons of *Apatosaurus* were collected the previous summer. Midway through the field season he was anxious to move to another locality. The DVP party, under Granger, was then busy working in Reed's Quarry (aka Quarry R), near Rock Creek, so Gilmore made a quick, opportunistic sweep of Bone Cabin Quarry, netting one ankle bone each from two different sauropod dinosaurs. For the latter half of the season, Gilmore removed to the Freezeout Mountains, where he had once collected successfully for Knight. There he opened up several new quarries, L, N, and O, and reopened the quarry worked previously by parties from the University of Kansas and the Field Columbian Museum (both in 1899), designating it Quarry M. Quarry L and Quarry O produced fragments of two caudal vertebrae of *Camarasaurus* and the right radius of a stegosaur, respectively. Quarry M, however, was a total bust. An annotated catalog of Carnegie Museum dinosaurs lists no catalogued fossils from Quarry M, although John S. McIntosh contends that several fine vertebrae were collected there before Gilmore abandoned the quarry as exhausted. Quarry N, on the other hand, was extremely productive. It yielded a partial skeleton of *Allosaurus*, parts of two tangled *Camarasaurus* skeletons, a bounty of *Apatosaurus* bones, and about a dozen bones of *Stegosaurus*. It also produced something identified only as "strange bone." Some fossils were still in the ground when Gilmore finished his field season. One possible reason for his haste and impatience that summer is that Gilmore was engaged to be married to his Laramie sweetheart, Laure Coutant, on October 22, 1902. He returned to Pittsburgh with his new bride sometime in the fall, bringing home forty boxes of Jurassic dinosaurs. He returned to the Freezeout Mountains the following summer, 1903, and cleaned out Quarry N, which netted seven more boxes of material. He also collected a few more bones from Quarry O. With nothing more promising in sight, Hatcher sent him to collect marine reptiles in western Kansas for the latter half of the 1903 field season.[26]

Utterback returned to the Powder River area in May 1903, resuming work on his *Diplodocus* quarry on the 12th. He met a gang of railroad workers on his train ride west, and persuaded one of them to join him at the quarry. This man proved to be a good worker, who could handle the team and do a majority of the heavy stripping and other hard labor. He was a novice at camp living, however, and showed no particular interest in the work. According to Utterback, "[He] cares no more for bones than a hog does for an acorn with a hole in it." The weather was ideal for bone digging, and together they made excellent progress. But by the first of June the *Diplodocus* quarry, now dubbed Quarry A, was beginning to show signs of exhaustion. Six dorsals, ten caudals, both sternal plates, and numerous foot bones and chevrons had already been taken up, and a few unidentified bones could still be seen leading into the bank. Nevertheless, Utterback was not very confident of continued success. While his young apprentice labored to strip off the overburden in order to take out the last few bones, Utterback located a new prospect on the slope above Quarry A and began to open it up. Quarry B, he predicted, would yield some very fine bones. He made one small stripping and uncovered a wealth of good, uncrushed bones. Pelvic bones, vertebrae, and ribs appeared in prolific numbers, so he suspended work on Quarry A in early June. Together with his hired hand he expanded Quarry B over the next several weeks, finding the bones so thickly commingled as to be almost impossible to work. They removed more than one hundred bones from a space roughly ten feet square, with more bones running into the hillside. Two skeletons, at least, seemed to be preserved here, although Utterback could not determine precisely what they were. One, he thought, might be a *Stegosaurus*. With so much work to do in Quarry B, they had no time to search the area for additional prospects. Nor did he think it very likely that he could find something better. Even so, when the heavy stripping was finished, sometime in July, he planned to take the team and wagon on an exploratory trip to the south.[27]

Local friends told Utterback that there were numerous finds to be had in a locality to the south and he was anxious to see this area and stake a claim to any significant specimens before someone else beat him to the punch. This was not an unreasonable fear. A businessman in nearby Buffalo, Wyoming, told him that a Mr. Ross of the University of Chicago had written a letter inquiring after the exact location of the Carnegie Museum's *Diplodocus* quarry (Utterback's Quarry A). Writing to Hatcher about his concern of being preempted, Utterback asked the curator to keep this matter in mind if he should happen to stop in Chicago on his way west. But Hatcher had no idea who this man could be. However, he did plan to visit the University of Chicago and the Field Columbian Museum very shortly, and promised to keep Utterback abreast of anything definite he might discover about the mysterious Mr. Ross. "In the mean time I suggest that . . . you . . . spend a few days looking the ground over and locating any specimens you may think of value, leaving your name and the name of the Museum which you represent thereon," Hatcher advised.[28] Presumably, rival collectors would feel honor bound to respect all such claims.

The inquisitive Chicago paleontologist was actually E. S. Riggs of the Field Columbian Museum. Riggs had read about the nearly perfect Carnegie Museum dinosaur prospect in Hatcher's *Science* article and was contemplating a trip to the same region to find more sauropod material with which to build a mounted dinosaur for Chicago—the specimen he collected near Fruita, Colorado, in 1901 was only half complete. Thanks to his budget-minded superiors, Riggs never made the trip. Possibly the businessman in Buffalo who advised Utterback about "Mr. Ross" was the same person who wrote to the Field Columbian Museum in 1897 or 1898 with information about a dinosaur prospect nearby, thus bringing the area to Riggs's attention in the first place. Moreover, someone in Buffalo, perhaps this same man, informed the American Museum's Barnum Brown of the whereabouts of Utterback's quarry when Brown was in town that same summer, 1903, following up yet another dinosaur tip. Late in July, Brown dropped by the quarry for a visit. He prospected for two days from the Carnegie Museum camp, then, finding nothing good, moved on to Montana.[29] Brown did not like to have company in the field and was anxious to find a promising new Jurassic locality for Osborn. But for Riggs's miserly superiors, and Brown's aloof impatience, there might have been a rush on dinosaurs in the Powder River country starting in 1903.

About the time of Brown's visit, Utterback's enthusiasm for the work at Quarry B began to wane. Perhaps he was influenced by Brown's disdain for the locality. "[W]hile I am taking out a great number of bones, and many good ones," he explained to Hatcher, "the results as a whole are entirely unsatisfactory. Have been thinking seriously of abandoning the quarry and [prospecting for a new locality]. One thing sure we can always return and work it as no one else will do so. . . . As I go into the hill many of the bones are badly crushed and broken."[30]

Hatcher, however, had a different plan in mind. While making a rapid reconnaissance of the geology of parts of Montana and Alberta on behalf of the U.S. Geological Survey, Hatcher located some Cretaceous dinosaur remains near Musselshell, Montana. After parting company with his survey party on August 1, 1903, he returned to gather them up. The site showed real promise, and Cretaceous dinosaurs were very much needed for the Carnegie Museum collections. Unfortunately, two more months of grueling fieldwork had taxed Hatcher's health yet again. Sick for a week, he summoned Utterback from Wyoming to take over the Montana dig and then he returned to Pittsburgh. Utterback closed up work at Quarry B on August 11, and then headed for Montana, stopping first in Buffalo to ship his twenty-nine boxes of dinosaur fossils, including a long, articulated series of thirty-eight posterior caudal vertebrae of *Diplodocus* and other miscellaneous sauropod specimens.[31] Hatcher had a reputation for doggedly pursuing fossils no matter what the personal cost, but his extra precautions in the summer of 1903 were surely warranted. Wilbur Clinton Knight, professor of geology at the University of

Wyoming and instigator of the Fossil Fields Expedition of 1899, took ill about the same time as Hatcher and then died unexpectedly of peritonitis on July 28. He was only forty-four years old.[32]

After a one-year hiatus, Utterback returned to the Jurassic beds of east-central Wyoming late in the summer of 1905. Heavy showers and flooded rivers made travel by wagon nearly impossible. For several weeks he made camp along the North Fork of the Powder River, fifty miles south of Buffalo. Between storms he had a chance to prospect and made two more or less promising finds. One yielded a partial hind limb of *Diplodocus* (although Utterback identified it as *Brontosaurus* in the field). The other was a small, partial skeleton of an ornithischian dinosaur called *Dryosaurus*, consisting of three dorsals, twenty-eight caudals, and most of the right hind limb. When these two specimens were up, and the bad weather had moved off, Utterback shifted operations to the Wyoming Cretaceous.[33]

"The Good Work Goes On"

For the 1904 field season, Hatcher planned to abandon fieldwork in the Jurassic completely. He assigned Peterson to collect fossil mammals in Nebraska and sent Utterback after more Cretaceous dinosaurs in Montana. Hatcher did no fieldwork at all. He devoted the spring and early summer of 1904 to research on ceratopsians, a group of horned dinosaurs from the Cretaceous. He inherited this sizable project from his former boss Marsh, via Osborn, the late Marsh's replacement as paleontologist for the U.S. Geological Survey. With a wealth of research specimens—most of them collected by Hatcher himself—split between Yale and the U.S. National Museum, the Carnegie curator had to do quite a bit of business travel. In the spring, he spent some ten days in Washington, D.C., where Gilmore had gone to do some fossil preparation on a *Triceratops* skull for Hatcher's project. Gilmore, who was then working in Washington on a contract basis, never came back. He accepted a job with the vertebrate paleontology staff at the National Museum and remained there for the rest of his career. Hatcher, however, returned to Pittsburgh early in June and resumed his frenetic work schedule.[34]

Late in June, Hatcher was feeling "somewhat run down," according to Director Holland, "having been exerting himself very strenuously."[35] Holland encouraged him to go home and recuperate, but Hatcher felt the pressure of too much work to do and stubbornly refused to leave. Later in the week Holland sent him home. But Hatcher returned to the museum, locked his office door, and worked late into the night, despite a high fever. "He joked about his illness," writes Tom Rea, "and ignored his doctor's orders." Finally, as he had with Prentice two summers previously, Holland ordered him out, even going so far as calling for an ambulance to take him to Mercy Hospital. Doctors diagnosed him with typhoid fever, and although he was extremely ill, he did not seem to be in imminent danger as of the first of July 1904. The disease was expected to plateau in four or five days and doctors were reluctant to make any long-term prognosis. Holland, however, aware of Hatcher's extraordinary constitution, expected "a full and speedy recovery." He sent a reassuring letter to Peterson with news of Hatcher's illness, and then left for a brief holiday excursion to St. Louis on July 2.[36]

Peterson was camped north of Harrison, Nebraska, late at night on the Fourth of July, when a telegram from his sister brought the stunning news that Hatcher was dead. He rode solemnly into town the next morning, wracked with sorrow for Hatcher's bereaved wife and their four children. He also felt keenly the abrupt and tragic end of his brother-in-law's noble work in science. Holland's reassuring letter, a few days delayed in the mail and now hopelessly irrelevant, waited for him in town. Peterson had been accomplishing great results in the field, but the news put his work in a new, somber perspective, completely extinguishing his enthusiasm for fossils (at least for the present). He wanted to close down his expedition and return to Pittsburgh as soon as possible, in order to be with his family. He wrote to Holland respectfully requesting permission to pull up stakes and come home.[37]

When a situation called for compassion, Holland was capable of writing a beautiful letter, but in his reply to Peterson's heartfelt request, he was all business. "I, as well as your sister," he wrote, "think the best thing you can do under the circumstances is to remain where you are and carry on the work that you are doing with energy. . . . I doubt very much whether it would be wise for you to . . . come back now. That would sacrifice the good results which in your letter you say you are just on the eve of achieving." This was a willful distortion of Peterson's report. In fact, he had already made some very important discoveries and was just on the point of writing to Hatcher with his exciting news when the tragic telegram arrived. "I should regret very much to have to report to Mr. Carnegie that [your] party had [not] succeeded in accomplishing anything this summer," Holland threatened, "which would certainly be the case if you were to come back here." He also clarified the new order of things for Peterson, explaining, "Until a successor to Prof. Hatcher is secured I shall act as the head of the Section of Paleontology in the Museum, and shall expect you to report to me everything . . . and to keep me fully advised of all your movements." He closed his letter with a perfunctory expression of kind regards and sympathy.[38]

Osborn, by contrast, who held Hatcher in very high esteem, seemed much more genuinely sympathetic. On vacation in Europe, he chanced to see the news about Hatcher's death in a Paris newspaper and wrote to Peterson to express his condolences.

> I hasten to write you of my deep sorrow and sympathy in learning of the death of Hatcher. . . . [I] was greatly shocked, for when I saw him last he was looking very well and was full of bright plans for the future. This is a great personal loss for I greatly admired Hatcher's scientific ability and enthusiasm, and always felt a fresh inspiration from talking to him. It is a hard blow to American paleontology, to which Mr. Hatcher was making such splendid contributions following his many years of magnificent work in the field—he was certainly our greatest collector. It is especially sad to think of his dying in the beginning of what promised to be the brightest and most satisfactory period of his life—when people could see his work and recognize his ability. . . . I return early in September to take up my work, and I shall always miss Hatcher.[39]

Carnegie was also in Europe when he heard the news about Hatcher. He wrote to Holland from his castle at Skibo, Scotland, with a touch of compassion and some inspiring marching orders. "Wire . . . tells me of our great loss, Professor Hatcher," he wrote. "I had not heard of his illness and was greatly shocked. I know how you valued him, and how difficult it will be to replace him. But when the flag drops from one, another steps forward and takes it up, and so the good work goes on."[40]

Holland took up the flag as acting curator, but made little or no attempt to fill the vacancy with a competent vertebrate paleontologist. "It is exceedingly difficult to find anyone in America who possesses the peculiar qualifications which belonged to [Hatcher]," Holland explained in his annual report. "He was undoubtedly the foremost paleontological field collector in the world."[41] Many prominent vertebrate paleontologists shared this opinion. For example, in a letter to Holland, Osborn wrote, "[Hatcher's] work as a collector was magnificent, probably the greatest on record."[42] W. B. Scott, who oversaw Hatcher's work at Princeton for several years, wrote:

> [Hatcher] may be said to have fairly revolutionized the methods of collecting vertebrate fossils, a work which before his time had been almost wholly in the hands of untrained and unskilled men, but which he converted into a fine art. The exquisitely preserved fossils in American museums . . . are, to a large extent . . . due to [his] energy and skill and to the large-minded help and advice as to methods and localities which

> were always at the service of any one who chose to ask for them. . . . No less than three great collections . . . owe their choicest treasures [directly] to the skill and devotion of Hatcher.[43]

Where would Holland ever find a man to take the place of someone with Hatcher's superlative skills and experience? He was not to look very far. "The Director of the Museum," Holland explained, referring to himself in the third person, "fortunately is in possession of much of the information possessed by Professor Hatcher in relation to the fossil fields of the West, and, aided by the skillful and intelligent labors of the . . . paleontological staff . . . he has felt justified in pushing forward the work already so brilliantly begun by Professor Hatcher, without immediately taking steps to appoint a successor."[44] It was a spectacular conceit that Holland sustained for years, despite the lack of confidence in his abilities among the paleontological assistants and in the broader scientific community.

ELEVEN

Last Days in the Jurassic

Osborn planned a busy summer field schedule for 1901. B. Brown explored unsuccessfully for promising new Jurassic localities in Colorado that summer. W. D. Matthew did the same in Utah, then joined Brown and other DVP collectors in the hunt for fossil mammals in eastern Colorado. Gidley, likewise, went to Texas to collect fossil mammals. Walter Granger and Albert Thomson began the 1901 field season searching fruitlessly for dinosaurs with George Wieland in the Jurassic beds east, north, and west of the Black Hills. With so many other DVP collectors afield, Peter Kaisen again took charge of the American Museum's Jurassic dinosaur fieldwork in southwestern Wyoming. Osborn had high expectations for this work, but with so many other costly field parties searching the West for fossils of all stripes, he could not afford to be too extravagant in the Jurassic. He gave Kaisen a strict budget that was "not to be exceeded in any case."[1]

Kaisen arrived in Wyoming in early May. He hired George Olsen, in Laramie, to serve as expedition cook. Olsen also pitched in at the quarry whenever extra hands were needed. They made camp at Bone Cabin Quarry on May 11, 1901, and began to do extensive horizontal stripping with plow and scraper at the northwest end of the excavation.[2] Two weeks later, on May 26, Osborn (along with E. Fraas) dropped in for a supervisory visit. Kaisen met them at the depot in Medicine Bow and drove them out to camp, arriving in time for a late supper and a good smoke. The weather was beautiful, so Osborn and his guest slept in the open on top of the bone-bearing layer. Osborn had purchased a mattress and a tarp in Denver and "slept like a top" in his Kenwood sleeping bag. The next morning, he made a careful study of the quarry with Fraas. He expressed himself as well pleased with Kaisen's progress and with the long-term prospects for well-preserved bones in the hard sandstone layer in which the party was then working. He was particularly anxious for Kaisen to find more skull material. He must have been somewhat dissatisfied with the food, however, for he consented to hire William Patton as camp cook and assistant. After lunch, they left for the Freezeout Mountains. Fraas drove the wagon and Osborn rode on Ute, a gentle black pony. They slept out again at Fale's Ranch. From the ranch they rode to Nine Mile Quarry, where Osborn gave directions to renew stripping on the east and south ends of the old excavation, hoping that Kaisen might recover more elements of the large *Brontosaurus* found there in 1899. They returned to Bone Cabin Quarry briefly, where Osborn inspected Kaisen's *Stegosaurus* prospect. Then they headed back to Medicine Bow and entrained for the Black Hills. Osborn delighted in his time on horseback. "I have hardly used my brain at all except in entirely new ways and upon new subjects," he rejoiced in a letter to his wife, "so I feel vastly refreshed mentally." From the saddle, he reflected long on the complexion of his department and especially on the giddy possibility of luring Hatcher away from the Carnegie Museum.[3] It must have been a very satisfying ride.

Osborn's departure heralded the arrival of bad weather, which slowed the work considerably. Kaisen, nevertheless, continued to make steady progress on his *Stegosaurus*. This prospect was remarkably complete and featured a pair of forelimbs with parts of both feet, a hind limb, a sacrum, five or six dorsal and more than a dozen caudal vertebrae, and at least three big dorsal plates. In addition, Osborn reported a short series of cervical vertebrae and a sheet of dermal ossicles. The sacrum, unfortunately, ran into an old prospect hole dug years before, and was soft and somewhat broken up. Another important discovery was the partial skeleton of an unidentified theropod dinosaur, including pelvis, numerous vertebrae, and the back part of a large skull. A long series of *Diplodocus* caudals and a single unidentified jaw fragment turned up, also.[4]

W. H. Reed, meanwhile, was working as a freelance collector in 1901. He was dug in at a new quarry near a place called Six Mile Gulch, a few miles from Rock Creek station. There he found a number of big, beautifully preserved bones and left them in situ, covered with canvas and a light layer of dirt. "You are probably aware that I am no longer in the employ of the Carnegie Museum but stil [*sic*] I am out for bones," he explained in a letter to Osborn. No longer beholden to any particular institution, Reed had a proposition for the DVP curator. "[I] have had good luck so far and have found two things that I think you may want. . . . [Y]ou can send some of your men to look them over and take them up if you wish. . . . [E]verything I get is for sale."[5] Reed's letter arrived in mid-July around the same time as Matthew's indifferent report about Jurassic dinosaurs in Utah, Brown's similar news about southeastern Colorado, and Wieland and Granger's failure in and around the Black Hills, so Osborn was just then very greedy for success. He forwarded Reed's letter to Granger with instructions to investigate the new quarry immediately with a view to making him an offer. "We may find it will pay to do some more work in [Reed's] . . . quarry and if we can buy his . . . bones at a bargain I think we had better take them," he reasoned.[6] Reed's prospect, apparently, was the best new thing in sight.

Reed had fallen on hard times since his clash with Hatcher and his subsequent resignation from the Carnegie Museum and had been unable to find a buyer for his latest discoveries. Granger assured Osborn that they could turn Reed's financial circumstances to their advantage. He went to see Reed's prospects during the first part of August, as instructed, and was suitably impressed by the quality of Reed's fossils. He found "first class bones" of *Allosaurus* and *Stegosaurus*, still lying in situ. Nearby, Reed had removed a number of bones from a second quarry consisting almost exclusively of large sauropods. Best of all was a complete forelimb of *Camarasaurus*, which was exceptionally large and well preserved. "Apparently there are many more bones in his quarry . . . perhaps enough to furnish work for a season or more. There might be some excellent things there," Granger reported to Osborn. Granger thought that Reed's asking price of $1,000 was "rather exorbitant," and was confident that he would accept any amount of money that Osborn might care to offer.[7] Two weeks later, with no word as yet from Osborn on the outcome of his negotiations with Reed, Granger was a lot less blasé. He wrote to the DVP curator with a strongly worded recommendation: "I earnestly hope to hear that you have made some arrangements with Reed for the purchase of his bones, for the more I have thought of the matter, lately, the more I am impressed that we should have the fossils."[8]

Bone Cabin Quarry, meanwhile, had been producing some fine material in the spring and early summer of 1901. Osborn sent the good news to Granger, while the latter was still laboring in vain near the Black Hills, implying that he could redeem his field season by teaming up with Kaisen in Wyoming. "Good work and good luck in the Bone Cabin Quarry may retrieve the failures of the early part of the season," Osborn hinted darkly. But *new* Jurassic localities remained the curator's chief obsession. "It is extremely important for us to find a new working place," he wrote. Then, repeating himself for emphasis, he stressed that "it is *absolutely necessary* for us to find a good locality. You must make inquiries everywhere and intimate that we will pay well for a good prospect." Nevertheless, Osborn wanted Granger to be as frugal as possible, hoping that through the strictest economy they could stay and work Bone Cabin Quarry until very late in the season. "You are liable to strike at any moment one or more fine skulls," he wrote hopefully.[9]

Oddly, this is exactly what happened. Granger's very first find in the quarry, when he arrived from the Black Hills late in July, was the better part of the large skull of a sauropod, probably *Brontosaurus*. It was not in top shape, however. Located near the edge of one of the old prospect holes, the skull was somewhat weathered as a result. Also, gophers had burrowed underneath it, doing more unspecified damage. Worst of all, the skull showed evidence of tool marks, and since it was found so close to the surface, Granger concluded that it had been scalped accidentally by Kaisen's eager scraper. But, to his credit, he did not name names. Kaisen, meanwhile, found a large *Brontosaurus* pelvis and partial hind limb, and an *Allosaurus* hind limb to boot, with foot and pelvic bones nearly complete. Granger was keeping camp expenses low by living "as economically as is consistent with health and good work." He was clearly happy to be back at Bone Cabin Quarry. "[T]he work goes merrily," he reported.[10] Osborn was partaking in another Colorado mountain holiday when he received Granger's news. He was pleased, but not entirely placated, cautioning Granger to "use the *scraper* very carefully, [because] a Brontosaurus skull is a grand thing. *You will surely find more*," he added confidently.[11]

In fact, they did find more. In the latter part of August, Kaisen found the lower jaw of a carnivorous dinosaur. Nearby was a reasonably good portion of the rest of the skull. A week or so later he found the posterior part of another sauropod skull with an atlas articulated to it. All this material was found well preserved in a hard layer of sandstone on the western perimeter of the dig. A few hundred square feet, at minimum, of this same productive layer remained untouched. As the field season extended into early September, Granger became more and more convinced that he and Kaisen would not be able to close out Bone Cabin Quarry by that fall and feel satisfied that it was completely exhausted. Assuming that he would be returning to the quarry the next summer, Granger concluded to close up work there for the season and move the outfit immediately to Nine Mile Quarry. He wrote to Osborn for his approval and for permission to remain in the field through the first half of October. This was not an entirely insignificant matter, as Granger required an extra $100 to cover the additional expenses.[12] But Osborn was particularly anxious for success. He wrote to Granger from Chicago on September 6 (the day President McKinley was shot) lamenting that 1901 "has been a very costly season—with many disappointments." Still not content with the outlook for the following season, Osborn advised Granger in the meantime to "keep a sharp lookout for new localities and prospects for 1902."[13]

By this time, Osborn had become fixated on Nine Mile Quarry and he eagerly approved Granger's plan to reopen it. He had settled on the partial *Brontosaurus* collected at Nine Mile Quarry in 1899 as the basis for a composite skeleton to be mounted for display at the museum and he was anxious to acquire every possible fragment. "It is . . . important to make an *extensive clearing* around the 9 Mile . . . ," Osborn advised. "Strip on *a large scale from the start* so that you can *be absolutely sure* that there is no more of this splendid animal on which so much of our success in making a complete mount depends." McKinley had died early that morning and Osborn added that he was "greatly depressed over the loss of our President."[14]

They moved camp to Nine Mile Quarry on September 18, 1901, and began working there in earnest on the following morning. As per Osborn's instructions, they did extensive stripping on two sides of the old excavation, extending it eight feet west and six feet south. The depth of the bone-bearing horizon was ten to twelve feet in some places and a hard concretionary layer slowed the digging significantly, so it took ten full days to reach pay dirt. Unfortunately, they found precious little new material. One well-preserved cervical vertebra turned up, along with two caudals with their neural spines broken off. They also found a small bone from the ankle, and an eighteen-inch long fragment of the distal end of a fibula. That was it. Kaisen and Granger worried that their results would not please Osborn. Nevertheless, they were both certain that the quarry was now thoroughly exhausted.[15]

Meanwhile, Osborn closed the deal for Reed's bones without incident. He offered Reed $400 for all the specimens, both quarries, and every prospect described in Granger's letter. Osborn claimed disingenuously that the DVP could not afford to pay even this modest amount. Although it was considerably less than the original asking price, Reed accepted Osborn's offer, apparently without making any fuss. Wary of any future "misunderstanding," Osborn ordered Granger to bring a check to Reed in person and secure a carefully detailed bill of sale specifying all of the fossil materials involved in their transaction. This proved impractical, however, as Reed was then working as an assayer at a mine some fifty miles from Laramie and farther still from Bone Cabin Quarry. Reed sent his own bill of sale to Granger, which, although less detailed, covered basically the same materials. Granger emphasized in a letter to Osborn that the land on which Reed's prospects were found was owned by a local rancher who had purchased it from the railroad, so that Reed's document covered a moral right to the prospect only, rather than a legal one.[16]

By the time that Granger and Kaisen had finished up their work at Nine Mile Quarry, there was very little time left for fair-weather work in Wyoming. Osborn wanted something done immediately with Reed's quarry, however, so he instructed Granger to take up the carnivore specimen right away and to leave the rest of his prospects "well protected against the winter and against visitors."[17] Accordingly, Granger and Kaisen removed to Six Mile Gulch on October 1, 1901, and took up residence in a cabin kindly furnished by Reed. Their first task was to pack the specimens Reed had already taken out, which included a large number of (mostly) unassociated bones of both theropods and sauropods. These filled some fourteen or fifteen boxes. Next they finished excavating the theropod bones Reed had exposed in situ. This specimen included an almost complete pelvis and a string of vertebrae reaching from just anterior to the sacrum through the first few caudals. Granger reported that a few of these bones had been slightly damaged earlier in the spring when the quarry was trampled by cattle. Osborn, naturally, was not amused, and demanded an explanation.[18]

Next they excavated Reed's sauropod, which included a sacrum, eight to ten presacral vertebrae, a few caudals, an ilium, and a scapula. Granger thought this specimen was very peculiar and Osborn was anxious to see it.[19] The weather held very favorably through early October. On the 7th, they shifted work to Reed's general quarry (Quarry R, hereafter), from whence the unassociated bones had come. In the time remaining, they took up two sacra, a scapula, and a few small limb bones. Then they covered the quarry with a cattle-proof layer of dirt and closed up work for the season on October 19, 1901.[20]

"Put as Much Heart into It as You Can"

Granger and Kaisen returned to Medicine Bow on June 22, 1902. They retrieved their outfit from storage and established camp at Reed's old cabin on Rock Creek, three miles south of Quarry R. Olsen joined the expedition again as cook, teamster, and general assistant. Each morning he drove the diggers from camp to quarry, which was situated on a knoll of dark clay standing some twelve feet above the level of a small, dry basin east of Medicine Bow. Dinosaur bones cropped out of the clay on two sides of the knoll, most often in a thin, hard concretionary layer. The bones were well preserved, but mostly disarticulated. Using a team, plow, and scraper, the party made a new stripping of about twenty-two hundred square feet to determine the extent of the fossil deposit.[21]

As usual, Osborn had high expectations for the field season and he saddled Granger with the burden of success. "I hope you will push your work this summer with great energy and persistence, and accomplish fine results," he wrote to his field-worker. "Put as much heart into it as you can, because that is the direct road to success in everything; and I have had the feeling during the last year that you have not put quite enough of this element into your work."[22] (Maybe the curator was also concerned about Granger's salvation, for hard work, in Osborn's worldview, was the key to material *and* spiritual success.[23]) Granger was undoubtedly much taken aback by Osborn's rebuke, but he responded with measured diplomacy. "We are all putting in our best energies . . . ," he protested meekly. "I want you to feel that I am always doing my best in the field."[24]

If Granger was optimistic about the collecting conditions at Quarry R, Kaisen was less so. They began taking out bones by the second week of July. There was a great profusion of vertebrae, ribs, and pelvic bones of sauropods and a few also of *Allosaurus*. Yet the bones were not as abundant as at Bone Cabin Quarry. Moreover, limb bones were relatively rare. Although very few bones were found articulated, Granger was certain that by carefully mapping the material in situ, many of the bones could be confidently associated back at the museum. With such an abundance of uncrushed bone, Granger felt that an important discovery was likely to happen at any time. But by early August, nothing particularly significant had come to light. Kaisen reported to Osborn that the quarry did not look "extra good." He was eager to resume operations at Bone Cabin Quarry. Granger, on the other hand, who was due to spend most of August surveying with Matthew in the Bridger Basin, wanted Kaisen to remain at Quarry R at least until the end of August.[25]

Granger departed August 6, 1902, leaving Kaisen and Olsen to continue the work at Reed's old prospect. They found that the bones were best-preserved and easiest to work along the very edges of the deposit, where the concretionary matrix was thinner and softer. So Kaisen worked the western side and Olsen the eastern side of the knoll. When Olsen turned up better results, Kaisen moved over and joined him. Kaisen soon found the lower jaw of a carnivorous dinosaur, the first (and only) skull material recovered from Quarry R. By the end of August the cache of bones had virtually dried up, although Kaisen and Granger both believed that there was more material to be found in the unexplored parts of the knoll. When Granger returned to camp on September 7, the party decided to remove at once to Bone Cabin Quarry.[26]

Some excellent new fossil material turned up in Bone Cabin Quarry almost right away, including a long series of forty-two caudal vertebrae; the upper jaw of a carnivorous dinosaur (possibly pertaining to the more or less complete *Allosaurus* skull Kaisen collected in 1901); several bones, including vertebrae, ribs, and limb bones of a small dinosaur (possibly referable to the small carnivorous dinosaur discovered in 1900); and a reasonably complete forelimb of *Morosaurus*, including foot bones and toes. A complete forelimb with foot was very high on Osborn's list of desiderata, although the specimen tended to confirm that Hatcher was right and Osborn wrong about the presence of a claw on the first digit.[27] "This quarry has certainly given us many surprises," Granger wrote to Osborn, "and just now, as we thought we were about to close it up for good, it looks very much as if another season or a part of one, anyhow, would be necessary to work out all of the good ground. We are getting into an awful mess of bones, some of them . . . first class."[28]

Gratified by the results from Wyoming, Osborn, nonetheless, was still not quite satisfied. "I am especially pleased with your recent discoveries," he wrote to Kaisen about the latest returns from Bone Cabin Quarry, "and I hope you will soon come across some more skull material. Altogether we are promised a very successful season at the quarry." In the same letter, he expressed concern about keeping up with his rivals. "I think we have the finest carnivorous Dinosaur material in the world; but I envy the Carnegie Museum their complete skeletons. Mr. Hatcher writes me that they have found a magnificent *Diplodocus* which seems to be almost perfect."[29] He relayed this same news in a letter to Granger, asking, in jealous frustration, "Where are we?"[30] He was tentatively planning a trip west to see the quarry and hoped to stop in Chicago and Pittsburgh en route to investigate his rivals' progress in person.

By this time, the quarry was so essential to the DVP's Jurassic dinosaur success that Osborn began to feel uneasy about its ambiguous legal status. Probably he learned, through Hatcher or otherwise, that Gilmore had made a hasty stop there over the summer, while Granger, Kaisen, and Olsen were working at Quarry R, and collected two small bones. "Who owns the Bone Cabin Quarry?" Osborn asked in a letter to Granger. "I think it would be well for us to stake out a claim on the basis of our continued residence there. Will you kindly ascertain whether this is possible . . . ? It is becoming so valuable that I do not want to trust to chance."[31] But Granger had a different perspective and he wrote persuasively to disabuse the curator of any notion of ownership. He explained that the quarry was located on a section of land once owned by the railroad, but later purchased by a ranching outfit—just like the situation at Quarry R. Whether or not a claim to the mineral rights on privately held land would stick, Granger had no idea. But at the very least, as he pointed out to Osborn, a claim for mineral rights would immediately draw unwanted attention to the value of their quarry. The landowners were already very favorably inclined toward the DVP party, he argued, and no local resident was likely to attempt to work the quarry during the cold winter months. W. C. Knight, Wyoming's state geologist, could probably explain the legal issues involved, if necessary. But, "it is my very strong opinion," he stressed, "that the very best way hold to these workings is to simply abandon them in the fall, as we have done, and then quietly open them up again in the spring."[32] Thereafter, Osborn let the matter rest.

The first snow of the season fell on the night of September 25, 1902, but the weather continued to be mostly favorable for fossil collecting for another month. Quality fossils turned up repeatedly, including a pair of lower jaws and a premaxilla of *Allosaurus* and a number of excellent limb and pelvic bones of *Brontosaurus*. By late October, work on the current stripping was essentially finished. Granger was reluctant to start a new stripping for fear of winter weather settling in before it could be finished. On November 2, the party abandoned work in the quarry for the season, packed their spoils into fifty-one boxes and headed for Medicine Bow. They waited ten idle days for a freight car before Granger and Kaisen gave it up, returning to the New York museum and leaving Olsen to supervise the loading by himself whenever the tardy boxcar should make an appearance.[33]

Barnum Brown, meanwhile, spent the summer of 1902 in Montana, searching for Cretaceous dinosaurs. He had a few useful fossil tips to follow up and a mandate from Osborn to use every resource available to locate promising exposures. "[T]here is every reason to think that by careful inquiry among the natives, by making friends wherever you can, and by energetic prospecting, you may find something of real value," Osborn advised.[34] At the same time, he was wary of attracting the competition. He warned his collector that "if you are striking it rich we will not say very much about it, because if we do we shall probably have some companions next year. I notice that our friends in other Museums do not hesitate to poach on our preserves."[35] In fact, Brown did strike it rich in the Cretaceous. But Osborn, although pleased by Brown's success, still pined for a huge, exhibit-quality Jurassic dinosaur. When a promising tip arrived at the museum in late September of a large dinosaur in South Dakota, Osborn detailed Brown to track it down "at once," but a series of miscues conspired to make the trip to investigate the specimen impossible for the present.[36]

Brown, however, had found another Jurassic dinosaur lead of his own. In the lobby of the Billings State Bank, a small number of miscellaneous fossil vertebrate specimens, including at least one Jurassic dinosaur limb bone, attracted his attention. The source of these bones, he soon learned, was a series of Jurassic exposures along Beauvais Creek on the Crow Indian Reservation. On October 21, 1902—the day before the Laramie wedding of his friend and rival Charles Gilmore—Brown ventured south to explore these beds, while Osborn waited anxiously for his report. "I am, as you know, extremely desirous of finding a new Jurassic locality, since, while we are getting splendid materials from the quarries in Wyoming, there is no prospect of complete skeletons or of completing the specimens already obtained," Osborn explained pessimistically.[37] But what Brown found was only somewhat promising. The exposures were limited in area and only moderately productive in vertebrate fossils. A special expedition to collect from these beds was not warranted, in Brown's estimation, but it would be worthwhile to take a longer look at these beds in conjunction with another Montana Cretaceous expedition in some future field season.[38]

"We Are Certainly Not Holding Our Own"

Osborn planned another two-pronged assault on the Jurassic for the summer of 1903. As in years past, a party of diggers, including Kaisen and P. Miller, would spend their season excavating at several known localities in southeastern Wyoming. Reed's Quarry R and especially Bone Cabin Quarry were still producing fossils by the carload, but Osborn's confidence in them as a likely source for an exhibit-quality skeleton was about played out. He therefore detailed Brown to search widely for fossil reptiles in Wyoming, South Dakota, and Montana. "I am especially desirous of finding a new region for Jurassic Dinosaurs," he instructed.[39] A galling letter from Elmer S. Riggs informed Osborn that his Field Columbian Museum rival was working on a complete restoration of *Brontosaurus*, based on the fossil material he and his party had collected in western Colorado in 1901. This news provoked a competitive outburst that Osborn relayed in a letter to Brown. "I do hope we shall be able to get some fine large dinosaur material this season. We are certainly not holding our own,"

Osborn complained, although with scant justification. "Riggs is going to make a new restoration of Brontosaurus, a thing we could not do from our material. In fact, both Carnegie and Chicago have done better than we have, with the single exception of Bone Cabin Quarry."[40] It was a subtle rebuke intended to spur Brown to great discoveries.

Brown left New York for the field on the evening of May 12, 1903. He stopped briefly in Chicago to visit with Riggs, his former KU colleague. Then he journeyed to Edgemont, South Dakota, where he rented an outfit and hired a field assistant. Together they headed into the Cretaceous beds south of the Black Hills and hunted down a number of good specimens of marine reptiles and pterosaurs. Brown anticipated at least two months of productive work at this locality.[41] Osborn delighted in Brown's report, but he was impatient for results with Jurassic dinosaurs. One storied specimen, in particular, was really tormenting him. Sometime after Brown's departure, a letter arrived at the museum from a man named E. A. Sparhawk, of Buffalo, Wyoming. Sparhawk claimed to know the location of an excellent specimen of *Brontosaurus*, and Osborn wanted Brown to make arrangements to examine it as soon as possible. But Osborn's instructions were sufficiently vague as to give Brown the flexibility to remain for an indefinite period collecting in the Cretaceous. "[I]t appears to be a good plan for you to make a number of prospects and locations, and then take the trip to the Big Horn country to search for [Sparhawk's] Brontosaurus," he wrote, then added ambiguously, "I do not like to postpone that too long for fear someone else may snap up the prospect."[42] So Brown continued working in South Dakota for another five weeks, while Osborn stewed. Meanwhile, welcome news arrived in New York that the Carnegie Museum's latest *Diplodocus* skeleton—the Powder River specimen found by W. H. Utterback in 1902, which was reported by Hatcher to be an excellent specimen—was not turning out as well as expected. But this did nothing to dampen Osborn's competitive spirit. To Brown he reiterated, "I still feel very strongly that one of the chief objects of our life at present must be the completion of our Jurassic Dinosaur, so I am earnestly hoping that your visit to Sparhawk . . . may be crowned with success."[43]

Instead, the eagerly anticipated dinosaur was a bust. Brown tracked Sparhawk down at the Occidental Hotel in Buffalo. He learned straightaway that the specimen was not nearly as good as the letter to Osborn implied. He went to see it anyway and found only a single, badly weathered limb of no value. As for the informant, "Sparhawk himself is a bar-room inhabitant and general wild horse scheme promoter with views of large sums of money in [his] future," Brown noted disapprovingly.[44] Osborn claimed that he suspected some sort of fraud, but he thought it best to run down the tip regardless. He also tried to dispel the disappointment by repeating his news about the failure of the Carnegie Museum's *Diplodocus*. "It is quite badly crushed," he gloated in a letter to Brown.[45]

In Buffalo, Brown learned the location of the Carnegie Museum's new *Diplodocus* quarry and he wanted to see it for himself. He hired a team and wagon and headed south along the eastern flank of the Big Horn Mountains. He traced exposures of Jurassic rock for a stretch of nearly forty miles between Buffalo and the Middle Fork of the Powder River, searching for dinosaurs. In some places, the dinosaur-bearing beds were about sixty feet thick, but nowhere did he find bones in great quantities. The best exposures, in Brown's judgment, were located east and north of Kaycee, Wyoming, along the Red Fork of the Powder River, about ninety miles over rough mountain road from the nearest railroad. But, to Brown's chagrin, a Carnegie Museum party led by Utterback was already there at work on a reasonably productive quarry. Brown stopped in and asked his rival's blessing to spend a day or more prospecting the area from his campsite. According to Brown, the Carnegie collector was very hospitable, giving him a frank and optimistic assessment of the local prospects. He spent just a single day there, however, finding only one prospect consisting of two limb bones in poor condition. Thinking that he might have better luck in Montana, Brown headed back north. If the pickings were poor in the north, Brown proposed to return to the Powder River country to set up shop near the Carnegie quarries.[46]

In Sheridan, Wyoming, Brown rented a costly outfit and saddle horse and hired a man to cook and manage his campsite. He headed northwest from Sheridan on July 30, 1903—the day after W. C. Knight's lamentable death in Laramie—again tracing the Jurassic beds through the foothills of the Big Horn Mountains. The chief advantage of this area over the Jurassic exposures to the south was their nearness to the railroad—nowhere would Brown be more than twenty miles away from the train. In a little over one week's time, he reached the banks of Beauvais Creek, east of Pryor, Montana. Along the route, he scouted a few, infrequent dinosaur prospects consisting of badly crushed bones. To the north, however, where the Jurassic beds assume a more horizontal position, he found fossils in a far better state of preservation. Beauvais Creek, as well as several other tributaries of the Bighorn and Yellowstone rivers, cuts through the Jurassic in this area, providing excellent sections in which to search for bones. Over the next week he worked three separate prospects. The first was a *Stegosaurus* tail, which ran promisingly into a bank but then played out abruptly, yielding only a number of small dorsal plates, a few caudal vertebrae, and a humerus. A second partial skeleton, unidentified, he abandoned due to poor preservation. His third prospect showed the greatest promise. He identified it tentatively as a *Diplodocus*, with all four limbs intact (but lacking the feet) and six or eight vertebrae. He also wanted to spend some time developing a hillside where numerous bones had been found eroded out, some of which had been collected and displayed in a bank in Billings, Montana.[47]

Brown went to Pryor, Montana, to get lumber to pack his specimens. During this trip he developed a debilitating fever that laid him low for days. He nursed himself back to health with the aid of sage tea and returned to work on August 22. His outfit was so expensive that he decided that it would be prudent to finish work on his best prospect, pack his things, and return to Sheridan. Along with the Cretaceous specimens he obtained south of Edgemont, Brown expected to ship about thirty boxes of fossils back to New York, including five boxes from Montana and two more from Wyoming consisted mainly of Jurassic dinosaurs. He sent this news to Osborn, explaining, "[A]s I understood, you desired Dinosaurs from the Jurassic above all others so thats what I am sending back. . . . I think it advisable to work [here] another season."[48]

Meanwhile, Kaisen and Miller left New York headed for Wyoming on June 1, 1903. Kaisen detrained at Laramie, tracked down his friend Olsen, and urged him to work for the expedition again as cook and teamster. But Olsen declined. So Kaisen convinced his brother, Martin, to give the work a try. He also stopped for a friendly chat with Reed, who had returned to work at the University of Wyoming by 1903. Reed planned to open up a rival quarry on Como Bluff within the month. The DVP party established their first camp along Rock Creek near Quarry R on June 7th. The creek was flooded, so they had to take a long detour to cross it. But by August it would be nearly dry and its water unpalatable. Cold and rainy weather hampered the work for three full days.[49]

Serious digging commenced on the 9th, at the west end of the quarry, but all that turned up were some useless fragments. The east end was just as unproductive. At the center of the quarry the party had somewhat better luck. A small pocket of large, disarticulated bones preserved in a hard layer of sandstone yielded a few dorsal vertebrae, scapulae, and pelvic bones, and then played out. Kaisen, who had been spending his winters since 1900 working in the DVP's fossil preparation lab, feared that the most delicate structures of the vertebrae would be difficult, if not impossible, to prepare for exhibit because of the extreme hardness of the matrix. Underneath, the party found a layer of numerous smaller bones, including humeri, femora, vertebrae, and more scapulae, all disarticulated.

Following Osborn's mandate, many of these bones were left behind due to their diminutive size. On Sundays, and even on the Fourth of July, members of the DVP party squandered their free time searching Como Bluff and environs for better prospects, but they found nothing. Kaisen concluded that it would not pay to do any more stripping at Quarry R and he decided to relocate to Bone Cabin Quarry on or about the first of August. Osborn approved Kaisen's plan.[50]

Meanwhile, Reed showed up in mid-June. He took Kaisen on a walking tour of Como Bluff and pointed out a prospect just east of Marsh's old Quarry 13. Together they poked around the prospect and uncovered a large cervical vertebra, four small caudals, and one claw. Not only were these bones fairly well preserved, but they were also the first articulated specimens of the season. Kaisen thought it would be best to spend a day or two at this site before committing to Bone Cabin Quarry. A few days digging might uncover something worthwhile. Reed's plan, according to Kaisen, was to reopen Marsh's mammal quarry. But it is possible that Kaisen, whose English was somewhat limited, misunderstood Reed's intentions, as nothing about the Como Bluff prospect was ever mentioned again.[51]

The DVP party abandoned work at Quarry R on August 1, 1903, as planned. They worked briefly at a prospect two miles to the west, where they recovered a femur and ilium of *Stegosaurus*. By August 3, they had established a new camp at Bone Cabin Quarry, and work was begun on a stripping of about twelve hundred square feet. Quality fossil material, however, was much scarcer than in previous seasons. The best bones obtained included a partial skull of *Allosaurus*, a fairly complete and articulated sauropod forelimb, and a partial associated skeleton of *Stegosaurus*. Osborn was very pleased with the *Allosaurus* remains.[52]

Toward the close of the field season, the DVP party discovered a pocket of well-preserved bones that seemed to continue into the bank, showing promise of future rewards. For the present, however, Kaisen had the good fortune to find a skull, atlas, and axis of *Diplodocus*. A few caudals of the same genus also turned up. Osborn had not taken the trouble to send many letters to Kaisen in the summer of 1903, but he did so now, writing, "I . . . congratulate you most heartily on the discovery of the skull of *Diplodocus*. . . . On the whole I think you have had a successful season; and if the . . . skull turns out well we shall have every reason to be satisfied." Finding the skull gave Kaisen renewed hope for more and better specimens, but a careful two-day search in the same area produced nothing further. In mid-September, a bad storm dumped more than five inches of snow on the quarry. The party lost two days shoveling and then drying the quarry out. By September 17, with early winter weather threatening, Kaisen closed up shop. Thirty-two boxes weighing about five tons was the total catch for the season. Kaisen and Miller returned to New York sometime during the first week of October.[53]

Last Days in the Jurassic

The DVP temporarily abandoned work in the Jurassic for 1904. Osborn had crafted an ambitious plan to explore systematically for fossil mammals through the whole of the North American Tertiary. Granger—who married Anna Dean on April 7, 1904—was assigned to work the Eocene (including what is now called the Paleocene) beds. Accordingly, a large party, under Granger's direction, worked all summer in Wyoming's Bridger Basin—which he had explored with Matthew in 1902—collecting fossil mammals. They were supported by a team and wagon driven out from Medicine Bow. Granger had spent his last days in the American Jurassic.[54]

Brown married a New York biology teacher named Marion Raymond on February 13, 1904, the day after his thirty-first birthday. He spent the following summer hunting Cretaceous dinosaurs. On the Crow Reservation, in south-central Montana, he found a small strip of Jurassic rocks and stopped for a closer look. He spent a few days scratching at the remains of a stegosaur skeleton on Beauvais Creek. He uncovered the entire specimen, hoping to find a skull, but was disappointed. As it was badly crushed, mangled, and headless, Brown left the skeleton in the ground. He believed it would pay to return to this locality for Jurassic dinosaurs at another time.[55]

Work resumed again at Bone Cabin Quarry on June 30, 1905 (see figure 29). Osborn particularly wanted to get the skulls of *Stegosaurus* and *Brontosaurus*, as well as those parts of the postcranial skeleton of *Stegosaurus* still lacking in the DVP collections.[56] But Kaisen, who took charge of the work, began the season with a run of very bad luck. Guy Gibson, a man he summoned from Laramie to serve as his assistant, took ill after only seven days of work. He left, never to return. Kaisen hired Clark Hader, of Medicine Bow, to replace him. Hader took well to the heavy labor, but was too careless to be trusted on the delicate work with the specimens. The weather was poor, and rain hampered the dig for two days. Then the wagon, which was getting well worn, suffered a breakdown. Kaisen lost several days in hauling it to Medicine Bow for repairs. Still, the work progressed. With the help of his assistants, Kaisen made a

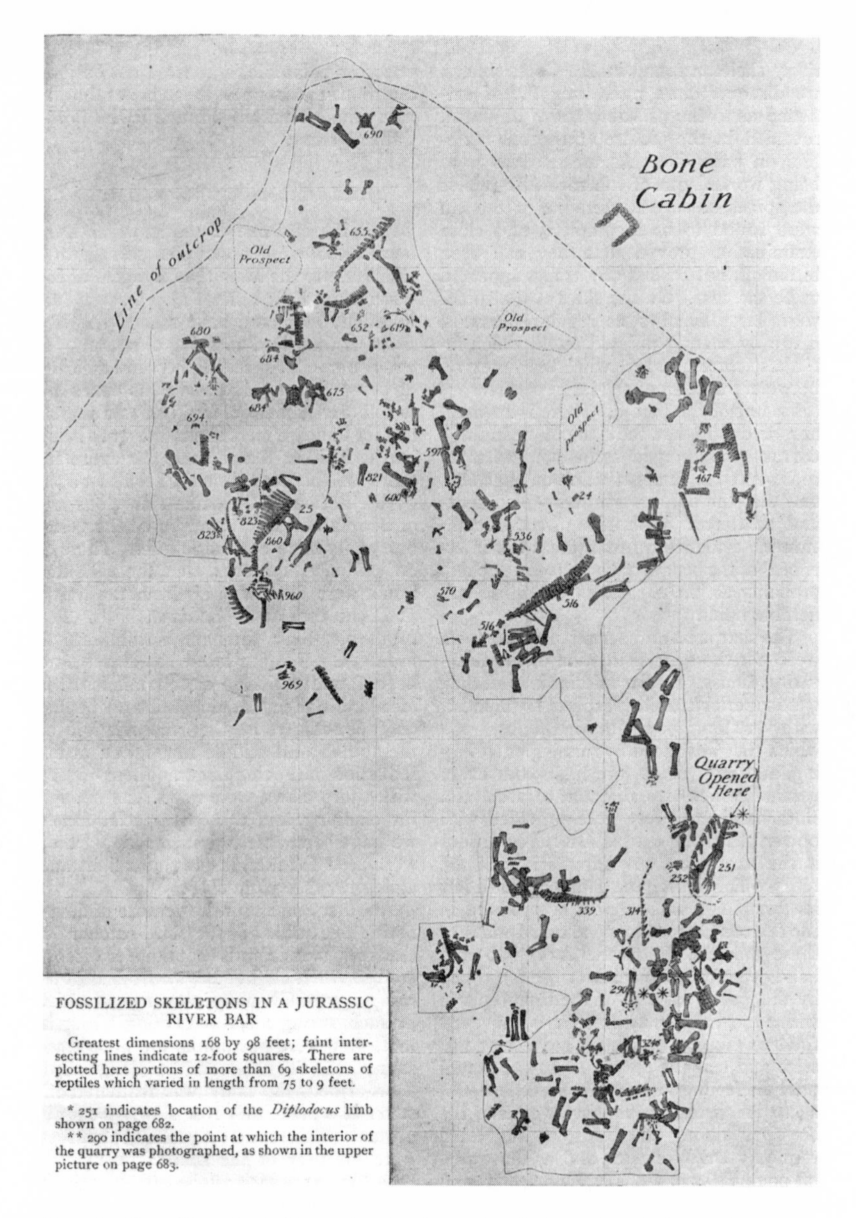
Bone
Cabin
Line of outcrop
Old
Prospect
Old
Prospect
Old
prospect
Quarry
Opened
Here
690
655
652
619
680
684
675
694
597
821
600
25
823
860
960
969
24
467
536
570
516
251
252
339
314
290
FOSSILIZED SKELETONS IN A JURASSIC RIVER BAR
Greatest dimensions 168 by 98 feet; faint intersecting lines indicate 12-foot squares. There are plotted here portions of more than 69 skeletons of reptiles which varied in length from 75 to 9 feet.
* 251 indicates location of the *Diplodocus* limb shown on page 682.
** 290 indicates the point at which the interior of the quarry was photographed, as shown in the upper picture on page 683.

Figure 29. Map of Bone Cabin Quarry showing the great profusion of Jurassic dinosaur bones excavated by the American Museum from 1898 to 1904. From Osborn, "Fossil Wonders."

stripping of approximately fifteen hundred square feet—large enough to last all summer. By mid-July, he was ready to hunt for bones, and in short order he uncovered a pocket of sauropod limbs. Some of these bones were well preserved, but most were badly crushed and broken. Kaisen had to take many of them out of the quarry, regardless, in order to make more room to work. What he wanted to know from Osborn is whether or not he should keep any of them.[57]

Osborn replied with a detailed set of instructions:

> I do not think we want to stop for limb bones of Sauropoda unless they are very good. Limb bones of carnivorous dinosaurs, of *Stegosaurus*, or limb bones of Sauropoda with foot bones associated we should keep. I feel it is important for you to supplement our *Stegosaurus* skeleton, and our carnivorous dinosaur skeleton, and to look out for anything new. . . . Cervical vertebrae, of course, might lead to a skull. . . . If you stop to take out things of little value your work will be limited, and you will not be able to get over so much ground. I feel, however, that I must trust largely to your own good judgment. You have had a lot of experience, and know just about what we need and what will be of greatest use to us.[58]

In mid-August, Kaisen collected his finest specimen of the season, a partial skeleton of a small dinosaur consisting of a skull, lower jaws, one hind limb, partial pelvis, and twenty-nine vertebrae running from the middle of the back to the middle of the tail. Judging from the size of the skull, Kaisen thought it might be a *Laosaurus*. "I would like to know if you think that little Dinosaur of any special value," he asked Osborn.[59] On August 26, Kaisen was joined by D. D. Streeter Jr., a volunteer laborer who had been with Granger in the Bridger Basin earlier in the summer. Streeter stayed for almost a month and made himself very useful. Sometime later Kaisen found a hind limb of *Allosaurus*, with femur, tibia, fibula, and many foot and toe bones. He also found another partial skull, but he could not identify the genus. With that, the pocket of bones that Kaisen had been working was exhausted, and the outlook for finding better material, he felt, was bad.[60]

A few badly crushed and broken bones turned up here and there in the quarry, but nothing worth saving. Weeks went by without finding anything worthy. A few poorly preserved bones could be seen running into the bank on the west side of the quarry, but they did not look promising. Moreover, there was now a layer of sandstone overburden ten or more feet thick atop the bone layer, thus rendering the digging both more difficult and more expensive. By mid-September, Kaisen had given up hope. "I don't know of any place to look for bones," he wrote despondently.[61]

Osborn was disappointed, of course, but his reply to Kaisen was surprisingly magnanimous. "I have such confidence in you that I know you have done your very best, and that we must be as philosophical as we can about the quarry playing out."[62]

Although Bone Cabin Quarry was not entirely exhausted, Osborn could afford to feel philosophical about giving it up for good. The DVP had already accumulated an enormous quantity of Jurassic dinosaurs of many different types, from gigantic sauropods to small, birdlike theropods. Many of these specimens were reasonably complete and suitable for display. Osborn never got the complete sauropod specimen he wanted, but there was sufficient material in the DVP collections with which to mount a composite skeleton of *Brontosaurus* early in 1905. More important, Osborn had a wandering eye, and his fancy was being diverted to the spectacular Cretaceous dinosaurs that Brown was then collecting in Montana and Wyoming, including the gigantic new theropod *Tyrannosaurus rex*, which Osborn first described in 1905. Osborn's competitors at the Field and Carnegie Museums had both abandoned the search for Jurassic dinosaurs in favor of other groups and other horizons, including the Cretaceous. So he was very keen on taking the lion's share of the best Cretaceous specimens before they fell into less deserving hands. "It is evident that we have a good field for next season," he wrote to Brown about his new locality in Montana. "In the meantime we must keep very quiet about it. I am tired of prospecting for the benefit of other Museums."[63]

So on October 2 the work in Wyoming was closed. Kaisen took six boxes weighing three thousand pounds to the freight office in Medicine Bow—the last shipment of Jurassic fossils from Bone Cabin Quarry. He sold the battered wagon to a local rancher for $15. Included in the fire sale was the gentle black saddle horse, Ute, which had done famous duty as Osborn's mount during a visit to the quarry in 1901. Osborn had decreed that this horse should be shot rather than sold to anyone who might not treat him kindly, but Kaisen could not pull the trigger. He felt the heavy weight of nostalgia for his years spent toiling at the quarry, and regret for the poor showing in this, his final season. To Osborn he wrote, "I will close up after working here Seven Years. I only wish that I had a bigger collection to close with."[64] Again, Osborn rose to the occasion with a generous reply. "I will welcome you back to the Museum, and I appreciate the grand work you have done in the Bone Cabin quarry," he wrote. "If anyone can find fossils there I know Kaison [*sic*] can."[65]

TWELVE

Putting Dinosaurs in Their Places

A decade of Jurassic fieldwork yielded massive quantities of unprepared fossil material for museum paleontologists in Chicago, Pittsburgh, and New York, where dinosaurs arrived by the ton from the American West. Though the effort expended to find them, dig them up, pack, and ship them was enormous, the work of making a useful museum collection of dinosaurs was only fairly begun in the field. Years of toil remained. First, all specimens arriving from the field had to undergo some amount—often an enormous amount—of painstaking fossil preparation. Next, certain specimens, like those that were completely new, or showed some novel feature, lent themselves to study, description, and publication. Finally, the best, the biggest, and the most complete specimens went on display, often in elaborate, lifelike mounts.

Fossil preparation, the costly and time-consuming process of freeing fossils from their rocky matrix and getting them into shape for study and display, posed a number of novel challenges for museum paleontologists. Finding adequate space to accommodate dinosaurs was often a serious problem. New fossil materials arriving from the field required room for temporary storage and dedicated laboratory space. Adapting a basic fossil preparation lab to the needs of dinosaur paleontology involved considerable extra investment in equipment and space. Finding, training, and retaining skilled fossil preparators and other support staff became increasingly expensive during the second Jurassic dinosaur rush. Paying modest wages to keep expenses down sometimes led to resentment or turnover. Savvy preparators used opportunities at rival museums to leverage better pay or better working conditions. The sheer volume of work, and its unique demands, led to increased specialization and professionalization among the science support staff. This, in turn, drove higher standards for the work, leading to important lab and fieldwork innovations. Preparators developed new techniques to handle the enormous workload, some of which required expensive new machinery, entirely new systems (e.g., electricity, or pneumatic apparatus), or new spaces in which to operate the equipment, some of which produced particularly noxious dust, noise, and/or fumes. Nevertheless, the essential task of fossil preparation, usually performed in backroom or basement labs by low-paid minions working in relative obscurity, was a vital prerequisite for the higher profile work of publishing original research and putting fossils on display.

If O. C. Marsh's seminal work on Jurassic dinosaurs served as a model for researchers who succeeded him, it also provided them with an irresistible target. The next generation of American vertebrate paleontologists, almost to a man, reviled their illustrious Yale predecessor, and many relished the idea of revising his body of work. New, better, or more complete specimens collected during the second Jurassic dinosaur rush provided the opportunity to correct some of Marsh's anatomical misinterpretations, many made on the basis of wholly inferior fossil materials. Moreover, Marsh, along with his rival, E. D. Cope, created hundreds of new taxa of dubious status, often from mere fragments, and it remained for their successors to straighten out the taxonomic confusion they created. Usually this involved combining several hitherto poorly known and apparently synonymous forms into a single, better known taxon. Paleontologists also discovered and described a number of entirely new dinosaurs during this period. New and better data provided fodder for original projects in biostratigraphy and systematics. Also, the greater number of active researchers brought a greater diversity of views to the table on the more subjective aspects of Jurassic dinosaur paleobiology, including new speculations on dinosaur posture and behavior. Cooperation on research projects made a feeble beginning during the second Jurassic dinosaur rush. Museum paleontologists sometimes shared resources, even swapping fossils or casts. Vital information about new fossil localities or important new fossil preparation techniques began appearing with ever-greater frequency in the scientific literature. Research topics and the fossil materials necessary to complete them properly were sometimes divided amiably among museum paleontologists. But cooperation was still the exception and not the rule.

Dinosaurs in the Rough

Developing an efficient system for storing and preparing fossils was an essential step in building a museum program in dinosaur paleontology. At the American Museum, a flourishing program in mammalian paleontology, established in 1891, lent the DVP a considerable competitive advantage over the upstart programs at the new museums in Pittsburgh and Chicago. Even so, the influx of Jurassic dinosaur specimens beginning in 1897 quickly overtaxed the DVP's handling capacity. Fortunately, Curator H. F. Osborn had the clout to get what he wanted from museum administrators. His program began in humble quarters, cramped and confined in the museum's basement. By 1898, its three storerooms were entirely filled with fossils. Osborn used this fact to leverage some new space. Late in 1899, the museum completely remodeled his department, assigning it to new offices on the uppermost floor of the east wing. Osborn was understandably pleased with his "very roomy" accommodations.[1]

The remodeled workspace for the DVP was a boon for fossil preparation. Better lighting and ventilation in the new top-floor fossil preparation lab made the work more pleasant and elevated its visibility and prestige. Rooms were retained in the basement, however, both for long-term storage of inferior fossils and to provide room for the dirtiest and noisiest lab work, which Osborn preferred to keep out of sight. The opportunity to upgrade the lab's systems and appliances was available in 1899, and it was probably taken, although it seems likely that improvements were continuously being made in the lab to keep it state of the art. The new lab featured an overhead trolley system, with chains and movable hoisting blocks attached to steel rails, which was used both to lift and move heavy blocks containing specimens and to suspend specimens while they were being fitted for mounting. The lab was wired for electricity, which provided power for reliable indoor illumination and to run certain tools, including the "indispensable" portable electric drill, a new invention. Small electric motors were useful for operating a multitude of essential tools. A two-horsepower motor operated a large lathe, which drove a rotary diamond saw used for cutting stone and fossil bone, wheels for grinding and sharpening hand tools, a drill for boring specimens, and a small saw for cutting and splitting metal. A smaller motor ran the blower on a miniature blast furnace used for heating and shaping metal armatures for mounting specimens, or for tempering or reshaping metal tools.[2]

At Pittsburgh's Carnegie Museum, Director W. J. Holland was a newcomer to vertebrate paleontology and he sometimes failed to anticipate

Figure 30. Hatcher (far right) visits the remodeled fossil preparation lab at the Carnegie Museum in 1903. © Carnegie Museum of Natural History.

fully the needs of his new department. It was not until October 1899, for example, when field-workers were already returning to Pittsburgh with their first assortment of Jurassic dinosaurs, that Holland appealed to the Committee on Buildings for space in the museum to establish a laboratory for fossil preparation and an office for J. L. Wortman, his new curator. The lab took shape rather quickly, with only a few start-up troubles, and preparators began slowly turning out specimens in early November. By January, only a few weeks before his resignation, Wortman was well satisfied with progress in the lab. Following J. B. Hatcher's first field season, in the summer of 1900, Holland provided a new, larger space for the preparation lab and storeroom. Hatcher and his staff spent a week arranging these rooms for maximum efficiency (see figure 30). Nevertheless, a growing preparation staff and a steady accumulation of Jurassic dinosaur fossils from the Sheep Creek, Garden Park, and Powder River localities ultimately swamped the available space. Preparators fitted up temporary quarters in the basement of a new museum building in 1906, but a lack of adequate space and proper appliances hampered their work. Until the new building was completed, and a permanent preparation lab established, finding adequate room for fossil storage and preparation would continue to be a problem that occasioned considerable inconvenience and loss of time.[3]

At the Field Columbian Museum, Geology Curator O. C. Farrington pressed forward with vertebrate paleontology, but he took a somewhat ad hoc approach to assimilating the new program within the structure of his Geology Department. Following the museum's inaugural paleontology expedition in 1898, room to accommodate the influx of fossil mammals had to be improvised somewhere within Geology's space (see figure 31). Somehow Farrington compressed the departmental library, in Hall 74, to half its original size. Once fitted with tables and a rack of storage trays, the space gained was just barely large enough to serve as the museum's first fossil preparation laboratory and storeroom (see figure 32). But when dinosaurs first arrived from Wyoming in 1899, the makeshift preparation lab proved too small for the work. Extra room was afforded by removing the remaining books and bookcases to the increasingly crowded curatorial office in Hall 73. The preparation lab, expanded to fill all of Hall 74, gained a turning lathe, a workbench, and a sink with running water. This, too, proved inadequate once work commenced on the enormous

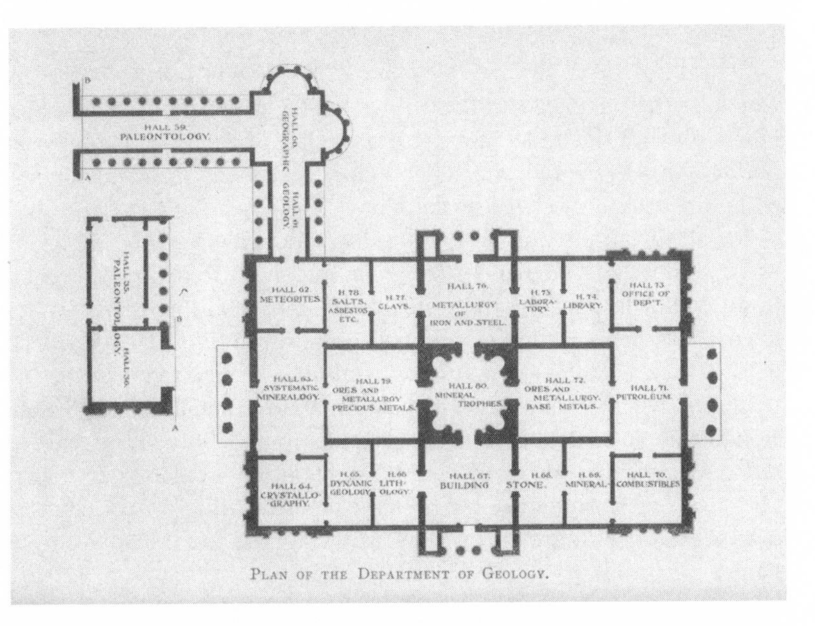

Figure 31. Floor plan of the Department of Geology in the West Pavilion at the Field Columbian Museum, ca. 1895. From Field Columbian Museum, "Historical."

Figure 32. Elmer S. Riggs (right) and an assistant work on Jurassic dinosaurs in the Field Columbian Museum's first, overcrowded, fossil preparation laboratory, ca. 1900. The enormous limb bone encased in plaster on the left is probably the femur of *Brachiosaurus*. Courtesy of the Field Museum, Negative #CSGEO3251c.

Jurassic dinosaurs from western Colorado. To provide more room for the work, Farrington agreed, in the spring of 1902, to swap his corner office in Hall 73 with the undersized preparation lab. The new lab included all the trappings of the old, and added a closet, revolving worktables, and a skylight with sliding overhead curtains.[4]

For all three museums, a staff of skilled and experienced technicians was the most vital ingredient for operating an efficient fossil preparation lab, but finding the right preparators and retaining their services for the long term could be a difficult proposition. Luring experienced but dissatisfied staffers from other institutions became a common practice. Osborn acquired his chief preparator, Adam Hermann, from Marsh. Holland, in turn, took Arthur Coggeshall from Osborn. Riggs bagged H. W. Menke from the American Museum, but then failed to get Albert Thomson or Charles Christman from the same institution, C. W. Gilmore from the Carnegie Museum, or even Charles Bunker from the University of Kansas.[5]

Seducing another institution's staff members was often interpreted as a hostile act. Osborn, for example, remarked bitterly about Hatcher's "absence of a clear feeling of right or wrong," when the latter allegedly co-opted his brother-in-law, O. A. Peterson, to accompany him to Patagonia in 1896. However, less than one month later, Osborn asked a Princeton collector in Hatcher's employ to make a special search for certain fossil mammals on behalf of the American Museum. Osborn declined to hire the Princeton collector outright, though, claiming that "no man's heart can be in two places at the same time."[6] Wortman wrote an apologetic letter to Osborn disavowing his own role in bringing Peterson from Princeton to the Carnegie Museum in 1900, but only because Osborn wanted Peterson to come back to work for the American Museum.[7] And S. W. Williston felt he owed Hatcher an apology and an explanation when Riggs tried to tempt his friend Sydney Prentice, the Carnegie Museum's scientific illustrator, with a position at the Field Columbian Museum.[8]

A higher salary, better working conditions, and greater opportunities to do certain kinds of preferred work, like research or fieldwork, were the chief inducements used to lure preparators to switch allegiances. The same were also sometimes used to try to persuade them to stay. Osborn was sometimes proactive in lobbying for his preparators. In 1900, for example, after instituting a new rule requiring his staff to work eight hours per day (instead of seven), he felt they deserved a "slight" raise. Preparators also had their own reasons for staying or leaving. Many of these men worked anonymously and some resented it. Peterson quit the American Museum because of a perceived lack of due credit. Those who stayed and did good work could sometimes negotiate for greater official acknowledgement. Paleontological assistant Arthur W. Slocum, for example, wanted a position title from the Field Museum "of sufficient merit to warrant publishing . . . in the Annual Reports as a member of the Scientific Staff." Some preparators used job offers at rival institutions to bargain for better terms. Still others, such as Norman Boss of the Carnegie Museum, tried this tactic and failed. Curators and administrators resented this practice very much and worked to suppress it. Farrington, for example, wrote a letter to an American Museum preparator warning that his museum "would not care to have its offer used to compel the payment of higher wages by a sister institution." Some, including Osborn, seemed to think that the gentlemanly thing to do was to deal preparators among themselves like baseball trading cards.[9]

Osborn expected unflagging loyalty from his subordinates, especially collectors and preparators, although he was sometimes reluctant or unwilling to meet the demands of workers who asked for more rewards, financial or otherwise, in return for their faithful service. He denied Princeton's J. W. Gidley a long-term opportunity with the DVP, for instance, because he felt it would be better to "train someone in [the work] whose sole interest is in the American Museum." Gidley stayed for years, anyway, always on a temporary basis, but he grew increasingly frustrated with his lot. In 1899 he complained, "It seems rather hard after all my years of experience . . . that I should be out here in the field working like a slave for . . . $50 per month, less than I was getting before I went to college."[10] Barnum Brown pleaded for years for a permanent position under Osborn, but did not get one until sometime after his return from Patagonia in 1900. He also negotiated repeatedly for better pay, but Osborn was exceedingly slow to raise his salary. Osborn seemed to think that the experience Brown was getting under his tutelage, the reputation he was winning, and the opportunity to publish some of his own results "ought to be sufficient reward" for the persistent low pay and lack of commitment on Osborn's part.[11] Riggs probably fell into permanent disfavor with Osborn after he cancelled a miserable arrangement he had made to work for the DVP for half pay, in 1898, in order to take a seemingly much more promising position in Chicago. After Wortman quit the DVP and joined the Carnegie Museum, taking Coggeshall with him, Osborn feared he would try to lure away more of his collectors. But Osborn expected them to feel honor bound to remain, writing in a thinly veiled warning to Walter Granger that "it would be a decided breach of faith for any man to leave the party before the close of the season."[12] Many of Osborn's subordinates, perhaps surprisingly, did remain loyal to the DVP. Rainger lists fourteen employees who stayed with Osborn for more than twenty years.[13]

Finding capable young men with little or no experience with fossils, but with reasonably good mechanical skills, and then training them to be excellent preparators was another common approach to staffing the preparation lab. Holland and Hatcher were especially keen to find and train their own preparators for the Carnegie Museum. But what were the qualities that suited a man for such a position? Hatcher felt that willing, interested, and modest young men were the best candidates to become well-trained workers. He also insisted on finding someone who would be agreeable. Holland, however, seemed not to get along well with anybody. He valued obedience most, and sought men who appeared to be pliant, modest, and willing to obey orders. He preferred to find a "college-bred" man "who has his way to work in the world." But he could be exceedingly picky. He turned one young man away for being "too sullen." Another was "too raw." Nor did he want a man with too much experience who might command an exorbitant price.[14] Osborn valued loyalty in his subordinates above all other virtues. He also seemed to take particularly well to men from the rural West. Over the long term, he seemed to get along much better with men who earned their reputations under him, men who owed him their careers. He had much poorer luck with Cope and Marsh castoffs, such as Hatcher, Peterson, and Wortman (Hermann was an important exception), or with men like O. P. Hay, who were accomplished scientists in their own right.[15]

Yet at the height of the second Jurassic dinosaur rush, no museum could afford to be too choosy about its preparators. Men of various skill levels and experience swelled the ranks of the preparation staffs at all three museums in the first few years of the twentieth century. Indeed, by 1900 the crush of dinosaurs coming in from the field created a terrible bottleneck in the DVP, despite efforts to mechanize and otherwise streamline the work. Osborn griped repeatedly that his preparation staff of seven men was too small. He added more and more men, and by 1903 the DVP boasted a preparation staff of fifteen.[16] When a similar crisis arrived at the Carnegie Museum, in 1903, Hatcher responded by contracting field operations. He kept Peterson in Pittsburgh for the summer to work on the backlog of unprepared fossil mammals. Later, in September, he recalled collector Earl Douglass from the field one month early, both because of a sudden and surprising drain of fieldwork funds and because of the abundance of work to do back at the lab.[17] Farrington urged the Field Columbian Museum to hire additional preparators in 1902 in order to keep abreast of the mounting Jurassic dinosaur workload. His request was denied, not because there was no need, but because the Geology Department already had seven employees.[18]

The high volume of work to be done during the second Jurassic dinosaur rush led to some increase in specialization and a sharper division of labor in museum paleontology departments. Osborn hired dedicated collectors and preparators from the very start. He would orchestrate the work of the department and reap most of the credit for its accomplishments, but he left the lower status manual labor to his staff of subordinates. He rarely participated in fieldwork, and seldom, if ever, involved himself with preparation.[19] Rainger has detailed how effectively the division of labor worked in the DVP and how Osborn profited by it. Yet it was sometimes a source of discord. Hatcher was particularly critical of Osborn's style of leadership. With Osborn in mind he wrote: "It seems to me that if some of the older workers in vertebrate paleontology would go to the trouble to go out into the field, do their own collecting, and familiarize themselves with the laboratory work, they would have a greater appreciation for the work and efforts of others."[20]

Hermann was the DVP's chief preparator during the second Jurassic dinosaur rush. He ran the departmental lab, supervised the other preparators, and, at Osborn's urging, developed new techniques for preparing and mounting fossils for display. He hardly ever participated in other departmental activities, however. Coggeshall, who trained under Hermann at the American Museum, later filled the same role of chief preparator for the Carnegie Museum. He also had a large staff of assistants to help with preparation. At the Field Columbian Museum, which had a much smaller paleontology staff than its Eastern rivals, the situation was very different. Riggs played the part of collector, chief preparator, researcher, and exhibit developer, and was the only vertebrate paleontologist of his era to make significant contributions in all four of these areas.

The need for greater speed and accuracy drove the development of a number of innovative fossil preparation techniques. Prior to the second Jurassic dinosaur rush, when the high volume of work first began to demand greater efficiency, fossil preparators worked exclusively with hand tools. Preparators removed the matrix from the bones by chipping it away with a tedious, repetitive tapping of light shoemaker's hammers on hardened steel chisels or awls. The work was exhausting for the preparator and sometimes too hard on the specimens. The jarring from the repeated blows caused much unwanted breakage in soft or brittle specimens, especially when the hardness of the matrix required a heavier stroke. A hardening agent of shellac or gum arabic prevented some breakage, but, other than exercising extreme caution, little could be done to protect thin edges or other delicate structures. Worse, a wide range of motion was required for wielding a hammer and chisel. On complicated bones with deep and intricate cavities, it was often impossible to find a place of purchase for the chisel and room to swing the hammer. But the greatest disadvantage of using hand tools was the slowness of the work.[21]

Preparators derived new techniques for speeding the work by adapting the technologies of other industries to fossil preparation. Hermann introduced the electric dental lathe and dental engine at the DVP laboratory. Both were useful for operating small grinding wheels, dental burs, or small rotary brushes. At some point, for matrix that was too hard to work effectively with tools, Hermann began experimenting with acid preparation. He had some success using hydrochloric acid and potash, both of which were useful for softening hard carbonate matrix. He had his most success using sandblasting equipment, which in trials was found to be very practical for cleaning matrix from large bone surfaces.[22]

The introduction of pneumatic tools was the most important innovation made in fossil preparation during the second Jurassic dinosaur rush. The complete pneumatic apparatus consisted of an air compressor, air tank, pressure gauge, piping and fixtures, and a suite of air tools, including pneumatic hammers and drills. The basic tool was the pneumatic hammer/chisel. This handheld, cylindrical device housed a hollow chamber where an air-driven hammer played lightly and rapidly upon the head of a chisel, causing it to vibrate. The vibrating chisel pulverized rock on contact. Work with the pneumatic hammer/chisel was faster, more accurate, more versatile, and easier on the fossils and the men who prepared them. Riggs developed this important new technique early in 1903 and was quick to share its specifications with colleagues at other institutions.[23]

Bully for Apatosaurus

Marsh's published work on Jurassic dinosaurs and especially his 1895 opus, *Dinosaurs of North America*, provided his successors with an established body of facts and ideas that aided them in their own researches, and, at the same time, it gave them a ready object for criticism and revision. Marsh was a particularly tempting target because he was so much disliked by the succeeding generation of American paleontologists. Harvard paleontologist Alfred S. Romer remembered that "all the old-timers I knew when young . . . all liked Cope and almost to a man they hated Marsh's guts."[24] It would be difficult to overstate just how eager they were to tear Marsh down. They scrutinized his work for mistakes, using these as an avenue of attack into his great scientific edifice.

Osborn was the ringleader in this effort, and he made no bones about his objective of completely revising Marsh's work on Jurassic dinosaurs. He spelled this objective out explicitly in letters to Wortman and Menke in 1898. Nor was he shy about sharing this motive with scientists outside of the DVP. W. C. Knight, for example, rightly suspected that Osborn was seeking to "break down Marsh's [dinosaur] work as far as possible."[25] Osborn began studying the Jurassic dinosaur literature in 1898, comparing Marsh's figures and descriptions against the two incomplete and only partially prepared sauropod dinosaurs his field party had collected at Como Bluff in 1897. By the spring of 1898, Brown's *Diplodocus* skeleton was turning out very well. So well, in fact, that Osborn determined to publish a complete restoration of the genus in short order, something Marsh had not yet attempted. The other skeleton was taking Osborn on great, giddy flights of enthusiasm for revising Marsh's work. "I am studying the literature very carefully," he wrote to Wortman.

> . . . Marsh's [*Brontosaurus excelsus*] restoration is incorrect in many particulars. I am convinced he has allowed too few dorso-lumbar [vertebrae], he has left 6 or 7 vertebrae out of the anterior part of the tail; he has given the anterior dorsal [vertebrae] spines, which is incorrect. Altogether I am more enthusiastic than ever in regard to the possibilities of our work in the Jurassic. I believe Marsh's Reptile work is as defective and full of faults as his Mammal work, and will crumble to pieces before our superior methods.[26]

It is hard not to see something deeply personal in Osborn's motives. He arrived at these conclusions very hastily, without having seen much fossil material and without previously having done any serious work on Jurassic dinosaurs. Wortman must have been surprised at his superior's findings, especially since Osborn had made them all after less than three weeks' worth of research and reflection, all undertaken after Wortman had left for the field in early April.[27]

Important revisions to Marsh's work did come out of the second Jurassic dinosaur rush. Much of this work was, of necessity, opportunistic in nature. A good example is Brown's *Diplodocus*, which, because it was more complete and better preserved than Marsh's original type specimen, provided a wealth of new data. Consequently, when Osborn described it, he was able to clear up a number of misconceptions about its osteology, especially in the pelvis and tail. Osborn was the first to argue that the sacrum was the highest point in the vertebral column, as well as the center of power and motion. Osborn went to great lengths describing the tail of Brown's *Diplodocus*, which was remarkably complete and showed a surprising heterogeneity of structure from one end to the other. He argued that the tail was a powerful locomotor organ for propelling the animal through water. On land, it served as a means of defense. It was also a counterbalance to the anterior portion of the body, enabling *Diplodocus* to assume, on occasion, a tripodal position, with the weight of the body supported by the hind limbs and tail. In contrast to other sauropod dinosaurs, he argued, *Diplodocus* was essentially long, light-limbed, and agile.[28] This view was an important break from the traditional interpretation of dinosaurs. Although it was grounded reasonably well in the factual evidence, Osborn's view had as much to do with contradicting Marsh's claim that sauropods were ponderous, sluggish creatures, as it did with skeletal anatomy.

At the Carnegie Museum, the *Diplodocus* skeleton collected under Wortman's supervision in 1899, supplemented by a second specimen collected in 1900, under Peterson, presented Hatcher with the opportunity to revise this genus further and to make a complete skeletal reconstruction (something Osborn, ultimately, was unable to do). Hatcher's reconstruction brought out the striking elongation of the neck and tail, and the "ridiculously" short body and tiny head. His description introduced a number of new skeletal features, including, according to Hatcher, the first known dinosaurian clavicle. He also revisited the contentious issue of the dorsal formula for *Diplodocus*, giving some ground to the disputed interpretation first published by Holland (see chapter 7). He discussed at some length the animal's probable habits. Nostrils on top of the head (as Marsh first picked out) and loosely articulated joints in the limbs and feet well-suited *Diplodocus* for an aquatic life, he argued. He also concluded, wrongly it now seems, that its lightly constructed skeleton, riddled with weight-saving vacuities, was an adaptation to living in water. Although capable of movement on land, he opined that *Diplodocus* must have been a "decidedly slow and clumsy" animal. This was a view derived from Marsh and opposed to Osborn that prevailed for many years. Hatcher believed he was justified in establishing a new species, *D. carnegii*, wisely named in honor of his patron. He chose two specific characters on which to hang his new species: first, relatively smaller cervical ribs (when compared with Marsh's *D. longus*); and, second, caudal spines inclined strongly backward. Finally, he reclaimed some credit for himself, his former colleagues, and his former boss Marsh for jump-starting the recent advances in American vertebrate paleontology:

> Where a generation ago the extinct vertebrate life of America was but poorly represented in our museums by imperfect series of teeth and isolated bones, we are now able to study many of these extinct animals from more or less complete skeletons. For these improved conditions we are mainly indebted to the late Professor Marsh, either directly by reason of the vast collections acquired by him, or indirectly through the improved laboratory and field methods developed by him and his assistants.[29]

Marsh and Cope did pioneering work on American Jurassic dinosaurs, establishing dozens of new taxa, many of which were deemed invalid by their successors. Because of their rivalry, they sometimes did hasty, slipshod work. The unseemly fossil feud, the rule of priority, and the circumstances of fossil preservation and discovery all fostered a profligate approach to the designation of new taxa. Cope and Marsh each rushed new names into print for fear of being preempted by his rival. The net result was a multitude of new names applied with scant justification to small morphological differences between specimens. More conservative taxonomists who followed Cope and Marsh derided this practice as "splitting."[30]

The goal of mounting a complete sauropod dinosaur for public display provided a practical incentive for re-joining split taxa. No amount of searching ever turned up a perfectly complete sauropod dinosaur. Therefore, to meet their exhibit goals, Osborn and others decided to build composite mounts from the accumulated parts of several incomplete but complementary specimens. This encouraged a broader view about variation within groups, as paleontologists looked for ways to simplify the dinosaur taxonomies they inherited from Cope and Marsh. There were often good scientific reasons for merging split taxa together, but the object of making a composite mount also played a significant part. It would not be appropriate, after all, to mount deliberately the bones of more than one legitimate taxon together and present it to the public as an animal from nature.

Osborn was the first to make a serious stab at revising the taxonomy of the various genera of American Jurassic sauropods, which had been muddled by Cope and Marsh during the first Jurassic dinosaur rush. Marsh had established the new genus and species *Titanosaurus montanus* in July 1877, based on a few bones found at Morrison, Colorado. Cope then followed in August with *Camarasaurus supremus*, based on a series of dorsal vertebrae from Cañon City. Because *Titanosaurus* was preoccupied, Marsh coined the new generic name *Atlantosaurus* in December. In the same paper, and with a brief description only, he introduced the new genus *Apatosaurus*, which was based on considerably more complete remains. Cope countered in the same month with *Amphicoelias*, based on another series of dorsals from Cañon City. In December 1879, Marsh briefly described a remarkably complete sauropod skeleton from Como Bluff, Wyoming, and named it *Brontosaurus excelsus*. This animal, which was later (in 1883) immortalized in Marsh's first dinosaur reconstruction, quickly became the best and most widely known American Jurassic dinosaur. Years later, Osborn argued that it was "*a priori* improbable that so many different genera of gigantic Saurians of similar size co-existed."[31] Indeed, he concluded as early as April 1898, after a quick study of the literature and a cursory look at only a few relevant fossils, that *all* these genera were synonymous. He was inclined to collapse them into a single genus, but which one? Marsh had published the earliest generic name, *Titanosaurus*, in July 1877. But since this name was unavailable, *Camarasaurus* was the name with the next longest pedigree. Osborn adopted it as the most appropriate name for the all-inclusive genus. And why not? He loathed Marsh, and was undoubtedly pleased to sink his enemy's several genera in favor of Cope's *Camarasaurus*. Moreover, he was then negotiating with Cope's executor to purchase the type specimen of this genus for the DVP collection and he had recently dispatched a field party to Como Bluff to search for a specimen to supplement the one obtained in 1897.[32]

In his first publication on this subject, issued in June 1898, Osborn seemed a lot less certain about this conclusion. He provided a literature review on the relevant sauropod genera and gave a brief description of the more robust of the two DVP sauropod specimens recently collected at Como Bluff, assigning it tentatively to the genus *Camarasaurus*. Although it does not explicitly state the case, Osborn's paper strongly implies that *Atlantosaurus*, *Camarasaurus*, *Apatosaurus*, *Amphicoelias*, and *Brontosaurus* are all synonymous, and that Cope's *Camarasaurus* enjoys rightful priority.[33] To toss so much of Marsh's Jurassic sauropod work into the waste bin of synonymy must have given Osborn great satisfaction.

The abundance of sauropod specimens arriving at the DVP from Bone Cabin Quarry later afforded Osborn the material basis for revising his own views on sauropod taxonomy. Early in 1900, he assigned Granger the task of measuring limb bones to determine their relative proportions. As a result of these studies, Osborn and Granger were able to distinguish three groups, which they assigned to Marsh's genera *Morosaurus* (established in 1878), *Brontosaurus*, and *Diplodocus* (also established in 1878). Osborn (with Granger) retreated from his earlier view, and instead equated Cope's *Camarasaurus* with *Morosaurus*, promising to provide a thorough description of the type of *Camarasaurus* in order to bolster his claim.[34] This tripartite division of the more common American Jurassic sauropod genera is still widely accepted today (although with the genus names *Camarasaurus* and *Apatosaurus* for the former two taxa).

At Chicago's Field Columbian Museum, Riggs published a notorious contribution to sauropod taxonomy in 1903. It appeared in his description of the partial sauropod skeleton he and his party collected near Fruita, Colorado, in 1901. He had made a sweep through several Eastern museums examining various dinosaurs. In New Haven, he had the opportunity to study Marsh's types. This is what he found:

> After examining the type specimens of these genera . . . the writer is convinced that [Marsh's] Apatosaur specimen is merely a young animal of the form represented in the adult of the Brontosaur specimen. . . . In fact, upon the one occasion that Professor Marsh compared these two genera he mentioned the similarity between . . . their respective types. In view of these facts the two genera may be regarded as synonymous. As the term "Apatosaurus" has priority, "Brontosaurus" will be regarded as a synonym.[35]

With that, Riggs rendered the name of America's best-known Jurassic dinosaur obsolete. His paper also resolved a number of puzzling anatomical problems.

Osborn, who was never one to withhold critical pronouncements on other paleontologists' work, liked Riggs's paper very much. But he was strangely reluctant to adopt the revised taxonomy. Osborn's opinion could have far-reaching effects. In a favorable review of the paper, which appeared in a February 1904 issue of *Science*, Osborn characterized the author's identification of *Brontosaurus* with *Apatosaurus* as an act of great "novelty," but without ever discussing the merits of the proposed synonymy. Even more damaging was Osborn's stubborn adherence to the name *Brontosaurus* in his closing paragraph. Osborn's review had a much wider readership than Riggs's original article and his opinion carried a great deal of weight. Riggs simply did not have the clout to carry off this revision on his own. On another East Coast swing in the spring of 1904, he reported back to Farrington, with obvious frustration, that paleontologists at the American and Carnegie Museums were still using *Brontosaurus* to refer to the sauropod skeletons they were then developing for exhibition. The following year, when the American Museum's prize sauropod went on display, the name *Brontosaurus* was affixed to its bones. Thereafter, thanks to museum displays, movies, and innumerable popular science books, the name gained tremendous currency with the science-literate public. Meanwhile, paleontologists slowly adopted Riggs's nomenclature. *Brontosaurus* lived on in the public's mind, while *Apatosaurus* gained favor in the technical literature. This situation, along with a strangely heated debate over the rightful title of a dinosaur stamp, inspired paleontologist and historian Stephen Jay Gould to write his famous essay "Bully for *Brontosaurus*," a semiserious pitch for a name that, despite its popularity, belongs in the limbo of synonymy.[36]

Paleontologists unearthed a number of entirely new forms during the second Jurassic dinosaur rush. At Bone Cabin Quarry, among the jumble of gigantic sauropod limbs and vertebrae, DVP collectors found a large part of the skeleton of a small, delicate-looking, carnivorous dinosaur. This remarkable specimen included a complete skull, pelvis, numerous vertebrae, and parts of both fore and hind limbs. A second specimen, consisting of a single, well-preserved hand, turned up later. Osborn briefly described these specimens in 1903, consigning them to a new genus and species, *Ornitholestes hermanni* (see figure 33). Narrow, elongated hands, with recurved claws, gave the animal rapid grasping ability. So did the "distinctively prehensile" and serrate teeth. The small, light construction of its skeleton, the cursorial structure of its hind limbs, and the balancing

Drawn by Charles R. Knight
RESTORATION OF THE "BIRD-CATCHING" DINOSAUR IN THE ACT OF CATCHING THE JURASSIC BIRD ARCHÆOPTERYX

Figure 33. Reconstruction of Osborn's *Ornitholestes hermanni* drawn by Charles R. Knight. From Osborn, "Fossil Wonders."

power of its long tail, were all adaptations for speed and agility. All this suggested to Osborn that his new theropod was well adapted for hunting Jurassic birds. The name he coined means Hermann's bird thief, an allusion to this supposed behavior and a grateful tribute to chief preparator Hermann's skill and success in mounting the skeleton for exhibit.[37] It is ironic that his determination to find a complete, exhibit-quality specimen of a gigantic sauropod dinosaur should yield this remarkable little carnivore, which was arguably Osborn's most important contribution to the American Jurassic dinosaur fauna.[38]

Hatcher, at the Carnegie Museum, used the two partial skeletons collected at Marsh's old quarry near Cañon City, Colorado, to establish the new genus *Haplocanthosaurus* and two new species, *H. priscus* and *H. utterbacki*, the latter named in honor of his intrepid collector William H. Utterback. He described these specimens in exhaustive detail in a memoir published late in 1903. He argued that *Haplocanthosaurus*, with its elongated torso (fourteen dorsals instead of the normal ten) and its simple neural spines, was the least specialized of all the sauropod dinosaurs. Other primitive features, shared with nonsauropod dinosaurs, seemed to point to a remote common ancestry, according to Hatcher. Always thinking ahead to the next research project, he advocated for a search for Triassic dinosaurs, with shared primitive features, to help resolve the questions of dinosaur phylogeny. This was motivated, perhaps, by his dashed hopes of collecting a panoply of dinosaurs from the long sequence of Jurassic exposures at Cañon City.[39] Hatcher's Carnegie Museum colleagues, O. A. Peterson and C. W. Gilmore, designated a new dinosaur *Elosaurus parvus* after a small, incomplete sauropod skeleton collected at one of the Sheep Creek quarries in 1901.[40] This specimen is now considered a juvenile *Apatosaurus*, however. Finally, Director Holland named a new species *Diplodocus hayi*, based on a partial skeleton recovered at Utterback's Quarry A on the Powder River.[41]

Riggs also described a completely new sauropod. In 1900, near Grand Junction, Colorado, his field assistant H. W. Menke had found a trail of bony fragments that he followed to the exposed and badly weathered distal end of a gigantic limb bone—probably a femur—protruding from the bedrock. Riggs tentatively identified the specimen as a poorly preserved *Brontosaurus*. When they developed the prospect further, however, they found a partial articulated sauropod skeleton of superlative dimensions. Its nine-foot-long ribs were the largest yet discovered. And there was something funny about the badly weathered femur. Once it was prepared back at the museum, Riggs realized that it was actually an enormous humerus, fully as long, in its weathered condition, as an associated femur found nearby. He estimated that its length, with its distal end complete, would exceed the femur by two or more inches. Not only was this specimen larger than any land animal discovered to date, but with forelimbs longer than its hind limbs, it also had proportions unique among Jurassic dinosaurs. Riggs penned a short article in *Science* describing the unusual features of this new dinosaur, and trumpeting its surprising size. He declined, however, to establish a new genus until more of the specimen had been worked out of the matrix. Two years later, after carefully studying the literature and examining numerous sauropod specimens in other museums, he coined the name *Brachiosaurus altithorax* to emphasize the long forelimbs and deep chest of his new dinosaur.[42]

Dinosaurs on Display

By 1901, Osborn had made the mounting of the Nine Mile *Brontosaurus* specimen for display at the American Museum one of his highest pri-

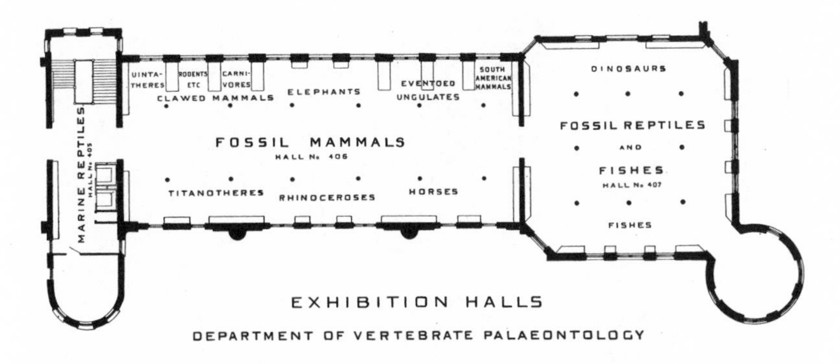

Figure 34. Floor plan of the exhibition halls for vertebrate paleontology at the American Museum of Natural History, ca. 1903. From Matthew, "Collection."

orities. But with nowhere to put such a monster as yet, work proceeded instead on finding more remains in the field to supplement their incomplete skeleton and on preparing and restoring the parts they already had. By the end of the year, this latter task was largely accomplished. So too was the construction of a new exhibit hall. Osborn held a reception there for museum members in November in order to give them a taste of what they might expect from the DVP in the future. Because the hall still lacked exhibit cases, Matthew arranged for the temporary display of a number of specimens. Especially noteworthy among these exhibits was the greater part of the skeleton of the Nine Mile *Brontosaurus*. "The bones . . . are laid out on a series of tables in as nearly as possible their natural relations," Matthew wrote to Osborn, "giving [visitors] a much more vivid idea than has heretofore been possible of the gigantic size of these animals."[43]

In 1903, the DVP staff began to implement the permanent installation of the new fossil reptile hall (see figure 34). A miscellaneous assortment of dinosaurs were moved into their new home with theropods arrayed against the east wall and an abundance of sauropods, including Brown's *Diplodocus*, spread around the north and west walls. In the center, work began in earnest on the permanent mount of the Nine Mile *Brontosaurus* (see figure 35). Two skilled ironworkers began working on a metal armature, while Hermann, along with most of his staff, worked to restore the missing bones. Mounting such a monster was a more formidable task than the chief preparator had anticipated. Hermann optimistically predicted that it would be completed by the following May. This was welcome news to Osborn, who hoped to open the new hall, or part of it

Figure 35. DVP staff members working on the mounted *Brontosaurus* ca. 1904. Image #17506. American Museum of Natural History Library.

Figure 36. The mounted *Brontosaurus* opened to the public at the American Museum in February 1905. From Matthew, "*Brontosaurus*."

at least, sometime in 1904.[44] The work, however, advanced much more slowly than expected. Still, by the end of the year, Hermann was pleased to report that it was nearly finished.[45]

Nearly, but not quite. Hermann spent the first few weeks of 1905 tinkering obsessively with the foot bones. Under pressure from Osborn, he finished by February 10. A proud curator, his DVP staff, and museum officials then hosted more than five hundred scientists and other special guests at a tea party with the new dinosaur on the 16th. Sixty-six feet eight inches in length and fifteen feet two inches high at the pelvis, *Brontosaurus* dominated the center of the DVP's new hall of fossil reptiles (see figure 36). It was the first permanently mounted sauropod dinosaur in the world, edging out a cast of Carnegie's *Diplodocus* at the British Museum by a scant few months. Approximately two thirds of the mounted skeleton consisted of bones from the Nine Mile *Brontosaurus*. Hermann supplemented the skeleton with a right forelimb found at Como Bluff. Likewise, the distal end of the tail, and the right hind foot, came from specimens collected at Bone Cabin Quarry. The remainder of the skeleton was modeled in tinted plaster. The exhibit was immensely popular, and it reawakened interest in the museum as a whole.[46]

But the reviews among rival paleontologists were not uniformly favorable. Riggs thought it looked too stiff. Coggeshall thought it was so technically unsophisticated that it did not even merit discussion. To be first to finish a mounted sauropod dinosaur was a great accomplishment, but it came at a price. The mounted dinosaurs that followed rapidly at other museums would benefit enormously by improving on Hermann's mistakes.[47]

Dinosaur exhibit development at the Carnegie Museum began rather humbly, in 1902, with a series of unimaginatively arrayed individual sauropod bones, both because chief preparator Coggeshall needed time to develop and perfect a system for mounting dinosaurs for display and especially because there was no space large enough in the original building to accommodate the enormous bulk of a fully mounted sauropod. Holland had ambitious plans to develop a Hall of Dinosaurs to house at least five complete skeletons, including no fewer than three sauropods. These plans would be held up for years, however, pending the completion of the new building. Meanwhile, an unexpected opportunity for Coggeshall and his staff to hone their skills occurred when Carnegie agreed to fund the making and mounting of a plaster replica of *Diplodocus carnegii* for the British Museum. Work began on this enormous project in the winter of 1902. In April 1903 the museum hired two skilled Italian modelers to sculpt some of the missing pieces, including the forelimbs, which were crafted on the basis of a smaller specimen, then scaled up slightly to match the size of the mounted replica. Molds then had to be made of all the bones and all the sculpted pieces. Plaster casts were then made from the molds. Because it was relatively easy to throw multiple casts from a single mold, plans were soon afoot to make several copies, one for each of several crowned heads of Europe. Soon the Carnegie Museum was a veritable *Diplodocus* factory, churning out facsimiles of its patron's namesake. The distribution of these copies to several European capitals, to Argentina, and to Mexico, helped make "dinosaur" a household word.[48]

When the time had come to execute a trial mount of the first replica, in May 1904, Holland negotiated for the free use of space in a massive exposition hall in downtown Pittsburgh. An ailing Hatcher had decided to forego fieldwork for the first part of the summer, at least, and he stayed in town to supervise the mount. He paid Coggeshall to moonlight on some foot bone models. He hired a pipefitter to help with the metal armature and a carpenter for the woodwork. Because there was little actual work for the curator to do, Hatcher made a ten-day research trip to Washington, D.C., in late May. By the time he returned in early June, the mounted replica was rapidly taking shape. It still lacked a skull. It was also missing a long series of chevrons, casts of which had been urgently requested from the American Museum, which was slow to fill the order. Nevertheless, Hatcher predicted confidently that the mounted *Diplodocus* would be completed by the first of July. Sadly, he took ill in the last days of June and did not live to see the skeleton finished. After Hatcher's death, the project lost some of its urgency. The work slowed to a more leisurely pace while plans for a formal gift ceremony at the British Museum took on ever more elaborate shape. The cast sailed for England in December. In March 1905, Holland and Coggeshall followed to supervise its construction.[49]

Mounting the original fossil bones of the Carnegie's *Diplodocus* in Pittsburgh presented a host of problems. The mount was a composite consisting of parts from both Sheep Creek skeletons, caudals from a specimen collected in the Powder River country, and sculpted or cast parts from a number of different specimens. The great weight of the fossils had to be supported without using an obtrusive metal framework that would detract from the splendor of the specimen. To accomplish this feat, Coggeshall, the chief preparator, designed a cast steel beam, conforming to the shape of the bones, which would support the vertebral column from below. In order to find the right shape for the beam, he articulated the vertebrae on their sides in a giant sandbox, placing them as close as possible to their natural positions. He then ran a generous bead of plaster along the ventral surface of the line of centra, approximating the desired thickness of the beam. This plaster cast was then taken to a mill where steel castings were made. He used this system from the seventh cervical to the seventeenth caudal, placing vertical rods for support at the pelvis, the first dorsal, and the ninth cervical. The vertebrae were then put in position on the frame and fixed firmly in place with plaster. He used metal rods of ½-inch to 1½-inch round steel to attach the limbs and ribs to the cast steel framework. To prevent the pelvis from tipping, he placed a stout 3½-inch flat iron bar, bent to conform to the bone, running crosswise underneath the sacrum. This was bolted to the frame in the middle and the two leg irons at either end. The whole ensemble, which was finished on time for the April 1907 reopening of the expanded Carnegie Museum, made for a beautiful and graceful mount (see figure 37). Admiring visitors could see very little of the metal framework. Coggeshall was justifiably proud of his accomplishment. In a letter to Riggs, he described his new system in detail, and crowed about the result. "Everybody here admits that it beats the Am[erican] Mus[eum] mount [by] a mile," he wrote.[50]

At Chicago's Field Columbian Museum, Farrington first floated the idea of mounting a sauropod dinosaur in May 1900, perhaps to compensate for an embarrassing lack of vertebrate fossils of all kinds. The museum committed a small amount of funding to the Geology Department for fossil fieldwork, suggesting a modicum of support, at least, for Farrington's ambitious plan. Riggs then spent the next two field seasons in the Jurassic beds of western Colorado searching for sufficient fossil

Figure 37. A fully mounted *Diplodocus* went on display at the Carnegie Museum in 1907. Casts of this specimen helped make "dinosaur" a household word. © Carnegie Museum of Natural History.

material with which to execute a mount. The partial, articulated *Apatosaurus* skeleton collected in 1901 held great promise. But the museum refused to fund a follow-up expedition in 1903, frustrating plans to find additional material to complete the mount. Still, the department possessed more than half of the skeleton of a single individual, with most of the bones magnificently preserved. Thanks to the labor-saving air tools introduced by Riggs, preparation of this specimen was completed by the summer of 1903. Most of the bones, including a thirty-foot articulated series of vertebrae, were placed in cases in an exhibit hall. In August 1903, Farrington submitted plans for mounting the dinosaur in a lifelike display and enclosing it in a gigantic glass case. Meticulous and fussy, Farrington worried about dust accumulating in the intricate cavities of the bones. Higinbotham, however, balked at the cost, denying Farrington's request.[51]

When the staff began considering plans for the new museum building, the idea of mounting a sauropod dinosaur seemed dead in the water. In a progress report, Farrington wrote:

> The probable removal of the Museum within a few years to a new and larger building makes it seem obvious that . . . attention need no longer be given extensively to installation in the present building and freedom from work of this character will afford time and opportunity for procuring and preparing material in readiness to occupy the larger space which the new building will afford.[52]

The museum banished fieldwork for dinosaurs indefinitely, largely because the great size of the specimens made their exhibition impractical. But plans for the new museum building stalled over the issue of where best to put it. For years, a series of legal battles delayed construction, and in this climate of uncertainty the idea of mounting the *Apatosaurus* for display revived. Farrington submitted detailed plans for the new exhibit—minus the glass enclosure—in February 1907.[53]

The project was enormous, and, after much delay getting the plan approved, it occupied Riggs and his small staff for more than a year and a half. Space for the mount was afforded by discarding the ceramic collections in Hall 33 and moving in the entire contents of Hall 35 (mostly Paleozoic invertebrate fossils). Jurassic dinosaurs would fill the vacated space in Hall 35. Wall cases were used to house miscellaneous specimens of *Diplodocus* and *Morosaurus*. The type specimen of *Brachiosaurus* found a new home there in 1908. At the center of the hall, Riggs began assembling the *Apatosaurus* skeleton. Though Coggeshall urged him to try the cast steel technique he had developed for Carnegie's *Diplodocus*, Riggs opted to mount his skeleton on a framework of his own design. He started with a rectangular base made from steel trusses eight by thirty feet. Four transverse I-beams and a single longitudinal I-beam acted as girders. Four vertical supports, firmly bolted and braced at the intersections of the girders, supported a T-beam, which in turn supported the spinal column. The T-beam was approximately thirty feet long, and bent to conform to the expected shape of the articulated vertebrae. It was four by four inches at one end, and tapered to two by two inches at the other—its size reduced in proportion to the load it was expected to bear. The bones were then attached individually to the framework, like jewels, by means of wrought iron clasps, which were bent to the shapes of the bones, and designed to be as inconspicuous as possible. Each bone needed to be custom fitted with iron clasps. So, for greater convenience, Riggs ordered a gas forge to do his own ironwork and had it installed on the west porch, just outside of Hall 35. A temporary sheet-iron shed erected over the forge allowed the men to work comfortably through the winter of 1907–08.[54]

There was trouble with the T-beam, which had to be sent back to the factory to be rebent at a cost of $12. Director Skiff wanted an explanation.

Figure 38. Riggs's *Apatosaurus* under construction in Hall 35. Note how the supporting beam does not conform to the curvature of the tail vertebrae. Riggs had to have the beam rebent. Courtesy of the Field Museum, Negative #CSGEO23972.

It seems that Riggs had carefully placed together a series of enlarged drawings of each of the bones, in order to approximate the desired curve of his T-beam, which was then bent to order. He followed the photographs of the Carnegie and American Museum mounts a little too carefully, however, and when he first attempted to mount his specimen he found that the tail would not assume that shape (see figure 38). Instead, it followed a more regular curve, which eliminated what Riggs called the "peculiar hump-back effect" of other mounted sauropods. Riggs believed that this was a consequence of the unusual completeness of his specimen, and the perfection of its articular surfaces. In short, he felt that his specimen was taking the natural shape of the animal in life, and that this mount, once completed, would be more accurate than those in Pittsburgh and New York.[55]

Riggs finished the work in the summer of 1908 and it opened to the public on August 1 (see figure 39). It was a beautiful mount, consisting of the bones from a single individual from the last cervical vertebra back to midtail. It stood fifteen feet above its base and stretched thirty feet from end to truncated end. A missing femur and ilium were modeled in plaster from their counterparts on the opposite side. Several ribs and chevrons were restored in plaster, also. A few other missing pieces, including bones from the right hind foot, the right fibula, and the left tibia, were supplied from casts of specimens at the Carnegie Museum. Holland had them made free of charge.[56] The head, neck, shoulder girdle, forelimbs, and the distal end of the tail were entirely wanting on the Field Columbian Museum specimen. Perhaps this was for the best, however, because the length of a complete *Apatosaurus* was about ten feet longer than the room in which it was first displayed.

Fifty years later, Riggs's "headless monster"[57] was mated with the bones of an additional specimen to make, at long last, a complete skeleton. By then, the museum had removed to its present location in a new building at the south end of Chicago's Grant Park. There, in 1921, the incomplete *Apatosaurus* was remounted in a cavernous new exhibit space designated Hall 38, where it continued to charm and confuse museum visitors. In 1954, Rainer Zangerl, then curator of fossil reptiles, put it thus: "[Sauropod dinosaurs] have a great deal of public appeal. . . . [But] in its present condition, [ours] does not convey a satisfactory idea of the animal as a

Figure 39. Riggs finished the mounted *Apatosaurus* as far as he was able. It opened to the public in August 1908. Thus it stood, truncated at both ends, for half a century. Courtesy of the Field Museum, Negative #CSGEO26576.

whole. It tends to confuse the visitor who often thinks of it as an animal having only two legs!"[58] An opportunity to solve this problem presented itself when geologist Edward Lee Holt found another specimen, which complemented Riggs's *Apatosaurus* almost perfectly, outside of Floy Junction, near Green River, Utah. The museum made arrangements with Holt to acquire the specimen, which was then carefully excavated by James H. Quinn and Orville L. Gilpin, two fossil preparators, in 1942. Gilpin then managed to incorporate the bones of the new specimen seamlessly with those of the old. The completed skeleton, seventy-two feet long, which finally debuted in April 1958, was a great success that attracted considerable attention from the local press. Riggs, who was then ninety years old and walking with the aid of two canes, stopped in Chicago in July 1959 to see the new mount. "By Joe, it was a pretty good job," he remarked.[59]

CONCLUSION

What's the Rush?

The second Jurassic dinosaur rush was a race among America's museum paleontologists to find the largest and finest sauropod dinosaur and to mount it for exhibit in a lifelike pose. Attendance at Philadelphia's Academy of Natural Sciences had spiked dramatically after the unveiling of Joseph Leidy's giant *Hadrosaurus*, the world's first mounted dinosaur.[1] Sauropod dinosaurs from the American Jurassic were the largest land animals then known, and when mounted for display would provide an even bigger drawing card for visitors. Othniel C. Marsh's restoration on paper of *Brontosaurus excelsus* suggested this possibility in a strikingly visual way. So compelling was Marsh's restoration that Andrew Carnegie used it to drum up enthusiasm for his proposed museum project in Pittsburgh.[2] Small in stature, Carnegie was fascinated by bigness.[3] He made sauropod dinosaurs an explicit goal for his museum, fronted the funding, and then entrusted Director William J. Holland to make this goal a reality. At the American Museum, in New York, several of Henry F. Osborn's subordinates urged him to expand his fossil mammal program to embrace dinosaurs. Initially reluctant to make this move, he became wildly enthusiastic for dinosaurs at the first sign of field success at Como Bluff, Wyoming. Jacob Wortman, Osborn's field foreman, had emphasized that dinosaurs would make spectacular exhibit objects. Osborn agreed, but he was probably motivated at least as much by the potential opportunity to attack Marsh's dinosaur work. Chicago's Field Columbian Museum blundered into the competition for Jurassic dinosaurs when it accepted an invitation to participate in the Fossil Fields

Expedition, largely because the offer presented an opportunity to do fossil vertebrate fieldwork on the cheap. Soon enough, however, Oliver C. Farrington and Elmer S. Riggs were both taken by the notion of attempting to mount the world's first sauropod dinosaur, and they successfully used this goal to generate a modicum of support for Jurassic fieldwork over several years. But the museum's conservative administration was never as committed to dinosaurs as its geology staff. Their lukewarm enthusiasm for paleontology quashed the Geology Department's pursuit of a complete, mounted sauropod. The result—a headless dinosaur rump, beautifully executed but incomplete—was a kick in the pants for Riggs and his fledgling program.

Why sauropod dinosaurs? Superlatives, in general, and superlative size, in particular, were fundamental virtues in American natural history museums around the turn of the twentieth century. The annual reports of the American, Carnegie, and Field Columbian museums are filled with them, especially "best," "most," "finest," "oldest," and "largest." This latter superlative was overused in ways that now seem ridiculous, as when Director Frederick J. V. Skiff crowed about the Field Museum's acquisition of the "largest specimen of mineral wax ever washed up on the Pacific Coast."[4] In late nineteenth-century America, there was a society-wide fascination with gigantic things, like the Krupp Gun or the Ferris Wheel at the World's Columbian Exposition, and museums exploited it to bring in visitors. James Secord termed a similar British phenomenon of the 1840s and '50s the "monstrous aesthetic," noting that exhibitions of prehistoric life at the Crystal Palace catered specifically to this taste.[5] If roadside attractions are any indication then there is something inherently appealing to Americans about the world's largest anything. The world's largest pencil draws visitors to Glen Burnie, Maryland, for example, while tourists flock to see the world's largest penguin in Cut Bank, Montana. Museums still exploit the popular appeal of monstrous things. Foot traffic in the Field Museum Library increased noticeably after the acquisition of a copy of *Bhutan*, the world's largest book.[6]

Great size was quite obviously the special appeal of sauropod dinosaurs. During the second Jurassic dinosaur rush, newspaper accounts of field discoveries routinely touted the extraordinary size—more often than not, greatly exaggerated—of the dinosaur specimens collected by museum paleontologists. This sensationalist newspaper coverage, coupled with the proliferation of mounted dinosaurs in public museums in the early twentieth century, ignited American dinomania. American paleontologists willingly participated in the hype, and thus helped to fuel the public's growing fascination with gigantic dinosaurs. Riggs, in several publications, emphasized the gigantic size of his new genus, *Brachiosaurus*, even though the crucial character that set this animal apart from other genera was its unique proportions. But he was not merely pandering to popular tastes, for great size also appealed to dinosaur paleontologists and patrons of science. When Charles W. Gilmore arrived at the University of Wyoming as a new student, Wilbur C. Knight showed him the bones in one of his Jurassic quarries. "They are immense," Gilmore wrote in his diary, "just the thing I would like to study."[7] Richard S. Lull, who collected for the DVP at Bone Cabin Quarry, "often remarked that he had a feeling of special affinity for large animals."[8] Osborn, too, according to Edwin H. Colbert, one of his many assistants, "liked to work on big things."[9] His preoccupation with fossil elephants, giant titanotheres, and especially sauropod dinosaurs seems to bear this out. Carnegie and Holland regularly acknowledged the appeal of the great size of sauropod dinosaurs in their correspondence. Carnegie, in fact, was so committed to mounting a giant sauropod that when Holland warned him that such a project would be enormously expensive and would require a large new hall at the museum in Pittsburgh he did not shrink from the challenge. "I should like to do the Colossal," was his fitting response.[10]

Mounting big, showpiece dinosaurs was a successful strategy for attracting the public's attention, but it was also a cause of some concern about content, both for museum paleontologists and their colleagues in other departments. In theory, the visitor who was lured inside the museum by dinosaurs would also become interested in other displays and in the systematic arrangement of the collections as a whole. In practice, however, the average museum visitor devoted most of his or her time to scrutinizing the "striking and showy objects" only. Farrington spent an afternoon studying the behavior of visitors to a paleontology exhibit at a rival museum (probably the American Museum) and "found that of the two hundred and fifty persons who visited the hall . . . all but eight devoted their full scrutiny to some life-size models of huge Tertiary mammals."[11] Most curators took the problems of museum education very seriously. At the American Museum, Osborn directed his DVP staff to make a special study of all "legitimate methods of attracting the attention and interest of visitors." Supplementary exhibits of fleshed-out models and full-color reconstructions of dinosaurs and other extinct vertebrates were an outcome of these studies.[12] Other museums soon copied these techniques. Nevertheless, museum paleontologists faced a continuous conflict between enticing visitors by catering to popular tastes and educating them with legitimate scientific content. This same conflict exists in museums today.[13]

Paleontologists raced to collect Jurassic sauropod dinosaurs in order to be the first to mount one for display—being first also was a virtue in American museums. Osborn's DVP won this race when its mounted "Brontosaurus" skeleton opened for public viewing in February 1905. Associate Curator William D. Matthew made the most of this unique feat in the *American Museum Journal*, writing that "no museum has ever before attempted to mount so large a fossil skeleton."[14] The glory was fleeting, however, as fully mounted sauropod dinosaurs soon appeared in London, Pittsburgh, Chicago, and elsewhere. Moreover, the drawback to being first was that other museums could and did improve upon the DVP's technical choices.

The race to collect sauropod dinosaurs was not about bragging rights alone, however. Paleontologists were also motivated to collect by a sense of competitive desperation: with so many rival museum parties actively seeking such a seemingly finite resource as vertebrate fossils, the urge to get them before they disappeared was keenly felt.[15] Riggs, Holland, and especially Osborn (and others) emphasized this point repeatedly in their urgent requests for funding and institutional support. A few others, including Samuel W. Williston, Wilbur C. Knight, and John B. Hatcher, believed that there were enough Jurassic dinosaurs in the American West for everybody. Yet these men clearly held the dissenting view about fossil resources.

Osborn's view about the scarcity of fossils and his intensely competitive nature were the driving forces behind the second Jurassic dinosaur rush. As curator of the DVP, Osborn set the agenda for vertebrate paleontology at the American Museum of Natural History. Wealthy and well connected in New York society, he subsidized the work of the DVP for many years, diverting his own salary to pay artists and other staffers, and providing additional funds to meet operating expenses and to make important fossil purchases. More important, Osborn had ready access to an almost limitless pool of potential funding through his close associations with the likes of J. P. Morgan. His social standing also smoothed the way for speedy approval of his grand departmental plans with museum trustees and other administrators. By 1901, thanks to his sterling connections, Osborn was himself a trustee. In 1908, he was appointed museum president. His DVP was among the most favored programs at the American Museum. According to Rainger, Osborn had the clout to dictate his own terms to museum directors Hermon C. Bumpus and Frederic A. Lucas.[16]

Osborn wanted to establish his DVP as the dominant institution in American vertebrate paleontology, and he had the wherewithal to make it happen. To accomplish this goal, he employed a stable of talented and ambitious upstarts in paleontology who would perform most of the routine work of the department and who would be completely dependent on him for their livelihoods. Osborn sent dozens of these men afield to collect fossil vertebrates—a good collection was a prerequisite for paleontological research and exhibit development. It was Osborn who set the stage for the second Jurassic dinosaur rush by sending his collectors into Marsh's classic locality at Como Bluff in 1897. Their successes attracted imitators and touched off a fierce competition to find and collect dinosaurs. Less successful was Osborn's frenetic search for productive Jurassic localities elsewhere, in Colorado, Utah, Montana, and South Dakota. No matter: the DVP maintained a presence in the Jurassic beds of Wyoming for all or part of nine field seasons, and accumulated an enormous collection of dinosaurs, even bigger and better than Marsh's storied collection at Yale.

From his lofty tower at the American Museum, Osborn's influence over American vertebrate paleontology was so pervasive as to set the tone for the discipline as a whole. When he drove his collectors to explore for new Jurassic exposures, to collect fossil vertebrates on a vast scale, and to put them on display in lifelike mounts, paleontologists at other institutions were usually quick to follow his lead. This is not to say that Osborn did everything first or best, or that the DVP always played a leading role in all aspects of American vertebrate paleontology. Hatcher, Riggs, and others all made important contributions in the field and developed important innovations of their own. Still, by 1905, Osborn's DVP had become the undisputed institutional leader in the field, supplanting Yale from its premier position in less than fifteen years. Moreover, Osborn was such a patriarchal figure at the American Museum of Natural History that most of the credit for the accomplishments of his talented staff devolved upon him. As patron and supervisor, Osborn robbed many of the laurels for Charles Knight's murals, Adam Hermann's preparation and exhibit techniques, and Barnum Brown's field discoveries. If American vertebrate paleontology was less competitive and more cooperative after 1905, it was because Osborn had by then assumed such a dominant position in the field. Osborn, curator and later president and trustee at the American Museum, Columbia University professor, and vertebrate paleontologist of the U.S. Geological Survey, had destroyed, assimilated, marginalized, outcompeted, and/or outlived all the serious competition. Most of those who remained in the field after the end of the second Jurassic dinosaur rush—including paleontologists at rival institutions—were more or less dependent on him for favors and professional opportunities.

Vertebrate paleontology also was a favored science at Pittsburgh's Carnegie Museum, where the mandate to collect and exhibit a sauropod dinosaur came directly from steel baron Andrew Carnegie, the museum's chief benefactor. Whether because of his friendship with Marsh, his fascination with big things, or his oversized vanity, Carnegie made dinosaur paleontology a top priority at his museum. He personally provided a generous annual fund to meet the expenses of the Vertebrate Paleontology Department. Holland, Carnegie's handpicked museum director, took a personal interest in seeing his patron's wishes fulfilled. After some naïve trial and error, he found and hired the ideal coterie of field and lab specialists to get the work done. Thanks to Carnegie's deep pockets, Holland's determination, and the skill, talent, and hard work of Hatcher, Jacob L. Wortman, Olof A. Peterson, William H. Reed, Gilmore, and others, the Carnegie Museum quickly established a thriving program in vertebrate paleontology, even besting Osborn's DVP in the acquisition of American Jurassic dinosaurs.

The specimens recovered at Sheep Creek, near Cañon City, and along the Powder River were as good as or better than anything the DVP collected at Bone Cabin Quarry. (The discovery and excavation of Jurassic dinosaurs at Carnegie Quarry, in Utah, beginning in 1909, would soon eclipse all of these earlier discoveries.) It would be difficult to overstate the popular significance of the Carnegie Museum's renowned *Diplodocus* skeleton. Duplicated some dozen times and exported to a number of populous European and Latin American capital cities, "Dippy" has probably been seen by more people than any other fossil. These casts were undoubtedly vectors for the international spread of dinomania. Moreover, duplicates and casts of important European fossils, which sometimes flowed back to Pittsburgh in grateful recognition of Carnegie's gift, were a boon for American vertebrate paleontology. This exchange of specimens facilitated comparative studies of American and European taxa. An ambitious fossil vertebrate research program, in which American Jurassic dinosaurs were only one part, was already well begun and yielding rich returns when it was cut short by Hatcher's tragic death. Had he lived longer, or had Holland found a worthy replacement, there is no telling how productive the Carnegie Museum might have been in terms of paleontological research. Carnegie's death, in 1919, was another blow to the cause. Carnegie, it seems, had already lost much of his earlier enthusiasm for dinosaurs. His generous appropriations dried up significantly in his later years, and he neglected to make any permanent provision for paleontology at the museum. After his death, money for dinosaur hunting was harder to come by. Consequently, the Pittsburgh party abandoned their fieldwork at the Carnegie Quarry in Utah, which was far from exhausted. Nevertheless, the Carnegie Museum had by then acquired "one of the greatest—perhaps the greatest—assemblage of Jurassic dinosaurs ever made."[17]

Vertebrate paleontology was not a favored science at Chicago's Field Columbian Museum. Instead, it suffered from a persistent lack of adequate resources and institutional support. The fundamental cause of this problem was the subordination of the vertebrate paleontology program within the Geology Department, which added an extra layer of administrative control and precluded the program from competing for museum resources as a full-fledged department. Nor was geology a particularly high-ranking member of the museum's departmental hierarchy, where anthropology and zoology ruled the roost. Geology Curator Farrington diverted the majority of his meager funding to mineralogy, economic geology, and meteoritics, while paleontology languished. A wealthy and well-connected patron could have turned the tide, but, in contrast to the situations in Pittsburgh and New York, no one stepped forward in support of paleontology in Chicago. Nor was Riggs very favorably situated to attract a patron. As an assistant curator, he had virtually no access to museum trustees or administrators, including Skiff. He was new to the city, poor, and—lacking a PhD or an impressive string of publications—he was not very prominent in his chosen field. With only a lowly assistant curator to plead meekly for its support, vertebrate paleontology at the Field Columbian Museum, more often than not, was starved for resources during the second Jurassic dinosaur rush. That Riggs enjoyed a number of extraordinary successes, even when measured against his better-funded rivals in Pittsburgh and New York, is a testament to his hard work, talent, and good fortune.[18]

Another crucial reason why the Field Columbian Museum was disinclined to support dinosaur paleontology as extravagantly as its Eastern rivals was the nagging problem of space. While founders debated the ideal size, style, and location for a permanent building, the museum temporarily occupied the former Fine Arts Palace, a ramshackle structure hastily built for the World's Columbian Exposition and abandoned when the fair closed in October 1893. Many exhibitors in the Mines and Mining building at the fair had donated their heavy, bulky displays of rocks, minerals, ores, and so forth to the museum, rather than pay the cost of hauling them away, such that when the museum opened in 1894, the Geology Department boasted a large number of exhibits that completely filled the West Pavilion of the building, and spilled over into a number of adjacent rooms in the Central Pavilion. Meeting the demands for increased exhibit space in the Central Pavilion for the growing collections of the Zoology and Anthropology departments meant that Geology could expand its space for exhibiting fossil vertebrates only by contracting other hard rock exhibits, which Farrington was often loath to do.

Meanwhile, administrators were reluctant to fund the expansion of vertebrate paleontology while the question of where and how they would construct a new building remained unanswered. Spending the money to mount gigantic Jurassic dinosaurs simply made no business sense so long as there was no ready space to put them, and when taking them down again and moving them would incur additional trouble and expense. Dinosaurs, according to Skiff, "occupy so much space as to make their exhibition impracticable."[19] This excessively cautious approach probably emanated directly from museum founder Marshall Field. In 1894, in response to a request to enlarge the scope and budget of the museum's Division of the Railway, Field wrote: "My judgment is that [we] should go slow in all expenditures from this time on, at least until [we] know definitely where the permanent home of the Museum is to be and where the money is to come from to maintain it."[20] The same conservative attitude toward funding and supporting the vertebrate paleontology program hindered its development for years, and relegated the museum to a distant third place finish in the race to acquire and display American Jurassic dinosaurs.

The practice of American vertebrate paleontology changed significantly during the second Jurassic dinosaur rush. In its new museum setting, paleontology found a rich new source of private philanthropic support, often solicited and controlled by entrepreneurial paleontologists like Osborn or administrators like Holland. Like their predecessors Cope and Marsh, these men continued to enjoy the lion's share of the credit for work accomplished, largely by virtue of maintaining tight control of the funding. Nevertheless, private support funded a groundswell of new work done largely by young, ambitious newcomers to paleontology, such as Riggs, Brown, and Walter Granger. Many of these men later became prominent vertebrate paleontologists in their own right; museums provided them with their first professional opportunities. Museums placed a premium on developing elaborate displays of mounted fossil vertebrates, often to meet the goals of their patrons. This required a battery of reasonably complete, exhibit-quality specimens, some of which could be acquired by purchase or exchange, but most of which would have to be collected in the field. To meet this need, museum paleontologists mounted a long-term, methodical field collecting campaign to search for new specimens and new fossil localities. They also developed improved field methods to meet the demand for better quality specimens and better contextual data. In the fossil preparation laboratory, technicians such as Riggs, Adam Hermann, and Arthur Coggeshall invented new labor-saving tools and techniques, including pneumatic hammers and sandblasting, to free specimens quickly and safely from their rocky matrix. Most of these techniques are still in use in modern fossil preparation labs. In terms of research, the second Jurassic dinosaur rush brought a surge of dinosaur revisionism. Much of Marsh's earlier body of work on American Jurassic dinosaurs was revised and rewritten in light of new and better specimens. Finally, after an initial period of intense competition to find and collect exhibit-quality dinosaurs, petty rivalry ultimately gave way to a relatively cooperative and congenial working relationship among museum paleontologists.

The second Jurassic dinosaur rush also transformed America's natural history museums. The emphasis in urban museums on education and popularization inspired an innovative period of fossil vertebrate exhibit development that spread quickly to other institutions and dramatically altered the public face of museums. Previously, it was common practice for museums to keep their vertebrate fossil collections in storage, strictly for the use of paleontologists. Before the turn of the twentieth century, Philadelphia's *Hadrosaurus* and a few duplicate casts were the only mounted dinosaurs in the United States. By 1920, there were dozens of skeletons of all kinds in museums, large and small, nationwide. These displays appealed to visitors, who came to museums in droves to see dinosaurs, and patrons, who shelled out money to fund such conspicuous displays (Chicago is a notable exception). Dinosaurs by then had become a status symbol; museums were (and still are) judged by the quality and quantity of their mounted dinosaurs.

Vertebrate paleontology thrived in its new museum setting. A specimen-based discipline, paleontology depends on access to fossils. By being concentrated in a small number of museums (and some universities) where the resources to make and maintain substantial fossil collections were (usually) available, paleontology found the ideal institutional setting. In museums, vertebrate paleontology was somewhat isolated from the universities and laboratories where pathbreaking developments in experimental biology and genetics were then taking place. Conceptually and methodologically, it remained wedded to what historian Ronald Rainger calls the morphological tradition.[21] But it was a peripheral field of inquiry only in the sense that the explosive growth in other subfields of biology and geology at American universities left it seeming small by comparison. The slow growth of paleontology was a consequence of the limits imposed by the size of collections and the difficulty of access to fossils, rather than by a diminishing level of interest on the part of biologists. On the contrary, vertebrate paleontology survived and flourished in most major museums, where a small group of dedicated practitioners continued to collect and study fossils, describe and name them, all with the intent of reconstructing the fragmented history of life on Earth.

Epilogue

Barnum Brown remained at the American Museum for the rest of his professional career, eventually succeeding to the position of curator of fossil reptiles. Although a modestly accomplished scholar, Brown is best known for his spectacular successes as a fossil hunter. He collected fossil vertebrates of all kinds, and from almost all corners of the globe, but was especially interested in dinosaurs. With funding provided by the Sinclair Oil Company, Brown opened up Howe Quarry, in Wyoming, another spectacular assemblage of late Jurassic dinosaurs. He died in New York City just short of his 90th birthday.

Arthur Sterry Coggeshall traveled throughout North and South America and Europe assembling casts of Carnegie's gift *Diplodocus*. For this work he received numerous medals and awards. Later he moved into museum education, becoming curator of public education at the Carnegie Museum and a popular public lecturer. He also enjoyed a very successful career as a museum administrator, including stints as director of the St. Paul Institute of Science in St. Paul, Minnesota, and director of the Illinois State Museum in Springfield. He was director of the Santa Barbara Museum of Natural History when he died in 1958.

Charles Whitney Gilmore moved to the National Museum late in 1903 and remained there for the rest of his professional career. With his wife and three daughters he made a happy home in Washington, D.C., that was a welcome refuge for other paleontologists traveling to the capital. He was a well-rounded paleontologist who led expeditions, collected, prepared, and curated important fossils, and published original research—including a still important monograph on *Apatosaurus*. He never shrank from his early fascination for gigantic fossil reptiles. He died in 1945 and was buried in Arlington National Cemetery.

Walter Granger spent his entire career as a paleontologist with the American Museum, working predominantly on fossil mammals. He excelled at fieldwork. Osborn took Granger as his assistant on an expedition to Egypt in 1907. Granger then spent many field seasons working in the early Tertiary beds of North America. In 1921, he was appointed chief paleontologist of the Central Asiatic Expeditions and spent the better part of the next ten years traveling between New York, China, and Mongolia. In 1927, he succeeded to the position of curator of fossil mammals. He died unexpectedly in 1941 in Lusk, Wyoming.

William Jacob Holland never found a successor to the late J. B. Hatcher and appointed himself to the position of curator of paleontology at the Carnegie Museum. To establish his credibility as a paleontologist, Holland published a monograph on *Diplodocus* that was somewhat error prone. Later he presided over Earl Douglass's excavations near Vernal, Utah, which netted three more Jurassic sauropod specimens and a wealth of other specimens for the Carnegie Museum. He feuded with O. P. Hay and others about the proper posture for sauropods and with H. F. Osborn about the proper head of *Apatosaurus*. To Holland's credit, subsequent science has vindicated many of his views. He died from a stroke in 1932.

Peter C. Kaisen worked for many years as a fossil collector and preparator at the American Museum. He worked for several field seasons under B. Brown's direction collecting Cretaceous dinosaurs in the Red Deer River badlands of Alberta. He also accompanied expeditions to Alaska and Mongolia. He remained with the museum until his death in 1936.

Richard Swann Lull studied under H. F. Osborn at Columbia University and received his PhD in 1903. In 1906 he was named assistant professor of vertebrate paleontology at Yale and associate curator at Yale's Peabody Museum. He remained there for fifty years. He died in 1957 in his 90th year.

William Diller Matthew succeeded H. F. Osborn as curator of the Department of Vertebrate Paleontology in 1911, and, like his predecessor, was largely interested in fossil mammals. In 1927 he accepted a position as professor of paleontology at the University of California, Berkeley. He died in 1930.

Paul Christian Miller spent six years working as a collector and preparator for the American Museum. In 1907, S. W. Williston invited him to become his chief assistant at the University of Chicago, where he remained until the end of his career. After Williston's death, he succeeded to the position of associate curator of paleontology at the Walker Museum.

Henry Fairfield Osborn succeeded to the position of president of the American Museum in 1908, gradually taking on more of the duties of an administrator and fewer of those of a scientist. During his tenure as president, the Department of Vertebrate Paleontology was a favored program that came to dominate American paleontology. He died of a heart attack in 1935.

Olof August Peterson spent the remainder of his professional career, which was devoted exclusively to fossil mammals, with the Carnegie Museum. After Hatcher's death, in 1904, he remained in the field as instructed and made a spectacular discovery of Tertiary fossils at Agate Springs, Nebraska. He named *Dinohyus hollandi*—the "terrible pig"—for his boss. He died in 1933.

William Harlow Reed was appointed assistant geology professor and museum curator at the University of Wyoming in 1904. There he taught classes, led field trips, and continued to collect fossil vertebrates in southeastern Wyoming for another decade. Like his eastern counterparts, Reed put a number of his specimens on display. Never much of an anatomist, some of Reed's reconstructions were somewhat fanciful. He was popular with his students, including W. C. Knight's son, Sam. Reed died in 1915.

Elmer Samuel Riggs spent his entire professional career with the Field Columbian Museum, where there was precious little institutional support for vertebrate paleontology. That all changed for the better in 1922, however, when Marshall Field III provided a windfall of research funding to the museum. Riggs enjoyed a late career renaissance, collecting dinosaurs in Alberta, Canada, and then leading a five-year expedition to South America to make a representative collection of its enigmatic fossil mammal fauna—a lifelong ambition. He retired from the Field Museum in 1942 and returned to Kansas. Riggs was widowed three times and had two sons. He died in Kansas in 1963. He was 94 years old.

Albert "Bill" Thomson remained with the American Museum as a lab and field worker for the rest of his career. He and Granger were close friends and frequent field companions, including two exciting trips to Mongolia in 1928 and 1930. Granger considered him the finest fossil preparator in the business.

William H. Utterback stayed with the Carnegie Museum as a fossil collector for several years after his friend Hatcher's death. Little is known about what became of him thereafter.

Samuel Wendell Williston remained at the University of Chicago for the rest of his career. He mentored a small number of graduate students there and informally collaborated with Riggs at the university and the museum. He worked predominantly on fossil reptiles of the American Permian and marine Cretaceous. Williston was also an accomplished entomologist specializing in diptera. He died in Chicago in 1918.

Jacob L. Wortman never found an agreeable position in American vertebrate paleontology. After he left the Carnegie Museum in 1900, he worked for two years at Yale's Peabody Museum, then quit paleontology altogether. He later relocated to his ranch in Long Pine, Nebraska. At some point he moved to Brownsville, Texas, and opened a drugstore. In 1912 he married Eugenie Brulay and they had two children. Wortman died in Texas in 1926.

Appendix

TABLE OF SELECTED AMERICAN JURASSIC DINOSAUR GENERA AND THEIR SYNONYMS

THEROPODS

Allosaurus (Marsh, 1877) [= *Creosaurus* (Marsh, 1878)]
Ceratosaurus (Marsh, 1884)
  C. nasicornis (Marsh, 1884)
Coelurus (Marsh, 1879)
Ornitholestes (Osborn, 1903)
  O. hermanni (Osborn, 1903)

SAUROPODS

Amphicoelias (Cope, 1877) [probably = *Diplodocus* (Marsh, 1878)]
Apatosaurus (Marsh, 1877) [= *Bronotosaurus* (Marsh, 1879), *Atlantosaurus* (Marsh, 1879), *Elosaurus* (Peterson and Gilmore, 1902)]
  A. excelsus [=*Brontosaurus excelsus* (Marsh, 1879)]
Barosaurus (Marsh, 1890)
Brachiosaurus (Riggs, 1903)
  B. altithorax (Riggs, 1903)
Camarasaurus (Cope, 1877) [=*Morosaurus* (Marsh, 1878)]
C. supremus (Cope, 1877)
Diplodocus (Marsh, 1878)
  D. carnegii (Hatcher, 1901)
  D. hayi (Holland, 1924)
  D. longus (Marsh, 1878)
Dystrophaeus (Cope, 1877)
  D. viaemalae (Cope, 1877)
Haplocanthosaurus (Hatcher, 1903)
  H. priscus (Hatcher, 1903) [= *H. utterbacki* (Hatcher, 1903)]

STEGOSAURS

Stegosaurus (Marsh, 1877)

S. ungulatus (Marsh, 1879)

ORNITHOPODS

Camptosaurus (Marsh, 1885)

Dryosaurus (Marsh, 1894)

D. altus [= *Laosaurus altus* (Marsh, 1878)]

Notes

INTRODUCTION

1. See "The Crystal Palace at Sydenham," *Illustrated London News*, December 31, 1853.
2. All three museums resumed Jurassic fieldwork in later years, some of it on an enormous scale. Collectors from the Carnegie Museum, for example, opened up Carnegie Quarry, near Vernal, Utah, in 1909. Productivity at this site eclipsed all previous Jurassic dinosaur fieldwork. Another important Jurassic locality was Howe Quarry, worked profitably by collectors from the American Museum in the 1930s. These expeditions, however, fall outside the scope of the present work.
3. "Dinosaur" appears an average of six times per year from 1901 to 1920, more than sixteen times per year in the 1920s, and almost twenty-four times per year in the 1930s. Similar patterns can be found by searching for this term in the *New York Times*.
4. There is a wide range of views on this issue, actually. British historian Hugh S. Torrens argues that dinomania began in Great Britain as early as 1853 ("The Dinosaurs and Dinomania over 150 Years," in *Vertebrate Fossils and the Evolution of Scientific Concepts: Writings in Tribute to Beverly Halstead, by Some of His Many Friends*, ed. W. A. S. Sarjeant [Amsterdam: Gordon and Breach Publishers, 1995], 255–84, on page 277). Paul Semonin (*American Monster: How the Nation's First Prehistoric Creature Became a Symbol of National Identity* [New York: New York University Press, 2000], 404), Robert V. Bruce (*The Launching of Modern American Science, 1846–1876* [Ithaca: Cornell University Press, 1988], 344), and Douglas J. Preston (*Dinosaurs in the Attic: An Excursion into the American*

Museum of Natural History [New York: St. Martin's Press, 1986], 64) place the start of American public interest in dinosaurs in the late 1870s, but with scant justification. Preston (64), for example, writes: "It goes without saying that . . . dinosaurs . . . riveted the public's attention [in the late 1870s]." W. J. Thomas Mitchell (*The Last Dinosaur Book: The Life and Times of a Cultural Icon* [Chicago: University of Chicago Press, 1998], 165) places this event as late as 1940. Stephen Jay Gould ("Dinomania," in his *Dinosaur in a Haystack: Reflections in Natural History* [New York: Crown, 1997], 221–37) notes that the popularity of dinosaurs has been fitful and episodic, but that modern dinomania dates only to sometime around 1975.

5. See John S. McIntosh, "The Second Jurassic Dinosaur Rush," *Earth Sciences History* 9, no. 1 (1990): 22–27; Ronald Rainger, *An Agenda for Antiquity: Henry Fairfield Osborn and Vertebrate Paleontology at the American Museum of Natural History, 1890–1935* (Tuscaloosa: University of Alabama Press, 1991); and Tom Rea, *Bone Wars: The Excavation and Celebrity of Andrew Carnegie's Dinosaur* (Pittsburgh: University of Pittsburgh Press, 2001).

CHAPTER ONE

1. The present account draws primarily from Michael F. Kohl and John S. McIntosh, eds., *Discovering Dinosaurs in the Old West: The Field Journals of Arthur Lakes* (Washington, D.C.: Smithsonian Institution Press, 1997); Elizabeth N. Shor, *The Fossil Feud between E. D. Cope and O. C. Marsh* (Hicksville, N.Y.: Exposition Press, 1974); and John H. Ostrom and John S. McIntosh, *Marsh's Dinosaurs: The Collections from Como Bluff* (New Haven: Yale University Press, 1966). See also Othniel C. Marsh, "Notice of a New and Gigantic Dinosaur," *American Journal of Science* 14 (1877): 87–88; and Edward D. Cope, "On a Gigantic Saurian from the Dakota Epoch of Colorado," *Paleontological Bulletin* no. 25 (1877): 5–10.
2. Henry F. Osborn, "The Recent Progress of Vertebrate Paleontology in America," *Science* n.s., 13, no. 315 (January 11, 1901): 45. Nathan Reingold argued that paleontology was one discipline in which Americans could, and did, best European science in the nineteenth century (*Science in Nineteenth Century America: A Documentary History* [New York: Hill and Wang, 1964], 239).
3. In order to retain the appropriate historical flavor of the story, the original names coined by the actors—including a number of antiquated synonyms such as *Brontosaurus*, *Atlantosaurus*, and *Morosaurus*—have been used throughout this book. An appendix at the end provides a list of synonyms for anyone interested in knowing the modern names.
4. Charles Schuchert and Clara M. LeVene, *O. C. Marsh: Pioneer in Paleontology* (New Haven: Yale University Press, 1940), 376.
5. Quoted in Henry F. Osborn, *Cope: Master Naturalist* (Princeton: Princeton University Press, 1931), 410.

6. Quoted in David R. Wallace, *The Bonehunters' Revenge: Dinosaurs, Greed, and the Greatest Scientific Feud of the Gilded Age* (Boston: Houghton Mifflin, 1999), 143.
7. Eric Buffetaut, *A Short History of Vertebrate Palaeontology* (London: Croon Helm, 1987), 129–33; Mark Jaffe, *The Gilded Dinosaur: The Fossil War between E. D. Cope and O. C. Marsh and the Rise of American Science* (New York: Crown, 2000); and Shor, *Fossil Feud.*
8. On Marsh, see Ostrom and McIntosh, *Marsh's Dinosaurs*; on Cope, see Osborn, *Cope.* Cope and some of his collectors worked also at the Como Bluff site.
9. Brent Breithaupt, "Biography of William Harlow Reed: The Story of a Frontier Fossil Collector," *Earth Sciences History* 9, no. 1 (1990): 9.
10. Letter, S. W. Williston to H. F. Osborn, January 24, 1895, Department of Vertebrate Paleontology Archives, American Museum of Natural History (hereafter DVP Arch., AMNH).
11. On Williston, see Elizabeth N. Shor, *Fossils and Flies: The Life of a Compleat Scientist; Samuel Wendell Williston (1851–1918)* (Norman: University of Oklahoma Press, 1971). For more on the expedition of 1895, see Michael F. Kohl, Larry D. Martin, and Paul Brinkman, eds., *A Triceratops Hunt in Pioneer Wyoming: The Journals of Barnum Brown and J. P. Sams; The University of Kansas Expedition of 1895* (Glendo, Wyo.: High Plains Press, 2004).
12. Charles E. Beecher, "Othniel Charles Marsh," *American Journal of Science* 7, Series 4 (June 1899): 412.
13. The two quotations come from Schuchert and LeVene, *Marsh*, 377; and Mark J. McCarren, *The Scientific Contributions of Othniel Charles Marsh: Birds, Bones and Brontotheres*, Peabody Museum of Natural History Special Publication Number 15 (New Haven: Peabody Museum of Natural History, 1993), 11, respectively. See also Othniel C. Marsh, "The Dinosaurs of North America," in *The Sixteenth Annual Report of the U.S. Geological Survey* (Washington, D.C.: Government Printing Office, 1896): 133–244; and Othniel C. Marsh, "Vertebrate Fossils [of the Denver Basin]," in S. F. Emmons, C. W. Cross, and G. H. Eldridge, *Geology of the Denver Basin in Colorado*, U.S. Geological Survey Monograph No. 27 (Washington, D.C.: Government Printing Office, 1896): 473–527.
14. The idea of fossil illustrations serving as proxies for specimens is further developed in Martin J. S. Rudwick, "George Cuvier's Paper Museum of Fossil Bones," *Archives of Natural History* 27, no. 1 (2000): 51–68.
15. *Brontosaurus* appeared first in Othniel C. Marsh, "Principle Characters of American Jurassic Dinosaurs, Part IV: Restoration of *Brontosaurus*," *American Journal of Science* 26 (ser. 3) (August 1883): 81–85. It was later revised and reprinted in Othniel C. Marsh, "Restoration of *Triceratops*," *American Journal of Science* 41 (ser. 3) (April 1891): 339–42. Other restorations appeared in Othniel C. Marsh, "Restoration of *Stegosaurus*," *American Journal of Science* 42 (ser. 3) (August 1891): 179–81; Marsh, "Restorations of *Claosaurus*

and *Ceratosaurus*," *American Journal of Science* 44 (ser. 3) (October 1892): 343–49; and Marsh, "Restoration of *Camptosaurus*," *American Journal of Science* 47 (ser. 3) (March 1894): 245–46. *Laosaurus* appeared for the first time in Marsh, "Dinosaurs of North America." See also Schuchert and LeVene, *Marsh*, 383.

16. On Marsh's paper reconstructions, see Richard Lydekker, "Some Recent Restorations of Dinosaurs," *Nature* 48, no. 1239 (July 27, 1893): 302–4; and Schuchert and LeVene, *Marsh*, 283-85. On earlier dinosaur restorations, see Martin J. S. Rudwick, *Scenes from Deep Time: Early Pictorial Representations of the Prehistoric World* (Chicago: University of Chicago Press, 1992), 78–80. See Adrian Desmond, *The Hot-blooded Dinosaurs* (London: Blond and Briggs, 1975); and James Secord, "Monsters at the Crystal Palace," in *Models: The Third Dimension of Science*, ed. S. de Chadarevian and N. Hopwood (Stanford: Stanford University Press, 2004), 138–69, for more on Richard Owen and the Crystal Palace models. On *Hadrosaurus*, see Richard C. Ryder, "Dusting off America's First Dinosaur," *American Heritage* 39, no. 2 (1988): 69–73; and Leonard Warren, *Joseph Leidy: The Last Man Who Knew Everything* (New Haven: Yale University Press, 1998). On Dollo's *Iguanodon*, see David B. Norman, "On the History of the Discovery of Fossils at Bernissart in Belgium," *Archives of Natural History* 14, no. 1 (1987): 59–75. A brief but excellent overview of the history of dinosaur reconstructions appears in William B. Ashworth, *Paper Dinosaurs, 1824–1969: An Exhibition of Original Publications from the Collections of the Linda Hall Library* (Kansas City, Mo.: Linda Hall Library, 1996).
17. See Marsh, "Principle Characters"; Marsh, "Restorations of Extinct Animals, Plate I," (New Haven: Privately printed, 1895); Marsh, "Restorations of Dinosaurian Reptiles, Plate II," (New Haven: Privately printed, 1895); and, Marsh, "Dinosaurs of North America." For more on Cope, see Osborn, *Cope*, 246 and 250. For Marsh, see Schuchert and LeVene, *Marsh*, 91 and 383–85. See also Clayton Hoagland, "They Gave Life to Bones," *Scientific Monthly* 56, no. 2 (February 1943): 114–33. A much reduced copy of Cope's *Camarasaurus* reconstruction eventually appeared in Charles C. Mook, "Notes on *Camarasaurus* Cope," *Annals of the New York Academy of Science* 24 (1914): 19–22; and Henry F. Osborn and Charles C. Mook, "*Camarasaurus, Amphicoelias*, and Other Sauropods of Cope," *Memoirs of the American Museum of Natural History* 3, no. 3 (1921): 247–387.
18. Carnegie's letter is quoted in Wallace, *Revenge*, 252–53. The legend of Carnegie's spoiled breakfast appears in several places, including Helen J. McGinnis, *Carnegie's Dinosaurs: A Comprehensive Guide to Dinosaur Hall at Carnegie Museum of Natural History, Carnegie Institute* (Pittsburgh: Carnegie Institute, 1982), 13; and Keith M. Parsons, *Drawing out Leviathan: Dinosaurs and the Science Wars* (Bloomington: Indiana University Press, 2001), 1. A more authoritative account of Carnegie's early interest in dinosaurs is

Rea, *Bone Wars*, 29–30 and 40–41. See also *New York Journal and Advertiser*, "Most Colossal Animal on Earth Just Found out West," December 11, 1898.

19. Letter, O. C. Farrington to F. J. V. Skiff, 10 May 1900, Director's General Correspondence, Field Museum Archives (hereafter DGC, FMA). *Atlantosaurus* is now commonly included with the genus *Apatosaurus*.
20. Letter, J. Wortman to H. F. Osborn, 8 or 9 June 1896, DVP Arch., AMNH.
21. Rainger, *Agenda*, 48 and 56. See also Douglas Sloan, "Science in New York City, 1867–1907," *Isis* 71 (1980): 35 and 59–60; and John Michael Kennedy, "Philanthropy and Science in New York City: The American Museum of Natural History, 1868–1968," (PhD dissertation, Yale University, 1968).
22. Andrew Carnegie, *Autobiography of Andrew Carnegie* (Boston: Houghton Mifflin Company, 1920), 348–49.
23. Frank C. Lockwood, *The Life of Edward E. Ayer* (Chicago: A. C. McClurg & Company, 1929), 189–90. This story, probably apocryphal, appears in a number of other sources as well.
24. For more on cultural philanthropy in Chicago, see Helen Lefkowitz Horowitz, *Culture and the City: Cultural Philanthropy in Chicago from the 1880s to 1917* (Chicago: University of Chicago Press, 1989 [1976]); and Kathleen D. McCarthy, *Noblesse Oblige: Charity and Cultural Philanthropy in Chicago, 1849–1929* (Chicago: University of Chicago Press, 1982). Henry Blake Fuller's *With the Procession, a Novel* (New York: Harper & Brothers, 1895) tells the story of David Marshall, a wealthy Chicago merchant who reluctantly turns to philanthropy. The fictional protagonist is based loosely on Marshall Field.
25. Thomas G. Manning, *Government in Science: The U.S. Geological Survey, 1867–1894* (Lexington: University of Kentucky Press, 1967), 205–11; A. Hunter Dupree, *Science in the Federal Government: A History of Policies and Activities* (Baltimore: Johns Hopkins University Press, 1986), 215–35; and Wallace, *Revenge*, 255–58. All seem to agree that Powell's alliance with Marsh was the survey's greatest weakness.
26. Steven Conn, *Museums and American Intellectual Life, 1876–1926* (Chicago: University of Chicago Press, 1998), 53; and Rainger, *Agenda*, 19–23.
27. Peter J. Bowler, *Life's Splendid Drama: Evolutionary Biology and the Reconstruction of Life's Ancestry, 1860–1940* (Chicago: University of Chicago Press, 1996), 31; Lynn K. Nyhart, "Natural History and the New Biology," in *Cultures of Natural History*, ed. N. Jardine, J. A. Secord, and E. C. Spary (Cambridge: Cambridge University Press, 1996), 426–43; Rainger, *Agenda*, 20; and, Ronald Rainger, "The Continuation of the Morphological Tradition: American Paleontology, 1880–1910," *Journal of the History of Biology* 14, no. 1 (Spring 1981): 158.
28. Henry F. Osborn, "Recent Zoopaleontology: Vertebrate Paleontology in the United States Geological Survey," *Science*, n.s., 18, no. 469 (December 25, 1903): 837. See also Osborn, "Recent Progress," 46.

29. Oliver C. Farrington, "The Museum as an Educational Institution," *Education* 17 (1897): 483.
30. Henry F. Osborn, "Models of Extinct Vertebrates," *Science* n.s., 7, no. 182 (June 24, 1898): 841. See also Frederick J. V. Skiff, "Uses of the Museum," *Chicago Times-Herald*, April 29, 1895; Oliver C. Farrington, "Dr. Frederick J. V. Skiff," *Proceedings of the American Association of Museums* 3, nos. 7 and 8 (April-May 1921): 197–98; and Henry F. Osborn, *Creative Education in School, College, University, and Museum: Personal Observation and Experience of the Half-Century, 1877-1927* (New York: Charles Scribner's Sons, 1927).
31. Conn (*Museums*, 6) makes this same point. So does Skiff, in "Uses": "There is a social problem met in the museum, when properly conducted. Good manners are taught. Boisterous people learn quiet ways, rude people learn orderliness, slovenly people learn cleanliness." For more on dinosaurs moralized, see Mitchell, *Last Dinosaur*, 145–46.
32. See Rainger, *Agenda*, 54–60. On the popularity of dinosaurs in museums, see Conn, *Museums*, 45.
33. See Joseph T. Gregory, "North American Vertebrate Paleontology, 1776–1976," in *Two Hundred Years of Geology in America: Proceedings of the New Hampshire Bicentennial Conference on the History of Geology*, ed. C. J. Schneer (Hanover: University of New Hampshire/University Press of New England, 1979), 317. The potato metaphor, which was originally Marsh's, can be found in Adam Hermann, "Modern Laboratory Methods in Vertebrate Paleontology," *Bulletin of the American Museum of Natural History* 26 (1909): 284.
34. Some of these issues are discussed further in Ronald Rainger, "Collectors and Entrepreneurs: Hatcher, Wortman, and the Structure of American Vertebrate Paleontology circa 1900," *Earth Sciences History* 9, no. 1 (1990): 14–21.
35. Letter, G. R. Wieland to J. B. Hatcher, March 22, 1904, John Bell Hatcher Papers, Section of Vertebrate Paleontology, Carnegie Museum of Natural History (hereafter Hatcher Papers, CMNH).
36. Osborn, "Recent Zoopaleontology: Vertebrate Paleontology," 837.
37. See Brian Regal, *Henry Fairfield Osborn: Race and the Search for the Origins of Man* (Burlington, Vt.: Ashgate Publishing, 2002), 26–46. The Morrison meeting with Lakes is mentioned in a letter (A. Lakes to H. F. Osborn, November 13, 1896, DVP Arch., AMNH), although Lakes's journal (Kohl and McIntosh, *Discovering Dinosaurs*, 61) makes it unclear whether or not they actually met.
38. Regal, *Osborn*, 46.
39. See Rainger, *Agenda*, 60–62.
40. Quoted in Wallace, *Revenge*, 230–31.
41. Schuchert and LeVene (*Marsh*, 319) allude disapprovingly to Osborn's role in ending Marsh's federal funding. Manning (*Government*, 209) provides more details.

42. For more on Osborn's ambitions vis-à-vis Marsh, see Rainger, *Agenda*, 80–84, and, Sloan, "Science," 63–64.
43. See H. F. Osborn's field notebook for 1897, DVP Arch., AMNH.
44. Rainger, *Agenda*, 83–84.

CHAPTER TWO

1. See letters, O. A. Peterson to H. F. Osborn, 14 February 1896, and H. F. Osborn to W. B. Scott, 15 February 1896, DVP Arch., AMNH. Osborn also wanted to send an expedition to Patagonia for the American Museum, and abandoned his plans (temporarily) when he learned about Princeton's effort.
2. With Scott and a number of other students, Osborn had organized and participated on the Princeton Scientific Expedition of 1877, his rite of passage as a geologist. But, whether because he was no good at it, or because he felt it was beneath him, he hardly ever participated in fieldwork again. His few excursions to the field with DVP parties were largely ceremonial. For more on Osborn and fieldwork, see Ronald Rainger, "Vertebrate Paleontology as Biology: Henry Fairfield Osborn and the American Museum of Natural History," in *The American Development of Biology*, ed. R. Rainger, K. R. Benson, and J. Maienschein (New Brunswick: Rutgers University Press, 1988), 225 and 249 n.24; and Edwin H. Colbert, *Digging into the Past: An Autobiography* (New York: Dembner Books, 1989), 171 and esp. 228.
3. See DVP annual report for 1896, esp. "Expedition into the San Juan Basin, 1896 (Osborn Report)," DVP Arch., AMNH.
4. Vincent L. Morgan and Spencer G. Lucas, "Walter Granger, 1872–1941, Paleontologist," *New Mexico Museum of Natural History and Science Bulletin* 19 (2002): 1–58. Unfortunately, as a museum insider, Granger moved into his new position in the DVP without leaving a paper trail recording Osborn's or Wortman's views concerning his attributes as a potential fossil collector.
5. This encounter and its significance are discussed in Kohl, Martin, and Brinkman, *Triceratops*, 29. For an account of the KU expedition in 1894, see Shor, *Fossils and Flies*, 141–42.
6. Letter, S. W. Williston to J. Wortman, 18 February 1896, DVP Arch., AMNH. Wortman's initial inquiry has not been located. Williston's response, however, suggests that he asked specifically about Brown.
7. See letters, J. Wortman to H. F. Osborn, especially those of 18 April and 30 April 1896, DVP Arch., AMNH.
8. Letters, H. F. Osborn to J. Wortman, 29 April 1896 and 27 May 1896, DVP Arch., AMNH. The sole basis for Osborn's opinion about Brown seems to be Williston's letter of recommendation. Osborn and Brown would not meet personally until the following summer.
9. Letter, E. S. Riggs to H. F. Osborn, 24 March 1896, DVP Arch., AMNH.

10. Letter, H. S. Osborn to E. S. Riggs, 1 April 1896, DVP Arch., AMNH. On the personal significance of the fossil horse project to Osborn, see Rainger, *Agenda*, 80–82.
11. Letter, E. S. Riggs to H. F. Osborn, 8 April 1896, DVP Arch., AMNH.
12. Letter, S. W. Williston to H. F. Osborn, 10 April 1896, DVP Arch., AMNH. *Dinictis* is an extinct genus of saber-toothed cat.
13. Williston to Osborn, 10 April 1896.
14. Letter, H. F. Osborn to J. Wortman, 14 April 1896, DVP Arch., AMNH.
15. Letter, J. Wortman to H. F. Osborn, 16 May 1896, DVP Arch., AMNH.
16. See Rainger, "Collectors," esp. 18, or Rainger, *Agenda*, 75.
17. Wortman to Osborn, 5 July 1896.
18. Letter, J. Wortman to H. F. Osborn, 21 July 1896, DVP Arch., AMNH. Wortman's original telegram and Osborn's reply, to which Wortman is here responding, are unfortunately lost.
19. Letter, J. Wortman to Mr. Winsor, 31 July 1896, DVP Arch., AMNH.
20. See Expedition into the Big Horn and Wind River basins, appended to the DVP annual report for 1896, and letter, J. Wortman to H. F. Osborn, 5 August 1896, DVP Arch., AMNH. See also American Museum of Natural History, *Annual Report of the President, Treasurer's Report, List of Accessions, Act of Incorporation, Constitution, By-Laws and List of Members for the Year 1896* (New York: American Museum of Natural History, 1897).
21. DVP annual report for 1896, DVP Arch., AMNH. Riggs was paid for two and two-thirds months.
22. Letter, B. Brown to J. Wortman, 15 November 1896, DVP Arch., AMNH.
23. Brown to Wortman, 15 November 1896. Brown first met Knight and Reed, curator and collector, respectively, for the University of Wyoming, while doing fieldwork under Williston in Wyoming in the summer of 1895 (see Kohl, Martin, and Brinkman, *Triceratops*, 73–74). He also likely met and conferred with Reed after the DVP expedition of 1896.
24. Letters, J. Wortman to H. F. Osborn, 16 May and 31 July 1896, DVP Arch., AMNH.
25. Letter, H. F. Osborn to B. Brown, 24 March 1897, DVP Arch., AMNH.
26. Letter, B. Brown to H. F. Osborn, 30 March 1897, DVP Arch., AMNH. A shameless self-promoter, Brown was not above taking a swipe at his predecessors. Another letter from S. W. Williston to Osborn (31 March 1897, DVP Arch., AMNH) provides additional details about the conditions at Como.
27. Brown to Osborn, 30 March 1897.
28. Williston to Osborn, 31 March 1897.
29. See, for example, Rainger, *Agenda*, 69, or Edwin H. Colbert, *Men and Dinosaurs* (New York: E. P. Dutton, 1968), 150.
30. Henry F. Osborn, "Fossil Wonders of the West: The Dinosaurs of the Bone-Cabin Quarry, Being the First Description of the Greatest 'Find' of Extinct Animals Ever Made," *Century Magazine* 68, no. 5 (1904): 681.

31. No dinosaur accessions are listed in American Museum of Natural History, *Annual Report of the President, Treasurer's Report, List of Accessions, Act of Incorporation, Constitution, By-Laws and List of Members for the Year 1892* (New York: American Museum of Natural History, 1893).
32. The quotation comes from a letter, H. F. Osborn to O. A. Peterson, 17 November 1893; see also letter, O. A. Peterson to H. F. Osborn, 27 October 1893, DVP Arch., AMNH. Peterson's unlucky locality was very close to the spot where Earl Douglass would open up the phenomenally productive Carnegie Quarry in 1909. More on Wortman's locality, which was revisited in 1900, appears below. For more details about Osborn's interest in this area, see Paul D. Brinkman, "Henry Fairfield Osborn and Jurassic Dinosaur Reconnaissance in the San Juan Basin, along the Colorado-Utah Border, 1893–1900," *Earth Sciences History* 24, no. 2 (2005): 159–74.
33. See DVP annual report for 1896; and, letters, J. Wortman to H. F. Osborn, 16 May 1896, H. F. Osborn to J. Wortman, 27 May 1896, and H. F. Osborn to E. S. Riggs, 27 May 1896, DVP Arch., AMNH.
34. Wortman to Osborn, 8 or 9 June 1896.
35. Letter, E. S. Riggs to H. F. Osborn, 12 June 1896, DVP Arch., AMNH. Riggs, like Brown, met Reed in the summer of 1895.
36. See Henry F. Osborn, *The American Museum of Natural History: Its Origin, Its History, the Growth of Its Departments to December 31, 1909*, 2nd ed. (New York: Irving Press, 1911), 78; American Museum of Natural History, *Annual Report of the President, Treasurer's Report, List of Accessions, Act of Incorporation, Constitution, By-Laws and List of Members for the Year 1895* (New York: American Museum of Natural History, 1896), 16; and Rainger, *Agenda*, 92–94. Although the asking price for Cope's collection of fossil mammals was $50,000, he only cleared about $32,000 on the sale. See Jane P. Davidson, *The Bone Sharp: The Life of Edward Drinker Cope* (Philadelphia: Academy of Natural Sciences of Philadelphia, 1997), 150.
37. H. F. Osborn's field notebook for 1897, DVP Arch., AMNH.
38. Letters, H. F. Osborn to G. Baur, 17 April and 30 April 1897, DVP Arch., AMNH.
39. Edwin H. Colbert, *William Diller Matthew, Paleontologist: The Splendid Drama Observed* (New York: Columbia University Press, 1992), 44–47 and 62–65.
40. Letter, B. Brown to H. F. Osborn, 8 May 1897, DVP Arch., AMNH.
41. Letter, B. Brown to H. F. Osborn, 6 May 1897, DVP Arch., AMNH.
42. The quotation comes from Brown to Osborn, 6 May. See also Brown to Osborn, 8 May.
43. Letter, B. Brown to H. F. Osborn, 2 May 1897, DVP Arch., AMNH.
44. Letters, H. F. Osborn to L. Osborn, 15 May, 17 May (two letters), 18 May (two letters), 20 May, 21 May, and 23 May 1897, Osborn Papers, New-York Historical Society (hereafter Osborn Papers, NYHS).
45. Letter, H. F. Osborn to L. Osborn, 30 May 1897, Osborn Papers, NYHS. For

more on Lakes and his work for Marsh, see Kohl and McIntosh, *Discovering Dinosaurs*.

46. Letter, H. F. Osborn to L. Osborn, 2 June 1897, Osborn Papers, NYHS; and Osborn's field notebook for 1897, DVP Arch., AMNH. Some versions of this seminal discovery, including one in the DVP Annual Report for 1897, claim that Osborn found the DVP's first dinosaur at Como Bluff, but the account in Osborn's own field notebook makes a strong case for Brown as the legitimate discoverer.
47. Letter, B. Brown to H. F. Osborn, 14 June 1897, DVP Arch., AMNH.
48. Brown to Osborn, 14 June 1897; letter, H. F. Osborn to B. Brown, 22 June 1897; and Osborn's field notebook for 1897, DVP Arch., AMNH.
49. Letters, J. Wortman to H. F. Osborn, 17 and 22 June 1897, DVP Arch., AMNH.
50. The quotation comes from a letter, J. Wortman to H. F. Osborn, 28 June 1897; see also letter, J. Wortman to H. F. Osborn, 24 June 1897, DVP Arch., AMNH.
51. Quoted in Wortman to Osborn, 24 June 1897; see also Wortman to Osborn, 28 June 1897.

CHAPTER THREE

1. For example, in a letter to Osborn (28 June 1897, DVP Arch, AMNH), Brown emphasized his own role on the expedition, writing, "I shall take out the finest specimen that has ever been found in the Jurassic, I think." In a later letter (B. Brown to H. F. Osborn, 15 August 1897, DVP Arch., AMNH), Brown wrote, "I have turned over the account [to Wortman] as you directed though I still manage affairs."
2. Wortman to Osborn, 24 June 1897.
3. Wortman's quotations come from a letter, J. Wortman to H. F. Osborn, 5 July 1897, DVP Arch., AMNH. See also Wortman to Osborn, 24 June 1897. Brown's quotation was taken from a letter, B. Brown to H. F. Osborn, 28 June 1897, DVP Arch., AMNH. Of course, it is impossible to know who conceived of the hydraulic scheme first, but Wortman was definitely the first to describe it on paper. By way of comparison, the DVP party spent approximately $1,600 at Como Bluff during the 1897 season.
4. Wortman to Osborn, 5 July 1897.
5. This paragraph is based on two letters, Wortman to Osborn, 5 July 1897, and B. Brown to H. F. Osborn, 15 August 1897, DVP Arch., AMNH; and esp. on Elmer S. Riggs, "The Discovery of the Use of Plaster of Paris in Bandaging Fossils," *Society of Vertebrate Paleontology News Bulletin* no. 34 (1952): 24–25. Brown's letter, incidentally, fails to credit Wortman's technique, and also differs slightly from Riggs's description, implying that Brown was applying what he considered to be his own procedure. Riggs's account is based on a story told to him in 1899 by Menke. Also, other collectors at

other times, including Williston and Charles Sternberg, developed similar fossil jacketing techniques independently. However, Wortman's innovation of 1897 has the strongest claim of continuity with the collecting routines followed by future paleontologists. On Williston's flour paste technique, see John N. Wilford, *The Riddle of the Dinosaur* (New York: Vintage Books, 1987), 126. On Sternberg's rice glue, see Charles H. Sternberg, *The Life of a Fossil Hunter* (Bloomington: Indiana University Press, 1990 [1909]), 88; and Katherine Rogers, *The Sternberg Fossil Hunters: A Dinosaur Dynasty* (Missoula, Mont.: Mountain Press Publishing Company, 1991), 45. Robert Plate (*The Dinosaur Hunters: Othniel C. Marsh and Edward D. Cope* [New York: David McKay Company, 1964], 194–95) credits Marsh with the idea of using strips of cloth soaked in plaster to make protective jackets. Norman, "History," 61, notes that Belgian collectors working at the Bernissart *Iguanodon* quarry independently invented plaster jackets in the late 1870s.

6. Letters, J. Wortman to H. F. Osborn, 13 July 1897; W. Granger to H. F. Osborn, 21 July 1897; and DVP annual report for 1897, DVP Arch., AMNH.
7. Letters, J. Wortman to H. F. Osborn, 20 and 25 July 1897, DVP Arch., AMNH.
8. Letters, J. Wortman to H. F. Osborn, 4 August 1897, and B. Brown to H. F. Osborn, 15 August 1897, DVP Arch., AMNH.
9. Letter, J. Wortman to H. F. Osborn, 4 August 1897, DVP Arch., AMNH. On Wortman and Patagonia, see Osborn to Scott, 15 February 1896.
10. Letter, B. Brown to H. F. Osborn, 15 August 1897, DVP Arch., AMNH.
11. Emphasis added. Wortman to Osborn, 4 August 1897. Osborn's original letter has not been located, but its contents can be gleaned from Wortman's reply.
12. Marsh had a reputation for not appreciating his field-workers. See chapter 1 for more on Osborn's "Marshiana."
13. Letter, J. Wortman to H. F. Osborn, 18 August 1897, DVP Arch., AMNH. Again, Osborn's original letter is missing.
14. Osborn recommended an increase in salary for Wortman in the DVP annual reports for 1897 and 1898 (DVP Arch., AMNH). In 1898 he wrote: "Although a highly trained and expert worker [Wortman] is receiving only one half as much as if he were professor in one of our colleges."
15. Wortman to Osborn, 18 August 1897.
16. Letter, J. Wortman to H. F. Osborn, 21 August 1897, DVP Arch., AMNH.
17. Letter, J. Wortman to H. F. Osborn, 8 September 1897, DVP Arch., AMNH.
18. Letter, J. Wortman to H. F. Osborn, 25 September 1897, DVP Arch., AMNH.
19. Letter, J. Wortman to H. F. Osborn, 1 October 1897, DVP Arch., AMNH.
20. Wortman to Osborn, 1 October 1897, and DVP annual report for 1897.
21. See letters, H. W. Menke to H. F. Osborn, 2 January, 16 February, 28 February, 1 April 1898; and H. F. Osborn to H. W. Menke, 9 March and 5 April 1898, DVP Arch., AMNH. See also Harold W. Menke, "From a Cabin Window," *Bird-Lore* 1, no. 1 (1899): 14–16.

22. Letters, H. F. Osborn to E. S. Riggs, 5 February 1898, and E. S. Riggs to H. F. Osborn, 28 February 1898, DVP Arch., AMNH.
23. Telegram, E. S. Riggs to H. F. Osborn, 7 April 1898, DVP Arch., AMNH. See also letters, A. Stewart to H. F. Osborn, 28 March 1898; O. C. Farrington to H. F. Osborn, 26 March 1898; and, H. F. Osborn to O. C. Farrington, 2 April 1898, DVP Arch., AMNH.
24. Letter, J. Wortman to H. F. Osborn, 1 May 1898, DVP Arch., AMNH.
25. Letter, J. Wortman to H. F. Osborn, 17 May 1898, DVP Arch., AMNH. See also letter, H. F. Osborn to J. Wortman, 13 May 1898, DVP Arch., AMNH.
26. DVP annual report for 1897.
27. Letter, H. F. Osborn to W. Granger, 25 April 1898, DVP Arch, AMNH. Osborn was already beginning to think that *Brontosaurus* and *Camarasaurus* were synonymous.
28. See letters, H. F. Osborn to J. Wortman, 28 April 1898, and H. F. Osborn to W. Granger, 28 April 1898, DVP Arch., AMNH.
29. Letter, H. F. Osborn to W. H. Menke, 30 April 1898, DVP Arch., AMNH.
30. Letter, W. Granger to H. F. Osborn, 21 April 1898, DVP Arch., AMNH.
31. Letter, H. F. Osborn to W. Granger, 25 April 1898, DVP Arch., AMNH. In fact, Riggs and the Field Columbian Museum party collected fossil mammals in South Dakota, Nebraska, Montana, and Wyoming. See Paul Brinkman, "Establishing Vertebrate Paleontology at Chicago's Field Columbian Museum, 1893–1898," *Archives of Natural History* 27, no. 1 (2000): 81–114.
32. DVP annual report for 1898; and letters, H. F. Osborn to J. Wortman, 23 April 1898, and J. Wortman to H. F. Osborn, 1 May and 17 May 1898, DVP Arch., AMNH.
33. Letters, W. Granger to H. F. Osborn, 28 April 1898; H. F. Osborn to H. W. Menke, 9 May 1898; and H. F. Osborn to J. Wortman, 10 May 1898, DVP Arch., AMNH.
34. Letter, J. Wortman to H. F. Osborn, 7 June 1898, DVP Arch., AMNH.
35. Wortman to Osborn, 7 June 1898.
36. Letters, J. L. Wortman to H. F. Osborn, 18 June and 10 July 1898; and DVP annual report for 1898, DVP Arch., AMNH.
37. The quarry map first appeared in Osborn, "Fossil Wonders," 691.
38. For details, see Ostrom and McIntosh, *Marsh's Dinosaurs*, 37; and esp. McIntosh, "Second Jurassic," 22.
39. See Breithaupt, "Biography," 11; and McIntosh, "Second Jurassic," 22.
40. This account is based primarily on Morgan and Lucas, "Walter Granger," 6–7. See also Osborn, "Fossil Wonders," 680; Edwin H. Colbert, *Dinosaurs: Their Discovery and Their World* (New York: E. P. Dutton & Company, 1961): 88–89; and Rainger, *Agenda*, 95.
41. Wortman to Osborn, 18 June 1898.
42. Letter, J. L. Wortman to H. F. Osborn, 18 June 1898, DVP Arch., AMNH.
43. The quotation comes from Wortman to Osborn, 10 July. See also letter, J. Wortman to H. F. Osborn, 24 June 1898, DVP Arch., AMNH. On the nature

of the matrix, see Frederick B. Loomis, "On the Jurassic Stratigraphy of Southeastern Wyoming," *Bulletin of the American Museum of Natural History* 14 (June 17, 1901): 195.

44. Letters, J. Wortman to H. F. Osborn, 23 July and 26 August 1898, DVP Arch., AMNH.
45. Letter, J. L. Wortman to H. F. Osborn, 20 August 1898, DVP Arch., AMNH.

CHAPTER FOUR

1. For details on the early history of vertebrate paleontology at the Field Columbian Museum, see Brinkman, "Establishing."
2. Both quotations were taken from Brinkman, "Establishing," 93. See also Farrington, "Museum."
3. For an account of Riggs's experiences on the expedition of 1895, see Shor, *Fossils and Flies*, or Kohl, Martin, and Brinkman, *Triceratops.*
4. For details, see Brinkman, "Establishing."
5. See letter, O. C. Farrington to F. J. V. Skiff, 26 May 1899, DGC, FMA. The late submission might have been Farrington's responsibility.
6. See Philip Gingerich, "History of Early Cenozoic Vertebrate Paleontology in the Bighorn Basin," in *Early Cenozoic Paleontology and Stratigraphy of the Bighorn Basin, Wyoming*, ed. P. Gingerich, University of Michigan Papers on Paleontology No. 24 (1980): 7–24.
7. Letters, O. C. Farrington to F. J. V. Skiff, 26 May 1899; F. J. V. Skiff to H. N. Higinbotham, 7 June 1899; and H. N. Higinbotham to F. J. V. Skiff, 22 June 1899, DGC (all filed under Farrington), FMA. There is no record of an Executive Committee meeting in June that can explain the late date of Higinbotham's reply. A special meeting of the Board of Trustees was held on 21 June, but Higinbotham was not present and no museum appropriations were discussed (see Ledgers, "Minutes of the Meetings of the Board of Trustees of the Field Columbian Museum, September 1893 to December 1912"; and "Record of Minutes of the Executive Committee of the Field Columbian Museum, May 1894 to December 1913," FMA).
8. Letter, W. J. Holland to A. Carnegie, 27 March 1899, Dr. William Jacob Holland Collection, Papers, 1880–1945, Carnegie Museum of Natural History Archives (Holland Papers, CMNH).
9. The exchange between Holland and Reed is detailed in a letter, W. J. Holland to A. Carnegie, 9 December 1898, Holland Papers, CMNH.
10. Letter, W. J. Holland to W. H. Reed, 19 December 1898, Holland Papers, CMNH.
11. Letter, W. J. Holland to A. Carnegie, 27 December 1898, Holland Papers, CMNH.
12. Letter, W. J. Holland to S. W. Downey, 10 May 1899, Holland Papers, CMNH. An excellent account of the political whirlwind surrounding this dinosaur can be found in Rea, *Bone Wars*, 52–67.

13. All quotations come from a letter, W. J. Holland to A. Carnegie, 5 May 1899, Holland Papers, CMNH.
14. Letter, W. J. Holland to A. Carnegie, 1 April 1899, Holland Papers, CMNH.
15. Letter, W. J. Holland to A. Carnegie, 27 March 1899, Holland Papers, CMNH. This was just overconfident bluster. Holland barely knew Reed and knew Osborn only by reputation. On the other hand, if he only meant to suggest that Reed was a better man for fieldwork than Osborn, he was certainly correct.
16. See letters, W. H. Holland to A. Carnegie, 1 April 1899; W. H. Holland to J. L. Wortman, 20 April 1899; and W. J. Holland to S. W. Downey, 10 May 1899, Holland Papers, CMNH. On Wortman's difficulties with Osborn, see Rainger, "Collectors."
17. The quotation comes from a letter, W. J. Holland to W. H. Reed, 12 May 1899, Holland Papers, CMNH. For more on Holland's competitive personality, see Rea, *Bone Wars*, 29, 41, and 51–58.
18. See Samuel W. Williston, "Wilbur Clinton Knight," *American Geologist*, 33, no. 1 (January 1904): 1-6. Williston describes Knight as "enthusiastic to a greater degree than is common among scientific men even" (1).
19. Quoted in Rea, *Bone Wars*, 105.
20. The quotation was taken from a letter, W. Granger to H. F. Osborn, 30 April 1899. See also letter, W. Granger to H. F. Osborn, 9 May 1899; and DVP annual report for 1899, DVP Arch., AMNH. On Granger's early years with the DVP, see Morgan and Lucas, "Walter Granger," 4–5.
21. Letter, R. S. Lull to H. F. Osborn, 2 February 1899, DVP Arch., AMNH.
22. Richard S. Lull, "Early Fossil Hunting in the Rocky Mountains," *Natural History* 26, no. 5 (1926): 457.
23. Letter, H. F. Osborn to R. S. Lull, 25 April 1899, DVP Arch., AMNH. For more on Lull, see George G. Simpson, "Memorial to Richard Swann Lull (1867–1957)," *Proceedings Volume of the Geological Society of America Annual Report for 1957* (1958): 127–34.
24. Letter, H. F. Osborn to W. Granger, 25 May 1899, DVP Arch., AMNH.
25. Letter, H. F. Osborn to W. Granger, 3 May 1899, DVP Arch., AMNH. Note the ambiguous language of Osborn's instructions.
26. Letter, Granger to Osborn, 9 May 1899. See also DVP annual report for 1899, DVP Arch., AMNH.
27. Letter, W. H. Reed to W. J. Holland, 9 May 1899, Holland Papers, CMNH. Gilmore enlisted in May 1898 and served as first sargeant of Torrey's Rough Riders, but saw no fighting. He received an honorable discharge on October 17, 1898, and returned to school in Laramie. See G. Edward Lewis, "Memorial to Charles Whitney Gilmore," *Proceedings Volume of the Geological Society of America Annual Report for 1945* (1946): 235–44.
28. The quotation comes from Granger to Osborn, 9 May 1899. See also Osborn to Granger, 15 May 1899.
29. W. J. Holland to S. W. Downey, 10 May 1899.

30. Letter, J. Wortman to W. J. Holland, 17 May 1899, Holland Papers, CMNH.
31. Letter, W. Granger to H. F. Osborn, 22 May 1899, DVP Arch., AMNH.
32. Letter, W. H. Reed to W. J. Holland, 26 May 1899, Holland Papers, CMNH.
33. Letter, J. Wortman to W. J. Holland, 6 June 1899, Holland Papers, CMNH; see also Rea, *Bone Wars*, 77–81.
34. Letter, H. F. Osborn to W. Granger, 15 May 1899, DVP Arch., AMNH. There are two letters in the DVP Archives from Osborn to Granger bearing this same date.
35. Osborn to Granger, 25 May 1899.
36. Letter, H. F. Osborn to W. Granger, 5 June 1899, DVP Arch., AMNH.
37. Letter, H. F. Osborn to W. Granger, 8 June 1899, DVP Arch., AMNH. In this letter, Osborn misquotes Marsh's maxim, which was quoted previously in a letter from Wortman.
38. See letter, H. F. Osborn to W. Granger, 18 May 1899, DVP Arch., AMNH. Osborn wanted to keep the contents of this letter secret, so he wrote it by hand rather than dictating it to a secretary. Across the top he wrote: "Destroy this letter." Wortman, for his part, likely held Osborn responsible for his failure to get the position as Marsh's replacement at Yale. See Rainger, "Collectors."
39. See letters, B. Brown to H. F. Osborn, 30 March and 8 May 1897; and J. Wortman to H. F. Osborn, 23 July and 26 August 1898, DVP Arch., AMNH. Details about Menke's activities in the field can be found in many letters in the DVP field correspondence. Although Menke's arrival date at the Field Museum is not known with certainty, a memo (O. C. Farrington to F. J. V. Skiff, 2 February 1899, DGC, FMA) establishes that he was in Chicago by February 2 at the latest.
40. Letter, W. Granger to H. F. Osborn, 30 April 1899, DVP Arch., AMNH.
41. Invitations were only mailed in early June, so late that many naturalists were unable to accept. See Wilbur C. Knight, "The Wyoming Fossil Fields Expedition of July, 1899," *National Geographic Magazine* 11, no. 12 (December 1900): 449.
42. Letter, O. C. Farrington to F. J. V. Skiff, 26 June 1899, Recorder's Office Accession Records—Geology (hereafter ROAR-Geology), No. 650, FMA. Although Farrington's plan was followed, no record of the museum's official approval could be located.
43. Recorder's Office, Expedition Vouchers (hereafter ROEV), Folder 1: Riggs, 1899, FMA. Collecting techniques for large fossil vertebrates remain largely unchanged.
44. See Rea, *Bone* Wars, 66; Arthur Coggeshall, "How 'Dippy' Came to Pittsburgh," *Carnegie Magazine* 25, no. 7 (July 1951): 239; Granger to Osborn, 9 May 1899; and E. C. Case MS, "Adventures in Memory," Ermine Cowles Case Papers, William L. Clements Library, University of Michigan (Case Papers, University of Michigan).
45. More on Butch and Sundance in Patagonia can be found in Anne Meadows,

Digging up Butch and Sundance (New York: St. Martin's Press, 1994). On Holland's safe fragments, see William J. Holland, "Bone Hunters Starting Well at Their Work," *Pittsburgh Dispatch*, July 25, 1899, 2. Henrika Kuklick and Robert E. Kohler (Introduction to H. Kuklick and R. E. Kohler, eds., *Science in the Field, Osiris* 11, second series [1996]: 1–14, on 11) make mention of the misidentification of naturalists as bandits on the margins of settled Western society.

46. See Shor, *Fossils and Flies*, 143–44.
47. Typescript, "Field Record Expeditions: 1898-1910, & 1922," Department of Geology Archives, Field Museum (Geol. Arch., FM). J. D. Dyer was a Wyoming state senator.
48. Knight, "Wyoming," 457; letters, E. S. Riggs to F. J. V. Skiff, 29 July and 4 August 1900, ROAR—Geology, No. 650, FMA.

CHAPTER FIVE

1. Letter, J. Wortman to W. J. Holland, 30 May 1899, Holland Papers, CMNH.
2. Coggeshall, "Dippy."
3. Wortman to Holland, 6 June 1899.
4. W. H. Reed to W. J. Holland, 10 June 1899, Holland Papers, CMNH.
5. J. Wortman to W. J. Holland, 18 June 1899, Holland Papers, CMNH.
6. Wortman to Holland, 18 June 1899.
7. J. Wortman to W. J. Holland, 28 June 1899, Holland Papers, CMNH.
8. Letter, W. J. Holland to J. Wortman, 3 July 1899, Holland Papers, CMNH.
9. Quoted in Rea, *Bone Wars*, 86.
10. Letter, W. Granger to H. F. Osborn, 1 June 1899, DVP Arch., AMNH.
11. Letter, W. Granger to H. F. Osborn, 18 June 1899, DVP Arch., AMNH.
12. Granger to Osborn, 18 June 1899.
13. H. F. Osborn to W. Granger, 29 June 1899, DVP Arch., AMNH.
14. Note that there was no talk of the Nine Mile skeleton being a new taxon, as there likely would have been during the *first* Jurassic dinosaur rush when Cope and Marsh competed to describe and name new taxa. The goal of mounting a sauropod dinosaur for display and the necessity of using multiple specimens to complete a composite mount likely encouraged "lumping" rather than "splitting" taxa during the second rush.
15. Letter, W. D. Matthew to H. F. Osborn, 2 July 1899, DVP Arch., AMNH.
16. Matthew to Osborn, 2 July 1899.
17. Matthew to Osborn, 2 July 1899.
18. Both quotations come from Matthew to Osborn, 2 July 1899.
19. Letter, H. F. Osborn to W. D. Matthew, 7 July 1899, DVP Arch., AMNH.
20. Letter, H. F. Osborn to W. Granger, 13 July 1899, DVP Arch., AMNH.
21. This account of the initial discovery of the Carnegie Museum's *Diplodocus* specimen is based on Rea, *Bone Wars*, 87. Many years after the events, Coggeshall wrote his own account of this same discovery, which is far

better known. Differing in several important details from the version recounted here, Coggeshall's story does not seem to square with the surviving evidence:

Perhaps *Diplodocus* should have been named the Star-Spangled Dinosaur, for it was discovered on the Fourth of July. The morning of the Fourth, Wortman and Reed mounted horses and pulled out to prospect an escarpment about two miles away, leaving me, as the youngest member, the writer, to prospect afoot.

The first indication of "Dippy" was a toe bone of the hind foot. After very close scanning of the ground, a few pieces of weathered bone were found. It was then that the heartbeats of the writer really became loud, for it was the best prospect any of us had discovered in over two months of hard and disappointing work, and we did so want to make good with a dinosaur for Mr. Carnegie. (Coggeshall, "Dippy," 240).

For a detailed discussion of the relative virtues of these two accounts, see Rea, *Bone Wars*, 234–35.

22. Letter, J. Wortman to W. J. Holland, 4 July 1899, Holland Papers, CMNH.
23. Rea, *Bone Wars*, 89.
24. Letter, J. Wortman to W. J. Holland, 19 July 1899, Holland Papers, CMNH.
25. See Rea, *Bone Wars*, 93–95, for a detailed (and humorous) account of Holland's suffering.
26. See Osborn to Matthew, 7 July 1899.
27. Letter, H. F. Osborn to L. P. Osborn, 21 July 1899, Osborn Papers, NYHS.
28. Letter, W. Granger to H. F. Osborn, 19 August 1899, DVP Arch., AMNH.
29. See Granger to Osborn, 19 August 1899, and DVP annual report for 1899, DVP Arch., AMNH.
30. Letter, W. D. Matthew to H. F. Osborn, n.d. [September 1899], DVP Arch., AMNH.
31. William D. Matthew, "Early Days of Fossil Hunting in the High Plains," *American Naturalist*, 26, no. 5 (1926): 451.
32. Matthew to Osborn, [September 1899]; DVP annual report for 1899, DVP Arch., AMNH.
33. Letter, W. J. Holland to J. Wortman, 22 August 1899, Holland Papers, CMNH.
34. Letter, W. J. Holland to J. Wortman, 2 September 1899, Holland Papers, CMNH.
35. Quoted in Rea, *Bone Wars*, 93.
36. See letter, J. Wortman to W. J. Holland, 11 September 1899, Holland Papers, CMNH.
37. Elmer S. Riggs, "Fossil-Hunting in Wyoming," *Science*, n.s., 11, no. 267 (February 9, 1900): 233; and Case, "Adventures," Case Papers, University of Michigan.
38. Letter, E. S. Riggs to F. J. V. Skiff, 27 August 1899, ROAR-Geology, No. 650, FMA.

39. Letter, E. S. Riggs to O. C. Farrington, 1 September 1899, ROAR-Geology, No. 650, FMA. If his time in the field could not be extended, Riggs asked to take his vacation time beginning 15 September—he may have wanted to remain in Wyoming and work the quarry for himself.
40. Letter, E. S. Riggs to O. C. Farrington, 5 September 1899, DGC, FMA. Ward was a paleobotanist, born in Joliet, Illinois, and then working for the U.S. Geological Survey. Schuchert had been working a quarry in the Freezeouts on behalf of the National Museum. Gilmore would return in 1902, reopening Riggs's abandoned quarry for the Carnegie Museum. He collected a few vertebrae before giving up the locality as exhausted (McIntosh, "Second," 23).
41. ROEV, Folder 1: Riggs 1899, FMA.
42. Letters, E. S. Riggs to O. C. Farrington, 5 September 1899, DGC; 11 September 1899, ROAR-Geology, No. 650; and E. S. Riggs to F. J. V. Skiff, 12 October 1899, ROAR-Geology, No. 650, FMA.
43. Letter, J. L. Wortman to W. J. Holland, 27 September 1899, Holland Papers, CMNH.
44. Letters, W. Granger to H. F. Osborn, 21 and 25 September 1899; H. F. Osborn to W. Granger, 21 September 1899; and DVP annual report for 1899, DVP Arch., AMNH.
45. See McIntosh, "Second," 23; accession card and inventory, ROAR-Geology, No. 650, FMA; and Field Columbian Museum, "Annual Report of the Director to the Board of Trustees for the Year 1898–1899," *Publications of the Field Columbian Museum, Report Series* 1, no. 5 (October 1899): 364–65. The *Stegosaurus* record was found in "Field Record Expeditions," Geol. Dept. Arch., FM.
46. Letter, H. F. Osborn to R. O. Johnson, 25 September 1899, DVP Arch., AMNH.
47. Letter, H. F. Osborn to W. B. Scott, 14 September 1899, DVP Arch., AMNH.
48. Letter, W. J. Holland to A. Darlow, 11 September 1899, Holland Papers, CMNH.
49. Letter, W. J. Holland to H. H. Smith, 8 November 1899, Holland Papers, CMNH.
50. Letter, H. F. Osborn to W. Granger, 11 September 1899, DVP Arch., AMNH.
51. No records have been located to show how Osborn came upon this information. Probably he heard about it informally, through a chance encounter with one of the many geologists traveling around the state. See Brinkman, "Osborn."
52. Letter, E. T. Jeffery to H. F. Osborn, 1 October 1899, DVP Arch., AMNH.
53. See C. Whitman Cross, "Stratigraphic Results of a Reconnaissance in Western Colorado and Eastern Utah," *Journal of Geology* 15 (1907): 639. The quotation comes from C. Whitman Cross, "Description of the Telluride Quadrangle," *U.S. Geological Survey, Geological Atlas, Folio 57* (1899): 3. This folio was published too late to have been the original source of Osborn's

dinosaur tip, unless he had seen it in manuscript. Given the timing, it seems likely that Cross and Osborn heard the same dinosaur rumors, probably independently.

54. Letter, A. L. Fellows to H. F. Osborn, 11 August 1899, DVP Arch., AMNH. Fellows might have discovered the dinosaur bones while surveying, although no reliable information on the initial discovery has been located. See Brinkman, "Osborn."
55. Letter, H. F. Osborn to J. Gidley, 11 October 1899, DVP Arch., AMNH. Nothing appeared in the local papers.
56. Letter, J. Gidley to H. F. Osborn, 21 October 1899, DVP Arch., AMNH.
57. The quotation comes from the DVP annual report for 1899, DVP Arch., AMNH; other information was taken from the DVP annual report for 1900. See also American Museum of Natural History, *Annual Report of the President, Treasurer's Report, List of Accessions, Act of Incorporation, Constitution, By-Laws and List of Members for the Year 1899* (New York: American Museum of Natural History, 1900).
58. Letter, H. F. Osborn to W. B. Scott, 14 September 1899, DVP Arch., AMNH.
59. Letter, W. J. Holland to C. C. Mellor, 31 January 1900, Holland Papers, CMNH. Mellor was then chairman of the Museum Committee of the Carnegie Institute.
60. For more on Wortman's conflict with Holland, see Rea, *Bone Wars*, 120–22; and Rainger, "Collectors." The controversial article appeared as Jacob L. Wortman, "The New Department of Vertebrate Paleontology of the Carnegie Museum," *Science*, n.s., 11, no. 266 (February 2, 1900): 163–66.
61. Howard W. Bell, "Fossil-Hunting in Wyoming," *Cosmopolitan Magazine* 28 (January 1900): 271–72.
62. Both quotations come from Riggs, "Fossil-Hunting," 233.

CHAPTER SIX

1. Letters, O. C. Farrington to F. J. V. Skiff, 10 October 1899, DGC; and F. J. V. Skiff to O. C. Farrington, 27 December 1899, Director's Letterbooks (hereafter DLB), FMA.
2. Field Columbian Museum, "Annual Report of the Director to the Board of Trustees for the Year 1899–1900," *Publications of the Field Columbian Museum, Report Series* 1, no. 6 (October 1900): 449.
3. Letter, O. C. Farrington to F. J. V. Skiff, 10 May 1900, ROAR-Geology, No. 723, FMA. Farrington wrote elsewhere that the museum was especially obligated to acquire objects of local or national significance. Other museum departments, especially Zoology, were then mounting very expensive international collecting expeditions.
4. Letter, H. N. Higinbotham to F. J. V. Skiff, 14 May 1900, ROAR-Geology, No. 723, FMA. As before, there is no record of an Executive Committee debate on this issue. Higinbotham apparently revised and approved the

expenditure personally. Note that the museum hierarchy was again followed to the letter. Riggs (presumably) devised the original plan, which was formally presented by Curator Farrington to Director Skiff, and by Skiff to President Higinbotham. Approval was passed back down the chain in exactly the reverse order.

5. Letter, F. J. V. Skiff to O. C. Farrington, 15 May 1900, DLB, FMA. The patronizing tone assumed by Skiff was common in correspondence of this type.
6. Letters, O. C. Farrington to F. J. V. Skiff, 16 May 1900, ROAR-Geology, No. 723; and F. J. V. Skiff to O. C. Farrington, 21 May 1900, DLB, FMA. Possibly Skiff and/or Higinbotham remembered that Riggs had gone slightly over budget the previous summer.
7. Sixty years later, Riggs was still annoyed. He recalled that the museum "provided . . . a rather small fund to cover such an undertaking." See letter, E. S. Riggs to Dept. of Interior, 5 October 1959, Elmer S. Riggs Interview and Correspondence, Colorado National Monument (hereafter Riggs Collection, CNM). A transcript of the Riggs interview and copies of some of this correspondence are kept by the Geology Department, Field Museum.
8. Letter, O. C. Farrington to F. J. V. Skiff, 10 May 1900, ROAR-Geology, No. 723, FMA. For more on Farrington in 1898, see Brinkman, "Establishing."
9. This specimen, collected in southeastern Utah by John Strong Newberry, was described by Cope in 1877 and assigned to the Triassic. Friedrich von Huene reassigned it to the Jurassic, but not until 1904. See John S. McIntosh, "The Saga of a Forgotten Sauropod Dinosaur," in *Dinosaur International: Proceedings of a Symposium Held at Arizona State University*, ed. D. L. Wolberg et al. (Philadelphia: Academy of Natural Sciences, 1997), 7. See also Brinkman, "Osborn."
10. For a discussion of the role of the railroad in stimulating scientific fieldwork, see Jeremy Vetter, "Science along the Railroad: Expanding Fieldwork in the US Central West," *Annals of Science* 61 (2004): 187–211.
11. See Othniel C. Marsh, "On the Geology of the Eastern Uintah Mountains," *American Journal of Science and Arts* 1, series 3 (1871): 191–98; and Wade E. Miller and Dee A. Hall, "Earliest History of Vertebrate Paleontology in Utah: Last Half of the 19th Century," *Earth Sciences History* 9, no. 1 (1990): 30.
12. See Albert C. Peale, "Geological Report on the Grand River District," in *Tenth Annual Report of the United States Geological and Geographical Survey of the Territories, Embracing Colorado and Parts of Adjacent Territories, Being a Report of Progress of the Exploration for the Year 1876*, ed. F. V. Hayden (Washington, D.C.: Government Printing Office, 1878): 163–85. Jurassic deposits are discussed on pages 178–80. On page 181, he writes: "At no point in the [Grand River] district did I obtain any fossils from its strata." Peale also reports on western Colorado geology in several other volumes of Hayden's *Reports*.

13. Elmer S. Riggs, "Dinosaur Hunting in Colorado," *Field Museum News* 10, no. 1 (1939): 4; "Dr. S. M. Bradbury Dies in San Diego Last Night," *Grand Junction Daily Sentinel*, September 19, 1913, 1 and 4; and, Riggs to Department of the Interior, 5 October 1959. Bradbury's original letter has not been located, and its contents can only be surmised from later accounts written by Riggs. It is not clear, for example, whether Bradbury identified or described the kinds of fossils he had collected.
14. Letter, O. C. Farrington to F. J. V. Skiff, 31 May 1900, ROAR-Geology, No. 723, FMA. Not much is known about Barnett. Riggs ultimately paid him $75 for three months' work (see ROEV, FMA).
15. Riggs, "Dinosaur Hunting in Colorado," 4. Not much else has been written about Riggs's fieldwork in western Colorado. One good source with excellent figures is Harley J. Armstrong and Michael L. Perry, "A Century of Dinosaurs from the Grand Valley," *Museum Journal* 2 (1985): 4–19. Geologist William Chenoweth has written a number of short articles on Riggs, the best of which is William J. Chenoweth, "The Riggs Hill and Dinosaur Hill Sites, Mesa County, Colorado," in *Paleontology and Geology of the Dinosaur Triangle: Guidebook for 1987 Field Trip Sept. 18-20, 1987*, ed. W. R. Averett (Grand Junction: Museum of Western Colorado, 1987), 97–100. Another good article is David Young, "*Brachiosaurus*: The Biggest Dinosaur of Them All," *Field Museum Bulletin* 46, no. 1 (1975): 3–9.
16. Elmer S. Riggs, "The Dinosaur Beds of the Grand River Valley of Colorado," *Field Columbian Museum Publication 60: Geology Series* 1, no. 9 (September 1901): 267–68. For a more modern discussion of the geology of this area, as well as an excellent account of Jurassic paleontology in the American West, see John Foster, *Jurassic West: The Dinosaurs of the Morrison Formation and Their World* (Bloomington: Indiana University Press, 2007).
17. Riggs, "Dinosaur Hunting in Colorado."
18. Letter, E. S. Riggs to H. Mosher, 24 June 1900, Private Collection. It is not known how or when Riggs and Mosher first met. Their correspondence provides a rich array of details about his fossil fieldwork. I am grateful to several of Riggs's descendants for making their private collection of letters and papers available to me.
19. Riggs to Mosher, 24 June 1900.
20. Riggs to Mosher, 24 June 1900.
21. The quotation comes from a letter, E. S. Riggs to F. J. V. Skiff, 26 July 1900, ROAR-Geology, No. 723, FMA. The poor conditions are detailed in another letter, E. S. Riggs to F. J. V. Skiff, 13 April 1903, DGC (filed under O. C. Farrington), FMA.
22. Riggs to Skiff, 26 July 1900, and Field Record Expeditions, Geol. Dept. Arch, FM.
23. See Al Look, *In My Back Yard* (Denver: University of Denver Press, 1951): 69. In 1988, staff from Grand Junction's Museum of Western Colorado

discovered a disturbed area littered with bits of plaster on the south slope of No Thoroughfare Canyon. A comparison of this site with historic photographs from the Field Museum confirmed that this was the location of Riggs's Quarry Twelve. See William L. Chenoweth, "Relocating Elmer Riggs' Quarry No. 12," in *Guidebook for Dinosaur Quarries and Tracksites Tour, Western Colorado and Eastern Utah*, ed. W. A. Averett (Grand Junction, Colo.: Grand Junction Geological Society, 1991): 17–18.

24. Riggs, "Dinosaur Hunting in Colorado," 4.
25. Elmer S. Riggs, "Structure and Relationships of Opisthocoelian Dinosaurs. Part II. The Brachiosauridae," *Field Columbian Museum Publication 94, Geological Series* 2, no. 6 (September 1, 1904): 230.
26. Riggs to Skiff, 26 July 1900.
27. See, for example, Riggs to Skiff, 26 July 1900, and E. S. Riggs to O. C. Farrington, 17 August 1900, DGC, FMA.
28. "Chicago Has the Largest Land Animal that Ever Lived," *Sunday Times-Herald* (Chicago), October 7, 1900; "Bones of the Largest Known Animal Found," *Sunday Tribune* (Chicago), October 7, 1900; "The Monster of All Ages," *Boston Journal*, October 14, 1900.
29. *Sunday Tribune* (Chicago), "Bones."
30. *Sunday Times-Herald* (Chicago), "Chicago."
31. Riggs, "Structure Part II," 230.
32. See Elmer S. Riggs, "The Largest Known Dinosaur," *Science* n.s., 13, no. 327 (1901): 549–550; and Elmer S. Riggs, "*Brachiosaurus altithorax*, the Largest Known Dinosaur," *American Journal of Science* 15 (April 1903): 299–306.
33. Riggs, "*Brachiosaurus*"; Riggs, "Structure Part II"; and Riggs, "Dinosaur Hunting in Colorado."
34. Letter, H. Mosher to E. S. Riggs, 2 August 1900, Private Collection.
35. "Animals of Past Ages: Prof. Briggs of Columbian Museum Lectured Last Night to a Select Audience," *Grand Junction Daily Sentinel*, July 28, 1900, p. 1. Riggs's own account of the lecture is given in a letter to Mosher, 16 August 1900, Private Collection.
36. Riggs to Mosher, 16 August 1900.
37. Riggs to Skiff, 26 July 1900.
38. "Some Prehistoric Specimens," *Grand Junction News*, August 4, 1900, p. 1. This information was almost certainly not from Riggs, leaving Bradbury as the likely source.
39. "Dinosaur Skeleton Found," *New York Times*, August 14, 1900, p. 2. In attempting to express the great size of the specimen, the author of this article blundered badly, writing: "One bone is nine feet long, and others are so large that it is almost beyond the strength of one man to lift them." The largest bones weighed almost seven hundred pounds, actually.
40. Riggs to Mosher, 16 August 1900.
41. Riggs, "Dinosaur Hunting in Colorado," 4.
42. The safety bicycle, with two identically sized wheels and pneumatic tires,

was a vast improvement on the penny-farthing bike. By the 1890s, bicycling had become a very fashionable fad.

43. Riggs to Mosher, 16 August 1900.
44. Riggs to Mosher, 16 August 1900.
45. Letter, E. S. Riggs to D. C. Davies, 13 August 1900, DGC, FMA.
46. Letter, E. S. Riggs to O. C. Farrington, 17 August 1900, DGC, FMA. This letter confirms that the party quit Quarry Thirteen on August 17, but leaves open the possibility that they left some bones still in the ground. No other records indicate that they ever returned to this site, however.
47. These specimens are described in Elmer S. Riggs, "The Fore Leg and Pectoral Girdle of *Morosaurus*, with a Note on the Genus *Camarosaurus*," *Field Columbian Museum Publication 63, Geological Series* 1, no. 10 (October 1901): 276–77. The localities of these discoveries were not recorded. According to Chenoweth, "Riggs Hill," the forelimb was found at a site two miles south of Fruita, north of the mouth of Wedding Canyon.
48. Letter, S. E. Meek to D. C. Davies, 12 November 1900, DGC, FMA.
49. "Now They Are Saying the Dinosaur Didn't Have Two Brains," *Inter Ocean* (Chicago), 1 March 1903, p. 3.
50. This rough itinerary was reconstructed from receipts preserved in ROEV, FMA. Riggs mentioned the promise in a letter to Mosher, 16 August 1900.
51. Steven F. Mehls, *The Valley of Opportunity: A History of West-Central Colorado* (Denver: Bureau of Land Management, Colorado State Office, 1982), 29.
52. Elmer S. Riggs Interview, Riggs Collection, CNM.
53. Letter, E. S. Riggs to F. J. V. Skiff, 6 February 1901, DGC, FMA.
54. Letter, W. Granger to H. F. Osborn, 13 June 1900, DVP Arch., AMNH.
55. DVP annual report for 1900, DVP Arch., AMNH.
56. Letter, W. Granger to H. F. Osborn, 25 June 1900, and DVP annual report for 1900, DVP Archives, AMNH. For a complete account of the DVP's Colorado dinosaur explorations, see Brinkman, "Osborn."
57. Hay, a professor of natural science at Butler College in Indianapolis with an interest in vertebrate paleontology, was president of the Indiana Academy of Science in 1891. Butler was secretary. Very likely Hay was in the audience when Butler presented his paper on Colorado fossils. Hay was also editor of the proceedings volume in which Butler's title was published. See Indiana Academy of Science, *Proceedings of the Indiana Academy of Science* 1 (1892). Butler's title appears on p. 73.

    No direct evidence of Hay's meeting with Butler exists. Both, however, participated in the 49th annual meeting of the American Association for the Advancement of Science. Butler served as vice-president of Section H–Anthropology; Hay presented a paper on fossil sharks and was elected to membership at the meeting. See American Association for the Advancement of Science, *Proceedings of the American Association for the Advancement of Science, Forty-Ninth Meeting Held at New York, N. Y., June 1900* 49 (1900), v, xlvii, and 243. On Hay's troubled history at the Field Columbian Museum, see Brinkman, "Establishing," 91–92.

58. See "A Scientific Society," *Grand Junction News*, May 30, 1891, p. 1; and "Academy of Sciences," *Grand Junction News*, April 9, 1892, p. 1.
59. Letter, O. P. Hay to W. Granger, [June 1900], DVP Archives, AMNH. Frank Kiefer was born near Brookville, Indiana, in 1863, and worked on a number of Indiana farms before relocating to Colorado. See *Progressive Men of Western Colorado* (Chicago: A. W. Bowen & Co., 1905), 559–60. Butler was born near Brookville, the son of a prosperous farmer. See William A. Daily and Fay K. Daily, *History of the Indiana Academy of Science, 1885-1984* (Indianapolis: Indiana Academy of Science, 1984), 181.

CHAPTER SEVEN

1. On Hatcher's difficulties working with others, see Rainger, "Collectors."
2. William B. Scott, *Some Memories of a Palaeontologist* (Princeton: Princeton University Press, 1939), 185. Biographical details on Hatcher can be found in William B. Scott, "John Bell Hatcher," *Science*, n.s., 20, no. 500 (July 29, 1904): 139–42; William J. Holland, "In Memoriam, John Bell Hatcher," *Annals of the Carnegie Museum* 2 (1904): 597–604; and George G. Simpson, *Discoverers of the Lost World: An Account of Some of Those Who Brought Back to Life South American Mammals Long Buried in the Abyss of Time* (New Haven: Yale University Press, 1984), chap. 7.
3. The best accounts of Hatcher's career under Marsh and Scott are found in Schuchert and LeVene, *Marsh*, chap. 8; and Url Lanham, *The Bone Hunters* (New York: Columbia University Press, 1973), chap. 17. See also Scott, "John Bell Hatcher."
4. See letters, H. F. Osborn to W. B. Scott, 12 January and 13 February 1900, DVP Arch., AMNH; and J. B. Hatcher to W. J. Holland, 12 February 1900, Hatcher Papers, CMNH.
5. See Rea, *Bone Wars*, 121–22.
6. See letters, W. J. Holland to S. H. Church, 1 February 1900; and, O. A. Peterson to W. J. Holland, 28 May 1900, Holland Papers, CMNH. Despite his significant experience as a field-worker under Marsh and W. C. Knight, Reed had been passed over the previous summer when Wortman was added to the Carnegie Museum staff.
7. Letter, W. J. Holland to A. Darlow, 1 March 1900, Holland Papers, CMNH.
8. Letter, W. J. Holland to C. C. Mellor, 8 February 1900, Holland Papers, CMNH.
9. Letter, W. C. Knight to J. B. Hatcher, 19 March 1900, Hatcher Papers, CMNH.
10. Letter, W. J. Holland to A. Carnegie, 19 March 1900, Holland Papers, CMNH.

11. Letter, W. J. Holland to J. B. Hatcher, 24 February 1900, Holland Papers, CMNH. The letter was mistakenly addressed to J. D. Hatcher.
12. Holland to Carnegie, 19 March 1900.
13. Letter, J. B. Hatcher to W. J. Holland, 31 March 1900, Hatcher Papers, CMNH.
14. Holland to Carnegie, 19 March 1900.
15. See letters, J. B. Hatcher to W. J. Holland, 17, 21, and 28 April, 9 and 31 May, and 4 June 1900, Hatcher Papers, CMNH; and, Rea, *Bone Wars*, 137–38. William Patten is likely the same man who worked for the Field Columbian Museum in the Freezeouts in 1899, although Riggs recorded his name as G. W. Patten (see chapter 5). He later worked for the DVP party (see chapter 11).
16. Letters, J. B. Hatcher to W. J. Holland, 9 and 31 May 1900, Hatcher Papers, CMNH. See also John B. Hatcher, "The Jurassic Dinosaur Deposits near Canyon City, Colorado," *Annals of the Carnegie Museum* 1, no. 11 (1901): 327–41.
17. Letter, W. H. Reed to J. B. Hatcher, 3 May 1900, Hatcher Papers, CMNH.
18. Letter, O. A. Peterson to J. B. Hatcher, 3 May 1900, Hatcher Papers, CMNH.
19. Letter, J. B. Hatcher to W. J. Holland, 9 May 1900, Hatcher Papers, CMNH.
20. Letter, O. A Peterson to J. B. Hatcher, 8 May 1900, Hatcher Papers, CMNH. It seems impossible that Peterson's letter, which was written on the evening of May 8, could have arrived in time to influence Hatcher's letter to Holland, which was written on the morning of the 9th, especially as it probably had to be forwarded from Medicine Bow. The similarity between the two letters suggests the possibility that Hatcher and Peterson had concluded that Reed's prospects were fictitious even before the expedition set out to find them.
21. Letter, W. J. Holland to J. B. Hatcher, 12 May 1900, Holland Papers, CMNH.
22. Letter, J. B. Hatcher to W. J. Holland, 18 May 1900, Hatcher Papers, CMNH.
23. Letter, J. B. Hatcher to W. H. Reed, 18 May 1900, Hatcher Papers, CMNH. This letter was enclosed with a letter to Peterson of a later date, and delivered by him in person on May 27.
24. Letter, O. A. Peterson to J. B. Hatcher, 25 May 1900, Hatcher Papers, CMNH.
25. Rea, *Bone Wars*, 140.
26. Letter, O. A. Peterson to J. B. Hatcher, 28 May 1900, Hatcher Papers, CMNH.
27. Letter, W. H. Reed to W. J. Holland, 27 May 1900, Hatcher Papers, CMNH.
28. Letter, W. J. Holland to W. H. Reed, 31 May 1900, Holland Papers, CMNH.
29. Letter, W. J. Holland to J. B. Hatcher, 31 May 1900, Holland Papers, CMNH.
30. Letter, J. B. Hatcher to W. J. Holland, 7 June 1900, Hatcher Papers, CMNH.

31. Letter, W. J. Holland to J. B. Hatcher, 12 June 1900, Holland Papers, CMNH.
32. Letters, W. H. Reed to W. J. Holland, n.d. [before 27 May 1900]; J. B. Hatcher to W. J. Holland, 7 June 1900; and O. A. Peterson to J. B. Hatcher, 26 June 1900, Hatcher Papers, CMNH.
33. Letters, O. A. Peterson to J. B. Hatcher, 16 June 1900; and O. A. Peterson to W. J. Holland, 16 June 1900, Hatcher Papers, CMNH.
34. Peterson to Hatcher, 26 June 1900.
35. Letter, O. A. Peterson to J. B. Hatcher, 23 August 1900, Hatcher Papers, CMNH.
36. Quoted in Schuchert and LeVene, *Marsh*, 220–21.
37. Letter, J. Wortman to H. F. Osborn, 4 June 1892, DVP Arch., AMNH.
38. Holland to Hatcher, 12 May 1900. See William J. Holland, "The Vertebral Formula in *Diplodocus*, Marsh," *Science*, n.s., 11, no. 282 (May 25, 1900): 816–18.
39. Hatcher to Holland, 18 May 1900.
40. Peterson to Hatcher, 26 June 1900.
41. Letters, J. B. Hatcher to W. J. Holland, 25 June 1900; O. A. Peterson to J. B. Hatcher, 4 July 1900; and O. A. Peterson to W. J. Holland, 15 July 1900, Hatcher Papers, CMNH. See chapter 5 for more on the fate of Holland's prized sauropods prospect.
42. Letter, O. A. Peterson to W. J. Holland, 2 August 1900, Hatcher Papers, CMNH. The skull, which Peterson initially referred to *Diplodocus*, is now believed to belong to *Camarasaurus*. See Parsons, *Drawing*, for an account of some of the controversy surrounding the identity of this skull.
43. Letters, O. A. Peterson to J. B. Hatcher, 9 and 15 August 1900, Hatcher Papers, CMNH.
44. Letters, O. A. Peterson to J. B. Hatcher, 15 and 23 August, and 10 and 26 September 1900; and O. A. Peterson to W. J. Holland, 26 September 1900, Hatcher Papers, CMNH. John S. McIntosh, "Annotated Catalogue of the Dinosaurs (Reptilia, Archosauria) in the Collections of Carnegie Museum of Natural History," *Bulletin of Carnegie Museum of Natural History* no. 18 (1981): 5–67, contains detailed information and modern identifications for the specimens taken out of these quarries. It is often difficult or impossible to reconcile some of the details found in the catalog with those found in the field correspondence.
45. The quotation comes from William J. Holland, "The Carnegie Museum Pittsburgh: Annual Report of the Director for the Year Ending March 31, 1901," *Publications of the Carnegie Museum* Serial No. 10 (1901): 19. See also letters, O. A. Peterson to J. B. Hatcher, 18 October 1900; and J. B. Hatcher to W. J. Holland, 3 October and 8 November 1900, Hatcher Papers, CMNH.

46. Letter, W. J. Holland to O. A. Peterson, 18 September 1900, Holland Papers, CMNH.
47. Letter, J. B. Hatcher to W. J. Holland, 25 July 1900, Hatcher Papers, CMNH.
48. The quotation comes from a letter, W. H. Utterback to J. B. Hatcher, 2 November 1900; see also letter, M. P. Felch to J. B. Hatcher, 28 October 1900, Hatcher Papers, CMNH.
49. Letters, W. H. Utterback to J. B. Hatcher, 12 November and 2 December 1900, Hatcher Papers, CMNH.
50. The quotation comes from Utterback to Hatcher, 2 December 1900; see also letters, W. H. Utterback to J. B. Hatcher, 12 and 17 November 1900, Hatcher Papers, CMNH.
51. William J. Holland, "The Carnegie Museum Pittsburgh: Annual Report of the Director for the Year Ending March 31, 1900," *Publications of the Carnegie Museum* Serial No. 7 (1900): 20–21. See also letters, W. H. Utterback to J. B. Hatcher, 31 January, 8 and 23 February, and 6 April, 1901, Hatcher Papers, CMNH.
52. Utterback to Hatcher, 31 January 1901.
53. See letters, J. Wortman to H. F. Osborn, 18 January 1901, and H. F. Osborn to J. Wortman, 21 January 1901, DVP Arch., AMNH.
54. Letter, W. J. Holland to J. B. Hatcher, 9 November 1900, Hatcher Papers, CMNH.
55. Letter, W. J. Holland to J. B. Hatcher, 21 March 1901, Hatcher Papers, CMNH.
56. See letters, J. B. Hatcher to W. J. Holland, 28 November and 3 December 1900, Hatcher Papers, CMNH.
57. Numerous letters in the Hatcher-Holland correspondence deal with Hatcher's Antarctic plans. The letters that bear directly on this point include J. B. Hatcher to W. J. Holland, 21 and 23 March 1901, and Holland to Hatcher, 21 March 1901, Hatcher Papers, CMNH.
58. See John B. Hatcher, "Vertebral Formula of *Diplodocus* (Marsh)," *Science*, n.s., 12, no. 309 (November 30, 1900): 828–30. See also John B. Hatcher, "*Diplodocus* Marsh, its Osteology, Taxonomy, and Probable Habits, with a Restoration of the Skeleton," *Memoirs of the Carnegie Museum* 1, no. 1 (1901): 29, which asserts that *Diplodocus* had exactly eleven dorsal vertebrae, including a highly modified last dorsal that functions as a sacral vertebra. Finally, John B. Hatcher, "Osteology of *Haplocanthosaurus*, with Description of a New Species, and Remarks on the Probable Habits of the Sauropoda and the Age and Origin of the Atlantosaurus Beds," *Memoirs of the Carnegie Museum* 2 (1903): 14, concedes that "whether or not all five of these vertebrae should be regarded as true sacrals must remain largely a matter of individual opinion." In other words, the distinction between Hatcher's dorsal number eleven and Holland's sacral number one was entirely subjective.

59. Letter, J. B. Hatcher to W. J. Holland, 30 November 1900, Hatcher Papers, CMNH. Simpson (*Discoverers*, 122) reads this letter as Hatcher's way of saying, "When you call me that, smile!" This famous quotation, incidentally, comes from Owen Wister's novel *The Virginian*, which takes place in and around Medicine Bow, Wyoming.
60. Letter, W. H. Holland to J. B. Hatcher, 20 December 1900, Hatcher Papers, CMNH.

CHAPTER EIGHT

1. Letter, H. F. Osborn to B. Brown, 25 July 1905, DVP Arch., AMNH.
2. For a colorful account of Miller's career, see Neil M. Clark, "Adventure, Here I Am; Come A-Shootin'!" *American Magazine* 104, no. 6 (December 1927): 56–57 and 163–66. The quotation comes from a letter, J. Wortman to W. J. Holland, 11 September 1899, Holland Papers, CMNH. See also DVP annual report for 1900, DVP Arch., AMNH. On Wortman as a campfire philosopher, see Henry F. Osborn, "J. L. Wortman—A Biographical Sketch," *Natural History* 26, no. 6 (1926): 652–53.
3. Walter Granger, "Memorial to Frederick Brewster Loomis," *Proceedings of the Geological Society of America for 1937* (June 1938): 173–82. See also letters, F. B. Loomis to H. F. Osborn, 21 January and 26 June 1900, DVP Arch., AMNH.
4. See DVP annual report for 1900; and letters, Loomis to Osborn, 26 June 1900, and W. Granger to H. F. Osborn, 17 July 1900, DVP Arch., AMNH.
5. Letter, H. F. Osborn to F. B. Loomis, 6 July 1900, DVP Arch., AMNH.
6. Loomis to Osborn, 26 June 1900; and DVP annual report for 1900.
7. The quotations come from a letter, H. F. Osborn to W. Granger, 6 July 1900, DVP Arch., AMNH. See also Osborn to Loomis, 6 July 1900. On Osborn's feelings with respect to the Carnegie Museum, see DVP annual report for 1900, DVP Arch., AMNH. See especially the subheading entitled "The Overdraft."
8. Letter, W. Granger to H. F. Osborn, 7 July 1900, DVP Arch., AMNH.
9. Letter, H. F. Osborn to W. Granger, 10 July 1900, DVP Arch., AMNH. See also Osborn to Loomis, 6 July 1900.
10. The quotations come from Granger to Osborn, 7 July and 17 July 1900.
11. DVP annual report for 1900, and, letter, W. Granger to H. F. Osborn, 13 September 1900, DVP Arch., AMNH.
12. DVP annual report for 1900.
13. Letter, O. A. Peterson to J. B. Hatcher, 23 August 1900, Hatcher Papers, CMNH.
14. Granger to Osborn, 13 September 1900. Brown's *Diplodocus*, collected at Como Bluff in 1897, lacked most of the spinal column from the pelvis forward.
15. Letter, H. F. Osborn to W. Granger, 17 September 1900, DVP Arch., AMNH.

16. DVP annual report for 1900, and, letter, W. Granger to H. F. Osborn, 21 September 1900. Granger's quotation was taken from a letter, W. Granger to H. F. Osborn, 12 October 1900. On Osborn's trouble acquiring a free freight car from Morgan, see letters, H. F. Osborn to W. Granger, 25 September, 27 September, and 4 October, DVP Arch., AMNH.
17. Letters, G. R. Wieland to H. F. Osborn, 16 September 1900; and H. F. Osborn to W. Granger, 17 September 1900, DVP Arch., AMNH. See also Joseph T. Gregory, "George Reber Wieland, 1865–1953," *Society of Vertebrate Paleontology News Bulletin* 39 (1953): 27–28. For more on Wieland's experience collecting *Barosaurus* for Marsh, see John S. McIntosh, "The Genus *Barosaurus* Marsh (Sauropoda, Diplodocidae)," in *Thunder-Lizards: The Sauropodomorph Dinosaurs*, ed. V. Tidwell and K. Carpenter (Bloomington: Indiana University Press, 2005), 38–77, especially pages 40–43.
18. Granger to Osborn, 21 September 1900.
19. Osborn to Granger, 27 September 1900.
20. Letters, G. R. Wieland to H. F. Osborn, 18 and 24 September; and H. F. Osborn to W. R. Wieland, 20 and 27 September, DVP Arch., AMNH. The value of Osborn's first offer is not known.
21. See James E. Todd, "The First and Second Biennial Reports on the Geology of South Dakota with Accompanying Papers, 1893–6," *South Dakota Geological Survey Bulletin* 2 (1898): 69–70.
22. Letter, G. R. Wieland to H. F. Osborn, 3 October 1900, DVP Arch., AMNH.
23. Letters, G. R. Wieland to H. F. Osborn, 6 and 10 October 1900, DVP Arch., AMNH.
24. Letter, G. R. Wieland to H. F. Osborn, 16 October 1900, DVP Arch., AMNH.
25. Letter, H. F. Osborn to W. Granger, 18 October 1900, DVP Arch., AMNH.
26. Letter, G. R. Wieland to H. F. Osborn, 21 October 1900, DVP Arch., AMNH.
27. Wieland to Osborn, 21 October 1900.
28. See letters, G. R. Wieland to H. F. Osborn, 26 and 30 October 1900, DVP Arch., AMNH. The quotation comes from the latter.
29. Letter, W. Granger to H. F. Osborn, 31 October 1900, DVP Arch., AMNH.
30. Letter, H. F. Osborn to G. R. Wieland, 5 November 1900, DVP Arch., AMNH.
31. Letter, G. R. Wieland to W. Granger, 10 November 1900, DVP Arch., AMNH. See also letters, G. R. Wieland to H. F. Osborn, 10 and 16 November, DVP Arch., AMNH.
32. See DVP annual report for 1900; postcard, G. R. Wieland to H. F. Osborn, 4 November 1900; and, letters, G. R. Wieland to H. F. Osborn, 26 November 1900; and W. Granger to H. F. Osborn, 19 and 29 November 1900, DVP Arch., AMNH.
33. Letter, H. F. Osborn to G. R. Wieland, 2 May 1901, DVP Arch., AMNH.
34. Letters, G. R. Wieland to H. F. Osborn, 16 and 19 May 1901, DVP Arch., AMNH; and, Albert Thomson's field notebook for 1901, Paul Miller/Albert Thomson Collection (Miller/Thomson Collection), FMA.

35. Thomson's field notebook for 1901, Miller/Thomson Collection, FMA.
36. All of Osborn's quotations come from two letters with the same date, H. F. Osborn to L. Osborn, 31 May 1901, Osborn Papers, NYHS. See also Osborn [MS] "Geological Tour," Osborn Papers, NYHS; and, Thomson field notebook for 1901, Miller/Thomson Collection, FMA.
37. Letter, H. F. Osborn to B. Brown, 25 July 1902, DVP Arch., AMNH.
38. Letter, H. F. Osborn to O. P. Hay, 8 July 1903, DVP Arch., AMNH.
39. Osborn to Osborn, 31 May 1901.
40. Osborn, [MS] "Geological Tour," Osborn Papers, NYHS.
41. Letter, H. F. Osborn to W. Granger, 4 June 1901, DVP Arch., AMNH.
42. DVP annual report for 1901; and letter, G. R. Wieland to H. F. Osborn, 9 June 1901, DVP Arch., AMNH.
43. Letter, G. R. Wieland to H. F. Osborn, 17 June 1901, DVP Arch., AMNH, and Thomson field notebook, Miller/Thomson Collection, FMA.
44. Letter, H. F. Osborn to W. Granger, 29 June 1901, DVP Arch., AMNH.
45. Wieland to Osborn, 9 June 1901.
46. Letter, W. Granger to H. F. Osborn, 30 June 1901, DVP Arch., AMNH.
47. Letter, W. Granger to H. F. Osborn, 22 July 1901, DVP Arch., AMNH, and Thomson field notebook for 1901, Miller/Thomson Collection, FMA.

CHAPTER NINE

1. Letter, E. S. Riggs to O. C. Farrington, 16 March 1901, ROAR-Geology, No. 790, FMA.
2. Letter, O. C. Farrington to F. J. V. Skiff, 18 March 1901, ROAR-Geology, No. 790, FMA.
3. Riggs to Farrington, 16 March 1901.
4. ROEV, FMA.
5. See letters, F. J. V. Skiff to G. B. Harris and E. T. Jeffery, 10 April 1901, DLB, FMA.
6. Letter, E. S. Riggs to H. Mosher, 30 April 1901, Private Collection.
7. Riggs to Mosher, 30 April 1901.
8. Riggs to Mosher, 30 April 1901.
9. Riggs to Mosher, 30 April 1901.
10. Letter, E. S. Riggs to P. Miller, 26 December 1959, Riggs Collection, CNM.
11. Riggs Interview, Riggs Collection, CNM. Additional information was taken from Riggs, "Dinosaur Hunting in Colorado"; and ROEV, FMA.
12. ROEV, FMA.
13. Elmer S. Riggs, "Hunting Fossils, Grand Valley, Colo. [1959]," manuscript, Riggs Collection, CNM.
14. Riggs, "Hunting Fossils," Riggs Collection, CNM.
15. See a brochure from the Museum of Western Colorado, "Dinosaur Hill: An Exhibit on the Tale of a Dinosaur, 90 Years from Beginning to End" (Grand Junction: Museum of Western Colorado, 1992).

16. Riggs, "Hunting Fossils," Riggs Collection, CNM.
17. Riggs, "Hunting Fossils," Riggs Collection, CNM.
18. *Grand Junction News*, 1 June 1901.
19. Riggs, "Hunting Fossils," Riggs Collection, CNM. See also ROEV, FMA.
20. ROEV, FMA.
21. Letters, W. C. Knight to H. F. Osborn, 1 April and 8 April 1901, DVP Arch., AMNH. The quotation comes from the latter.
22. See Henry F. Osborn [MS],"Geological tour, titanotheres & sauropoda, May 17–June 8, 1900 [1901]," Osborn Papers, NYHS; and, letter, J. B. Hatcher to W. J. Holland, 24 May 1901, John Bell Hatcher Papers, Section of Vertebrate Paleontology, Carnegie Museum of Natural History (Hatcher Papers, CMNH). The date on Osborn's MS is a mistake; it should be 1901.
23. Letters, W. D. Matthew to H. F. Osborn, 8 July and 18 July 1901, DVP Arch., AMNH.
24. Letters, B. Brown to H. F. Osborn, 5 July and 13 July 1901, DVP Arch., AMNH. Incidentally, Lee later became a prominent U.S. Geological Survey geologist.
25. See, for example, *Chicago Daily News* articles of June 19 ("Greatest 'U' of All: Immense Endowment, Beyond Dreams of Anybody but Harper, Hinted at") and June 20 ("Proposed Expansion of the University").
26. See letter, F. J. V. Skiff to M. Field, 4 October 1901; and Frederick J. V. Skiff and William R. Harper [MS], "Basis for a Plan of Association," Recorder, Historical Documents (hereafter RHD), Box 3, University of Chicago and Field Columbian Museum, 1901–1904, FMA.
27. Cartoon, *Chicago Daily News*, "Watch the Dinosaur Shrink!" 21 June 1901.
28. For a detailed discussion of the administrative history of vertebrate paleontology at the University of Chicago, see Ronald Rainger, "Biology, Geology, or Neither, or Both: Vertebrate Paleontology at the University of Chicago, 1892–1950," *Perspectives on Science* 1, no. 1 (1993): 478–519.
29. Letters, F. J. V. Skiff to H. Higinbotham, and F. J. V. Skiff to W. R. Harper, 26 February 1902, DLB, FMA.
30. Letters, F. J. V. Skiff to W. R. Harper, 22 March 1902; and W. R. Harper to S. W. Williston, 13 May 1902, Presidents' Papers, Department of Special Collections, Joseph Regenstein Library, University of Chicago (Presidents' Papers, Univ. of Chicago). Williston's annual salary would be augmented by $2,000 from the university.
31. Letter, F. J. V. Skiff to S. W. Williston, [26 May 1902], DLB, FMA.
32. Memorandum to Director, 28 July 1902, RHD, Box 8, Board of Trustees—Misc. Memos and Resolutions, FMA. On Williston's title, see letter, O. C. Farrington to F. J. V. Skiff, 1 December 1902, DGC, FMA.
33. Letter, S. W. Williston to J. B. Hatcher, 25 February 1903, Hatcher Papers, CMNH.
34. Riggs took a brief trip to North Dakota in July to investigate a tip from Williston about a Cretaceous fossil locality, but he collected very little (see

Field Columbian Museum, "Annual Report of the Director to the Board of Trustees for the Year 1901–1902," *Publications of the Field Columbian Museum, Report Series* 2, no. 2 (October 1902): 98.

35. Letter, W. C. Knight to E. S. Riggs, 26 January 1903, Elmer Samuel Riggs Correspondence (Riggs Correspondence), Geol. Dept. Arch., FM.
36. Letter, E. S. Riggs to O. C. Farrington, 11 March 1903, DGC (filed under Farrington), FMA.
37. Letter, O. C. Farrington to F. J. V. Skiff, 13 March 1903, DGC, FMA.
38. Letter, F. J. V. Skiff to D. C. Davies, 16 March 1903, DGC (filed under Farrington), FMA.
39. Letter, E. S. Riggs to F. J. V. Skiff, 13 April 1903, DGC (filed under Farrington), FMA.
40. Letter, F. J. V. Skiff to D. C. Davies, 15 April 1903, DGC (filed under Farrington), FMA.
41. Letter, F. J. V. Skiff to O. C. Farrington, 28 April 1903, DLB, FMA.
42. For more of Williston's financial difficulties in Chicago, see Shor, *Fossils and Flies*, 207–9.
43. Letter, S. W. Williston to J. B. Hatcher, 9 October 1903, Hatcher Papers, CMNH.
44. Letter, W. R. Harper to S. W. Williston, 17 December 1903, Presidents' Papers, Univ. of Chicago.
45. All quotations come from a letter, S. W. Williston to H. N. Higinbotham, 28 December 1903, DGC, FMA.
46. Letter, O. C. Farrington to H. N. Higinbotham, 7 January 1904, DGC, FMA.
47. Letter, E. S. Riggs to O. C. Farrington, 7 December 1904, DGC, FMA.
48. Letter, S. W. Williston to W. R. Harper, 30 May 1904, Presidents' Papers, Univ. of Chicago.
49. See letters, A. K. Parker to E. S. Riggs [filed under Farrington], 2 July 1904; and E. S. Riggs to O. C. Farrington, 3 September 1904, Riggs Correspondence, Geol. Dept. Arch., FM.

CHAPTER TEN

1. Letter, J. B. Hatcher to W. J. Holland, 12 May 1901, Hatcher Papers, CMNH.
2. Letters, J. B. Hatcher to W. J. Holland, 13 and 24 May 1901, Hatcher Papers, CMNH.
3. See Osborn [MS], "Geological tour," Osborn Papers, NYHS. See also Hatcher to Holland, 24 May 1901. For more on Fraas in East Africa, see Gerhard Maier, *African Dinosaurs Unearthed: The Tendaguru Expeditions* (Bloomington: Indiana University Press, 2003).
4. Letter, W. J. Holland to J. B. Hatcher, 29 May 1901, Hatcher Papers, CMNH.
5. The quotation comes from a letter, W. H. Utterback to J. B. Hatcher, 11 July 1901, Hatcher Papers, CMNH. See also letters, J. B. Hatcher to W. J.

Holland, 1 June 1900; and W. H. Utterback to J. B. Hatcher, 7 and 14 June 1900, Hatcher Papers, CMNH.

6. A letter, J. B. Hatcher to W. J. Holland, 1 June 1901, Hatcher Papers, CMNH, mentions that Gilmore needed money to get started in the field. Letters, J. B. Hatcher to W. J. Holland, 2 July 1901; and W. J. Holland to J. B. Hatcher, 13 July 1901, Hatcher Papers, CMNH (both complain about Gilmore's reluctance to write). See also Lewis, "Gilmore."
7. Letter, C. W. Gilmore to J. B. Hatcher, 26 December 1901, Hatcher Papers, CMNH. See also McIntosh, "Annotated Catalogue."
8. See letters, J. B. Hatcher to W. J. Holland, 2, 8, and 22 July 1901; and W. J. Holland to J. B. Hatcher, 5 July 1901, Hatcher Papers, CMNH.
9. Letter, J. B. Hatcher to W. J. Holland, 12 August 1901, Hatcher Papers, CMNH. See Hatcher, *Haplocanthosaurus*, 5–6, for Hatcher's views with respect to the relative age of the bone-bearing horizon in Utterback's quarry.
10. Letters, Hatcher to Holland, 12 August 1901; and J. B. Hatcher to O. A. Peterson, 31 August 1901, Hatcher Papers, CMNH.
11. Letters, G. F. Axtell to J. B. Hatcher, 3, 13, and 20 September, and 12 October 1901, Hatcher Papers, CMNH.
12. See Hatcher, "Jurassic Dinosaur Deposits," 338.
13. Letters, C. W. Gilmore to O. A. Peterson, 9 September 1901; O. A. Peterson to J. B. Hatcher, 14 September 1901; C. W. Gilmore to J. B. Hatcher, 26 December 1901; J. B. Hatcher to W. J. Holland, 31 October and 30 November 1901, and 31 January 1902, Hatcher Papers, CMNH. See also McIntosh, "Annotated Catalogue."
14. Letter, W. H. Utterback to J. B. Hatcher, 15 October 1901, Hatcher Papers, CMNH.
15. Letter, W. H. Utterback to J. B. Hatcher, 31 October 1901, Hatcher Papers, CMNH. See also letter, W. H. Utterback to J. B. Hatcher, 24 October 1901, Hatcher Papers, CMNH.
16. Letter, W. H. Utterback to J. B. Hatcher, 8 November 1901, Hatcher Papers, CMNH.
17. Letter, W. H. Utterback to J. B. Hatcher, 25 November 1901, Hatcher Papers, CMNH.
18. Letters, W. H. Utterback to J. B. Hatcher, 23 December 1901; and J. B. Hatcher to W. H. Holland, 31 December 1901, Hatcher Papers, CMNH.
19. Letter, J. B. Hatcher to W. J. Holland, 30 April 1902, Hatcher Papers, CMNH.
20. Letter, W. H. Utterback to J. B. Hatcher, 28 June 1902, Hatcher Papers, CMNH. See also letters, W. H. Utterback to J. B. Hatcher, 24 May, 5 and 12 June 1902, Hatcher Papers, CMNH.
21. John B. Hatcher, "Field Work in Vertebrate Paleontology at the Carnegie Museum for 1902," *Science,* n.s., 16, no. 410 (November 7, 1902): 752. See also letter, W. H. Utterback to J. B. Hatcher, 3 July 1902, Hatcher Papers, CMNH.

22. See Letter, H. F. Osborn to P. Kaisen, 3 October 1902, DVP Arch., AMNH. Hatcher's letter could not be located.
23. Letter, W. J. Holland to J. B. Hatcher, 13 August 1902, Hatcher Papers, CMNH.
24. Letters, W. H. Utterback to J. B. Hatcher, 20 September, 1, 7, 14, and 27 October 1902, Hatcher Papers, CMNH.
25. Letter, J. B. Hatcher to W. J. Holland, 31 May 1902, Hatcher Papers, CMNH.
26. Hatcher, "Field Work," 752, and John B. Hatcher, "Vertebrate Paleontology at the Carnegie Museum," *Science*, n.s., 18, no. 461 (October 30, 1903): 569–70, provide some scarce information about Gilmore's 1902-03 field seasons. On Quarry M, see also McIntosh, "Second," 23. On the "strange bone," see McIntosh, "Annotated Catalogue," 42. See also Lewis, "Gilmore."
27. The quotation comes from a letter, W. H. Utterback to J. B. Hatcher, 31 June 1903, Hatcher Papers, CMNH. See also letters, W. H. Utterback to J. B. Hatcher, 19 May, and 1 and 4 June 1903, Hatcher Papers, CMNH.
28. Letter, J. B. Hatcher to W. H. Utterback, 26 May 1903, Hatcher Papers, CMNH. See also letter, W. H. Utterback to J. B. Hatcher, 26 May 1903, Hatcher Papers, CMNH.
29. Letter, W. H. Utterback to J. B. Hatcher, 28 July 1903, Hatcher Papers, CMNH.
30. Utterback to Hatcher, 28 July 1903.
31. See McIntosh, "Annotated Catalogue"; and letters, W. H. Utterback to J. B. Hatcher, 12 August 1903; and J. B. Hatcher to W. J. Holland, 13 August 1903, Hatcher Papers, CMNH.
32. See Williston, "Knight."
33. See McIntosh, "Annotated Catalogue"; and letters, W. H. Utterback to W. J. Holland, 4 and 24 July and 14 August 1905, Hatcher Papers, CMNH. Utterback's Cretaceous fieldwork in 1903 and 1905 is beyond the scope of the present account.
34. See William J. Holland, "The Carnegie Museum Pittsburgh: Annual Report of the Director for the Year Ending March 31, 1905," *Publications of the Carnegie Museum* Serial No. 36 (1905): 23–24; and, letter, J. B. Hatcher to W. H. Utterback, 4 June 1904, Hatcher Papers, CMNH.
35. Letter, W. J. Holland to O. A. Peterson, 1 July 1904, Hatcher Papers, CMNH.
36. Rea, *Bone Wars*, 175; and Holland to Peterson, 1 July 1904.
37. Letter, O. A. Peterson to W. J. Holland, 5 July 1904, Hatcher Papers, CMNH.
38. Letters, W. J. Holland to O. A. Peterson, 8 July 1904; and Peterson to Holland, 5 July 1904, Hatcher Papers, CMNH.
39. Letter, H. F. Osborn to O. A. Peterson, 8 July 1904, Hatcher Papers, CMNH.
40. Quoted in Rea, *Bone Wars*, 175.
41. Holland, "Annual Report for 1905," 10.
42. Quoted in Holland, *In Memoriam*, 603.
43. Scott, "John Bell Hatcher," 139 and 141.

44. Holland, "Annual Report for 1905," 21–22. For several years, Holland listed himself as acting curator. By 1908, he had dropped the act and promoted himself to full curator of the Section of Paleontology. He was also in charge of the sections of Vertebrate Zoology, Entomology, Mineralogy, Comparative Anatomy and Osteology, Numismatics, Ceramics, Textiles, Graphic Arts, Transportation, and Carvings in Wood and Ivory. This left only the sections of Botany (which Holland gave up in 1906), Invertebrate Zoology, Archeology, and Historical Collections for others. See William J. Holland, "The Carnegie Museum Pittsburgh: Annual Report of the Director for the Year Ending March 31, 1908," *Publications of the Carnegie Museum* Serial No. 51 (1908).

CHAPTER ELEVEN

1. DVP annual report for 1901; and, letter, H. F. Osborn to P. Kaisen, 3 May 1901, DVP Arch., AMNH.
2. DVP annual report for 1901; and Osborn to Kaisen, 3 May 1901, DVP Arch., AMNH.
3. The quotations come from Osborn to Osborn, 31 May 1901; other information was taken from Osborn [MS], "Geological Tour," Osborn Papers, NYHS. Incidentally, Matthew, "Early Days," 451, notes derisively that "newcomers" often brought sleeping bags in the field.
4. Osborn [MS], "Geological Tour," Osborn Papers, NYHS; and, letters, P. Kaisen to H. F. Osborn, 15 and 27 June and 12 and 22 July 1901, DVP Arch., AMNH.
5. Letter, W. H. Reed to H. F. Osborn, 13 July 1901, DVP Arch., AMNH.
6. Letter, H. F. Osborn to W. Granger, [ca. 13 July 1901], DVP Arch., AMNH.
7. The quotations come from a letter, W. Granger to H. F. Osborn, 10 August 1901; see also letter, W. Granger to H. F. Osborn, 1 August 1901, DVP Arch., AMNH.
8. Letter, W. Granger to H. F. Osborn, 28 August 1901, DVP Arch., AMNH.
9. Letter, H. F. Osborn to W. Granger, 18 July 1901, DVP Arch., AMNH.
10. Letters, W. Granger to H. F. Osborn, 1 August 1901, and P. Kaisen to H. F. Osborn, 9 August 1901, DVP Arch., AMNH.
11. Letter, H. F. Osborn to W. Granger, 7 August 1901, DVP Arch., AMNH.
12. Letters, W. Granger to H. F. Osborn, 23 August and 9 September 1901, and P. Kaisen to H. F. Osborn, 8 and 27 August 1901, DVP Arch., AMNH.
13. Letter, H. F. Osborn to W. Granger, 6 September 1901, DVP Arch., AMNH.
14. Letter, H. F. Osborn to W. Granger, 14 September 1901, DVP Arch., AMNH.
15. DVP annual report for 1901; and, letters, P. Kaisen to H. F. Osborn, 29 September 1901, and W. Granger to H. F. Osborn, 6 October 1901, DVP Arch., AMNH.
16. Letters, H. F. Osborn to W. Granger, 11 August and 6 September 1901, and W. Granger to H. F. Osborn, 22 October 1901, DVP Arch., AMNH.
17. Osborn to Granger, 6 September 1901.

18. Letters, H. F. Osborn to W. Granger, 18 October 1901, and W. Granger to H. F. Osborn, 22 October 1901, DVP Arch., AMNH.
19. Letters, W. Granger to H. F. Osborn, 6 October 1901, and H. F. Osborn to W. Granger, 9 October 1901, DVP Arch., AMNH.
20. DVP annual report for 1901, and telegram, W. Granger to H. F. Osborn, 19 October 1901, DVP Arch., AMNH.
21. DVP annual report for 1902, and letter, W. Granger to H. F. Osborn, 23 June 1902, DVP Arch., AMNH.
22. Letter, H. F. Osborn to W. Granger, 1 July 1902, DVP Arch., AMNH.
23. Regal, *Osborn*, 55.
24. Letter, W. Granger to H. F. Osborn, 6 July 1902, DVP Arch., AMNH.
25. Letters, P. Kaisen to H. F. Osborn, 11 July 1902; and W. Granger to H. F. Osborn, 6 and 12 July and 3 and 8 August, 1902, DVP Arch., AMNH.
26. Letters, P. Kaisen to H. F. Osborn, 12 and 19 August and 7 September 1902, DVP Arch., AMNH.
27. Letters, W. Granger to H. F. Osborn, 15 and 20 September 1902, and H. F. Osborn to W. Granger, 2 October 1902, DVP Arch., AMNH.
28. Letter, W. Granger to H. F. Osborn, 20 September 1902, DVP Arch., AMNH.
29. Letter, H. F. Osborn to P. Kaisen, 3 October 1902, DVP Arch., AMNH.
30. Letter, H. F. Osborn to W. Granger, 18 September 1902, DVP Arch., AMNH.
31. Osborn to Granger, 18 September 1902, DVP Arch., AMNH.
32. Letter, W. Granger to H. F. Osborn, 8 October 1902, DVP Arch., AMNH.
33. DVP annual report for 1902, and letters, P. Kaisen to H. F. Osborn, 26 September 1902, and W. Granger to H. F. Osborn, 8 October 1902, DVP Arch., AMNH.
34. Letter, H. F. Osborn to B. Brown, 23 June 1902, DVP Arch., AMNH.
35. Letter, H. F. Osborn to B. Brown, 3 August 1902, DVP Arch., AMNH. This was almost certainly a reference to Gilmore's brief encroachment at Bone Cabin Quarry.
36. Letter, H. F. Osborn to B. Brown, 1 October 1902; and telegram, H. F. Osborn to B. Brown, 21 October 1902, DVP Arch., AMNH.
37. Letter, H. F. Osborn to B. Brown, 28 October 1902, DVP Arch., AMNH.
38. Letter, B. Brown to H. F. Osborn, 4 November 1902, DVP Arch., AMNH.
39. Letter, H. F. Osborn to B. Brown, 8 May 1903, DVP Arch., AMNH.
40. Letter, H. F. Osborn to B. Brown, 26 May 1903, DVP Arch., AMNH.
41. Letter, B. Brown to H. F. Osborn, 31 May 1903, DVP Arch., AMNH. See also Brown's field notebook for 1903, DVP Arch., AMNH.
42. Letter, H. F. Osborn to B. Brown, 11 June 1903, DVP Arch., AMNH.
43. Letter, H. F. Osborn to B. Brown, 8 July 1903, DVP Arch., AMNH.
44. Letter, B. Brown to H. F. Osborn, 23 July 1903, DVP Arch., AMNH.
45. Letter, H. F. Osborn to B. Brown, 28 July 1903, DVP Arch., AMNH.
46. Letter, B. Brown to H. F. Osborn, 29 July 1903, DVP Arch., AMNH. See also letter, W. H. Utterback to J. B. Hatcher, 28 July 1903, Hatcher Papers, CMNH.

47. Letter, B. Brown to H. F. Osborn, 15 August 1903, DVP Arch., AMNH.
48. Letters, B. Brown to H. F. Osborn, 21 August and 7 September 1903, DVP Arch., AMNH.
49. DVP annual report for 1903, and letter, P. Kaisen to H. F. Osborn, 11 June 1903, DVP Arch., AMNH.
50. Letters, P. Kaisen to H. F. Osborn, 25 June and 12 July 1903; and H. F. Osborn to P. Kaisen, 17 July 1903, DVP Arch., AMNH.
51. Letter, P. Kaisen to H. F. Osborn, 21 July 1903, DVP Arch., AMNH.
52. Letter, P. Kaisen to H. F. Osborn, 4 August 1903, DVP Arch., AMNH. This date almost certainly should be September 4. See also letter, P. Kaisen to H. F. Osborn, 15 September 1903, DVP Arch., AMNH.
53. Letters, P. Kaisen to H. F. Osborn, 15 and 17 September 1903, DVP Arch., AMNH. Osborn's quote comes from a letter, H. F. Osborn to P. Kaisen, 22 September 1903, DVP Arch., AMNH.
54. DVP annual report for 1904; George G. Simpson, "Memorial to Walter Granger," *Proceedings Volume of the Geological Society of America Annual Report for 1941* (1942): 161.
55. DVP annual report for 1904; letter, B. Brown to W. D. Matthew, 16 July 1904, DVP Arch., AMNH. See also G. Edward Lewis, "Memorial to Barnum Brown (1873–1963)," *Geological Society of America Bulletin* 75, no. 2: 19–28.
56. DVP annual report for 1905; letter, H. F. Osborn to P. Kaisen, 6 July 1905, DVP Arch., AMNH.
57. Letters, P. Kaisen to H. F. Osborn, 15 and 22 July 1905, DVP Arch., AMNH.
58. Letter, H. F. Osborn to P. Kaisen, 25 July 1905, DVP Arch., AMNH.
59. Letter, P. Kaisen to H. F. Osborn, 14 August 1905, DVP Arch., AMNH.
60. Letter, P. Kaisen to H. F. Osborn, 1 September 1905, DVP Arch., AMNH.
61. Letter, P. Kaisen to H. F. Osborn, 16 September 1905, DVP Arch., AMNH. See also DVP annual report for 1905, DVP Arch., AMNH.
62. Letter, H. F. Osborn to P. Kaisen, 22 September 1905, DVP Arch., AMNH.
63. Letter, H. F. Osborn to B. Brown, 3 October 1905, DVP Arch., AMNH.
64. Letter, P. Kaisen to H. F. Osborn, 28 September 1905, DVP Arch., AMNH.
65. Letter, H. F. Osborn to P. Kaisen, 3 October 1905, DVP Arch., AMNH.

CHAPTER TWELVE

1. On cramped quarters and planned improvements, see Rainger, *Agenda*, 90, and DVP annual reports for 1898 and 1899. See also letters, H. F. Osborn to J. Wortman (on the commodious new office spaces), 10 November 1899; H. F. Osborn to B. Brown (on basement storage), 25 July 1902; and A. Hermann to H. F. Osborn (on basement lab work), 22 December 1903, DVP Arch., AMNH.
2. See Adam Hermann, "Modern Methods of Excavating, Preparing and Mounting Fossil Skeletons," *American Naturalist* 42, no. 493 (1908): 46–47; and, Hermann, "Modern Laboratory Methods," 330–31.

3. See William J. Holland, "The Carnegie Museum Pittsburgh: Annual Report of the Director for the Year Ending March 31, 1904," *Publications of the Carnegie Museum* Serial No. 28 (1904): 24; William J. Holland, "The Carnegie Museum Pittsburgh: Annual Report of the Director for the Year Ending March 31, 1906," *Publications of the Carnegie Museum* Serial No. 43 (1906): 29; and, letters, W. J. Holland to T. G. McClure, 10 October 1899, Holland Papers, CMNH; J. B. Hatcher to W. J. Holland, 8 November 1900, Hatcher Papers, CMNH; and J. Wortman to H. F. Osborn, 4 November 1899, and 6 January [1900], DVP Arch., AMNH.
4. See Field Columbian Museum, "Annual Report of the Director to the Board of Trustees for the Year 1899–1900," *Publications of the Field Columbian Museum, Report Series* 1, no. 6 (October 1900): 447 and 449; and Field Columbian Museum, "Annual Report 1901–1902," 104.
5. Letters, O. C. Farrington to F. J. V. Skiff, 11 November 1905, DGC, FMA; and A. Thomson to E. S. Riggs, 11 January 1906, Riggs Correspondence, Geol. Dept. Arch., FM.
6. The quotations come from two letters, H. F. Osborn to W. B. Scott, 15 February 1896, and H. F. Osborn to J. W. Gidley, 9 March 1896, DVP Arch., AMNH.
7. Letter, J. Wortman to H. F. Osborn, 6 January [1900], DVP Arch., AMNH.
8. Letter, S. W. Williston to J. B. Hatcher, 25 February 1903, Hatcher Papers, CMNH.
9. Osborn's quotation comes from DVP annual report for 1900. On A. W. Slocum, see letter, O. C. Farrington to F. J. V. Skiff, 9 January 1906, DGC, FMA. On N. Boss, see letter, J. B. Hatcher to W. J. Holland, 16 January 1904, Hatcher Papers, CMNH. On Farrington, see letter, O. C. Farrington to C. Christman, 26 January 1906, DGC, FMA. For an example of Osborn dealing a preparator, see letter, H. F. Osborn to W. B. Scott, 12 January 1900, DVP Arch., AMNH.
10. On Gidley, see letters, H. F. Osborn to J. W. Gidley, 18 March 1896; and J. W. Gidley to H. F. Osborn, 1 August 1899, DVP Arch., AMNH.
11. Letter, H. F. Osborn to B. Brown, n.d. [May 1899], DVP Arch., AMNH. Other letters express the same ideas. See especially H. F. Osborn to B. Brown, 12 January 1900, DVP Arch., AMNH.
12. Letter, H. F. Osborn to W. Granger, 5 June 1899, DVP Arch., AMNH.
13. Rainger, *Agenda*, 80.
14. See letters, W. J. Holland to J. B. Hatcher, 12 June, 6 July, and 17 July 1900, Holland Papers, CMNH.
15. More on Osborn's working relationships appears in Rainger, "Collectors." Insightful firsthand accounts of Osborn's imperiousness can be found in George G. Simpson, *Concession to the Improbable: An Unconventional Autobiography* (New Haven: Yale University Press, 1978), 40; and, Colbert, *Digging*, 168–71. See Robert W. Howard, *The Dawnseekers: The First History of Ameri-*

can Paleontology (New York: Harcourt Brace Jovanovich, 1975), 270–71, for some less sympathetic accounts.

16. DVP annual reports for 1900, 1901, 1903, and 1904.
17. Letters, J. B. Hatcher to O. A. Peterson, 26 May 1903, and J. B. Hatcher to E. Douglass, 4 September 1903, Hatcher Papers, CMNH.
18. Letters, O. C. Farrington to F. J. V. Skiff, 14 November 1902, and H. N. Higinbotham to F. J. V. Skiff, 29 November 1902, DGC, FMA.
19. The abstract of a paper Osborn read before a meeting of the (short-lived) Society of the Vertebrate Paleontologists of America claims that "the writer has recently been experimenting with a sandblast, driven by a compressed air engine, with admirable results." It is difficult to take this claim literally, however. See Henry F. Osborn [abstract], "On the Use of the Sandblast in Cleaning Fossils," *Science*, n.s., 19, no. 476 (February 12, 1904): 256.
20. Letter, J. B. Hatcher to T. W. Stanton, 6 January [1904], Hatcher Papers, CMNH. See also Rainger, *Agenda*, especially chap. 4. On specialization in the preparation lab, see DVP annual report for 1903.
21. Elmer S. Riggs, "The Use of Pneumatic Tools in the Preparation of Fossils," *Science,*, n.s., 17, no. 436 (May 8, 1903): 747–49; and, Elmer S. Riggs [MS], "Hunting Fossils, Grand Valley, Colo.," Riggs Collection, CNM.
22. Hermann, "Modern Laboratory Methods."
23. Riggs, "Pneumatic Tools"; and, letter [draft], E. S. Riggs to A. Hermann, 30 June 1903, Riggs Correspondence, Geol. Dept. Arch., FM. Incidentally, this technology continues to be the bedrock of modern fossil preparation labs.
24. Quoted in Shor, *Fossil Feud*, 218.
25. Quoted in Rainger, *Agenda*, 94.
26. Letter, H. F. Osborn to J. Wortman, 28 April 1898, DVP Arch., AMNH.
27. Osborn to Wortman, 28 April 1898, DVP Arch., AMNH.
28. Henry F. Osborn, "A Skeleton of *Diplodocus*," *Memoirs of the American Museum of Natural History* 1, no. 5 (1899): 191–214.
29. Hatcher, *Diplodocus*, 1.
30. Osborn and Holland both railed against "splitters." See, for example, Osborn, "Recent Progress," 48 (where the author refers derisively to "species makers"), and William J. Holland, *The Moth Book: A Popular Guide to a Knowledge of the Moths of North America* (New York: Doubleday, Page & Company, 1903), 112–13. Ronald Rainger argues that other factors, including scientific training with an emphasis on description, caused scientists such as Cope and Marsh to neglect variation within species ("Paleontology and Philosophy: A Critique," *Journal of the History of Biology* 18, no. 2 [summer 1985]: 267–87).
31. Henry F. Osborn, "Additional Characters of the Great Herbivorous Dinosaur *Camarasaurus*," *Bulletin of the American Museum of Natural History* 10 (June 4, 1898): 229.

32. Osborn to Granger, 2 May 1898; see also Osborn to Wortman, 28 April 1898, DVP Arch., AMNH.
33. Osborn, "Additional Characters."
34. Henry F. Osborn and Walter Granger, "Fore and Hind Limbs of Sauropoda from the Bone Cabin Quarry," *Bulletin of the American Museum of Natural History* 14 (July 9, 1901): 200. Osborn must have known that *Camarasaurus* was the older of these two names, so his choice is puzzling.
35. Elmer S. Riggs, "Structure and Relationships of Opisthoceolian Dinosaurs, Part I, *Apatosaurus* Marsh," *Field Columbian Museum Publication 82, Geological Series* 2, no. 4 (August 1, 1903): 170.
36. See Henry F. Osborn, "Recent Zoopaleontology, the Sauropoda," *Science*, n.s., 19, no. 476 (February 12, 1904): 272; letter, E. S. Riggs to O. C. Farrington, 26 March 1904, Riggs Correspondence, Geol. Dept. Arch., FM; and Stephen J. Gould, *Bully for Brontosaurus* (New York: W. W. Norton, 1991), 79–93.
37. Henry F. Osborn, "*Ornitholestes hermanni*, a New Compsagnathoid Dinosaur from the Upper Jurassic," *Bulletin of the American Museum of Natural History* 19 (July 23, 1903): 459–60.
38. Another ironic discovery by the DVP party was the tiny, incomplete jaw of a new Jurassic mammal designated *Araeodon intermissus* in 1937. This specimen was found in Marsh's Quarry Nine at Como Bluff in 1897. See George G. Simpson, "A New Jurassic Mammal," *American Museum Novitates* 943: 1–6.
39. See Hatcher, "*Haplocanthosaurus*."
40. See Olof A. Peterson and Charles W. Gilmore, "*Elosaurus parvus*; a New Genus and Species of the Sauropoda," *Carnegie Museum Annals* 1: 490–99.
41. See William J. Holland, "The Skull of *Diplodocus*," *Memoirs of the Carnegie Museum* 9: 379–403.
42. Riggs, "Largest Known Dinosaur"; and Riggs, "*Brachiosaurus*."
43. See DVP annual report for 1901.
44. DVP annual report for 1903.
45. DVP annual report for 1904; and, letter, W. D. Matthew to B. Brown, 8 July 1904, DVP Arch., AMNH.
46. See William D. Matthew, "The Mounted Skeleton of *Brontosaurus*," *American Museum Journal* 5, no. 2 (April 1905): 63–70; "500 to Drink Tea under Big Dinosaur," *New York Times*, February 15, 1905, p. 9; "Old and Young Call to See the Dinosaur," *New York Times*, February 20, 1905, p. 12.
47. See letter, A. S. Coggeshall to E. S. Riggs, 2 December 1906, Riggs Correspondence, FM.
48. See Rea, *Bone Wars*, 164–67, 173–74. See also letter, J. B. Hatcher to W. J. Holland, 30 April 1902, Hatcher Papers, CMNH. On the celebrity status of *Diplodocus*, see Rea, *Bone Wars*, chap. 16.
49. Rea, *Bone Wars*, 169, 174–77; and, letters, J. B. Hatcher to W. H. Utterback,

4 June 1904, and W. D. Matthew to W. J. Holland, 1 July 1904, Hatcher Papers, CMNH.

50. See letter, A. S. Coggeshall to E. S. Riggs, 2 December 1906, Riggs Correspondence, Geol. Dept. Arch., FM. See also Coggeshall, "Dippy."
51. Letters, O. C. Farrington to F. J. V. Skiff, 10 May 1900, and 4 August 1903; and F. J. V. Skiff to O. C. Farrington, 6 October 1903, FMA.
52. Letter, O. C. Farrington to H. N. Higinbotham, 7 January 1904, DGC, FMA.
53. Field Columbian Museum, "Annual Report of the Director to the Board of Trustees for the Year 1904–1905," *Publications of the Field Columbian Museum, Report Series* 2, no. 5 (October 1905): 337 and 357; and, letters, O. C. Farrington to F. J. V. Skiff, 10 December 1906, and 8 March 1907, FMA. On the battle to keep the Field Museum out of Grant Park, see Lois Wille, *Forever Open, Clear and Free: The Historic Struggle for Chicago's Lakefront* (Chicago: Henry Regnery Company, 1972), 77–81.
54. Field Museum of Natural History, "Annual Report of the Director to the Board of Trustees for the Year 1907," *Publications of the Field Museum of Natural History, Report Series* 3, no. 2 (January 1908): 135–36.
55. Letter, H. W. Nichols to F. J. V. Skiff, 10 September 1907, DGC, FMA.
56. Letter, W. J. Holland to O. C. Farrington, 8 January 1908, DGC, FMA. See also Field Museum of Natural History, "Annual Report for 1907," 136.
57. Letter, C. C. Gregg to B. Patterson, 27 May 1954, FMA.
58. Rainer Zangerl [1954]. "Recommendation concerning letter from Mr. C. C. Ogle, of Salt Lake City, Utah (enclosed), relating to his discovery of a 40 ft. sauropod dinosaur skeleton." Unpublished memorandum, FMA.
59. The quotation comes from Riggs Interview, Riggs Collection, CNM. See also letter, E. S. Riggs to Department of the Interior, 5 October 1959, Riggs Collection, CNM; and Rainer Zangerl, "*Brontosaurus*—a Bulky Lump of Ancient Protoplasm," *Chicago Natural History Museum Bulletin* 29, no. 4 (April 1958): 6. On publicity, see Chicago Natural History Museum, *Report of the Director to the Board of Trustees for the Year 1958* (Chicago: Chicago Natural History Museum Press, 1959), 95.

CONCLUSION

1. Conn, *Museums*, 45.
2. Years later, in 1902, John Bell Hatcher's restoration of *Diplodocus* launched Carnegie's dinosaur gifting campaign when the king of England saw it hanging at Carnegie's Scottish castle and expressed an interest in getting a skeleton for the British Museum. See letter, A. Carnegie to W. J. Holland, 2 October 1902, Hatcher Papers, CMNH.
3. Rea, *Bone Wars*, 32.
4. Field Museum of Natural History, "Annual Report of the Director to the Board of Trustees for the Year 1912," *Publications of the Field Museum of Natural History, Report Series* 4, no. 3 (January 1913): 205.

5. Secord, "Monsters," 143.
6. The second largest anything, on the other hand, is infinitely less appealing. How else to explain what is so funny about the Kansas detour of Clark Griswold (Chevy Chase) to see the world's second largest ball of twine in the road trip movie *National Lampoon's Vacation*?
7. Lewis, "Gilmore," 236.
8. Simpson, "Lull," 129.
9. Edwin H. Colbert, *A Fossil-Hunter's Notebook: My Life with Dinosaurs and Other Friends* (New York: E. P. Dutton, 1980), 72. Simpson, "Lull," 129 makes this same point about Osborn.
10. See Rea, *Bone Wars*, 42–43.
11. Farrington, "Museum," 482–483.
12. Osborn, "Models," 841.
13. Gould, "Dinomania," discusses this conflict in the modern museum.
14. Matthew, "Mounted," 64.
15. A competition for other objects was taking place among American museums at this same time. Douglas Cole (*Captured Heritage: The Scramble for Northwest Coast Artifacts* [Seattle: University of Washington Press, 1985]) has documented the museum competition for Native American art and artifacts along the Pacific Coast of North America. The threatened extinction of the land mammals in Africa (and elsewhere) was a motivator for the competitive collecting of zoological specimens, also. See Penelope Bodry-Sanders, *Carl Akeley: Africa's Collector, Africa's Savior* (New York: Paragon House, 1991), 141–42.
16. Rainger, *Agenda*, 74.
17. Colbert, *Men*, 154. See also Rea, *Bone Wars*, esp. 215–16.
18. Brinkman, "Establishing," 107.
19. The quotation comes from Field Columbian Museum, "Annual Report 1904–1905," 357. For more on the museum's troubled occupation of the Fine Arts Palace and its effort to build a new building, see Sally G. Kohlstedt and Paul Brinkman, "Framing Nature: The Formative Years of Natural History Museum Development in the United States," *Proceedings of the California Academy of Sciences* 55, supplement 1 (2004): 23–26.
20. Letter, M. Field to J. G. Pangborn, 16 June 1894, DGC, FMA.
21. Rainger, "Continuation."

Bibliography

ARCHIVAL SOURCES

American Museum of Natural History (AMNH)
Department of Vertebrate Paleontology Archives (DVP Arch.)
Annual Reports
Barnum Brown Papers
Departmental Correspondence
Edward Drinker Cope Papers
Field Correspondence
Field Diaries
Henry Fairfield Osborn Papers
Carnegie Museum of Natural History (CMNH)
Archives
Dr. William Jacob Holland Collection, Papers, 1880–1945
Section of Vertebrate Paleontology
John Bell Hatcher Papers, 1899–1908
Olof August Peterson Papers, 1900–1934, 1953, 1955
Colorado National Monument (CNM)
Elmer Samuel Riggs Interview and Correspondence
Field Museum (FM)
Archives (FMA)
Director's General Correspondence (DGC)
Director's Letterbooks (DLB)
"Minutes of the Meetings of the Board of Trustees of the Field Columbian Museum, September 1893 to December 1912"
Paul Miller/Albert Thomson Collection
"Record of Minutes of the Executive Committee of the Field Columbian Museum, May 1894 to December 1913"
Recorder, Historical Documents (RHD)

Recorder's Office Accession Records—Geology (ROAR-Geology)
Recorder's Office, Expedition Vouchers (ROEV)
Department of Geology Archives (Geol. Dept. Arch.)
Elmer Samuel Riggs Correspondence
"Field Record Expeditions: 1898–1910, & 1922"
New-York Historical Society (NYHS)
Osborne Family Papers, 1832–1936
Private Collection
Elmer Samuel Riggs letters and papers
University of Chicago
Joseph Regenstein Library, Department of Special Collections
Presidents' Papers
University of Michigan
William L. Clements Library
Ermine Cowles Case Papers

PUBLISHED SOURCES

American Association for the Advancement of Science. *Proceedings of the American Association for the Advancement of Science, Forty-Ninth Meeting Held at New York, N.Y., June 1900* 49 (1900).

American Museum of Natural History. *Annual Report of the President, Treasurer's Report, List of Accessions, Act of Incorporation, Constitution, By-Laws and List of Members for the Year 1892*. New York: American Museum of Natural History, 1893.

———. *Annual Report of the President, Treasurer's Report, List of Accessions, Act of Incorporation, Constitution, By-Laws and List of Members for the Year 1895*. New York: American Museum of Natural History, 1896.

———. *Annual Report of the President, Treasurer's Report, List of Accessions, Act of Incorporation, Constitution, By-Laws and List of Members for the Year 1896*. New York: American Museum of Natural History, 1897.

———. *Annual Report of the President, Treasurer's Report, List of Accessions, Act of Incorporation, Constitution, By-Laws and List of Members for the Year 1899*. New York: American Museum of Natural History, 1900.

Armstrong, Harley J., and Michael L. Perry. "A Century of Dinosaurs from the Grand Valley." *Museum Journal* 2 (1985): 4–19.

Ashworth, William B. *Paper Dinosaurs, 1824–1969: An Exhibition of Original Publications from the Collections of the Linda Hall Library*. Kansas City, Mo.: Linda Hall Library, 1996.

Beecher, Charles E. "Othniel Charles Marsh." *American Journal of Science* 7, Series 4 (June 1899): 402–28.

Bell, Howard W. "Fossil-Hunting in Wyoming." *Cosmopolitan Magazine* 28 (January 1900): 265–75.

Bodry-Sanders, Penelope. *Carl Akeley: Africa's Collector, Africa's Savior*. New York: Paragon House, 1991.

Boston Journal. "The Monster of All Ages." October 14, 1900.

Bowler, Peter J. *Life's Splendid Drama: Evolutionary Biology and the Reconstruction of Life's Ancestry, 1860–1940*. Chicago: University of Chicago Press, 1996.

Breithaupt, Brent H. "Biography of William Harlow Reed: The Story of a Frontier Fossil Collector." *Earth Sciences History* 9, no. 1 (1990): 6–13.

Brinkman, Paul. "Establishing Vertebrate Paleontology at Chicago's Field Columbian Museum, 1893–1898." *Archives of Natural History* 27, no. 1 (2000): 81–114.

———. "Henry Fairfield Osborn and Jurassic Dinosaur Reconnaissance in the San Juan Basin, along the Colorado-Utah Border, 1893–1900." *Earth Sciences History* 24, no. 2 (2005): 159–74.

Bruce, Robert V. *The Launching of Modern American Science, 1846–1876*. Ithaca: Cornell University Press, 1988.

Buffetaut, Eric. *A Short History of Vertebrate Palaeontology*. London: Croon Helm, 1987.

Carnegie, Andrew. *Autobiography of Andrew Carnegie*. Boston: Houghton Mifflin Company, 1920.

Chenoweth, William J. "Relocating Elmer Riggs' Quarry No. 12." In *Guidebook for Dinosaur Quarries and Tracksites Tour, Western Colorado and Eastern Utah*, ed. W. A. Averett, 17-18. Grand Junction, Colo.: Grand Junction Geological Society, 1991.

———. "The Riggs Hill and Dinosaur Hill Sites, Mesa County, Colorado." In *Paleontology and Geology of the Dinosaur Triangle: Guidebook for 1987 Field Trip Sept. 18–20, 1987*, ed. W. R. Averett, 97-100. Grand Junction: Museum of Western Colorado, 1987.

Chicago Daily News. "Greatest 'U' of All: Immense Endowment, Beyond Dreams of Anybody but Harper, Hinted at." June 19, 1901.

———. "Proposed Expansion of the University." June 20, 1901.

———. "Watch the Dinosaur Shrink! [cartoon]." June 21, 1901.

Chicago Natural History Museum. *Report of the Director to the Board of Trustees for the Year 1958*. Chicago: Chicago Natural History Museum Press, 1959.

Clark, Neil M. "Adventure, Here I Am; Come A-Shootin'!" *American Magazine* 104, no. 6 (December 1927): 56–57 and 163–66.

Coggeshall, Arthur. "How 'Dippy' Came to Pittsburgh." *Carnegie Magazine* 25, no. 7 (July 1951): 238–41.

Colbert, Edwin H. *Digging into the Past: An Autobiography*. New York: Dembner Books, 1989.

———. *Dinosaurs: Their Discovery and Their World*. New York: E. P. Dutton & Company, 1961.

———. *A Fossil-Hunter's Notebook: My Life with Dinosaurs and Other Friends*. New York: E. P. Dutton, 1980.

———. *Men and Dinosaurs: The Search in Field and Laboratory*. New York: E. P. Dutton, 1968.

———. *William Diller Matthew, Paleontologist: The Splendid Drama Observed*. New York: Columbia University Press, 1992.

Cole, Douglas. *Captured Heritage: The Scramble for Northwest Coast Artifacts*. Seattle: University of Washington Press, 1985.

Conn, Steven. *Museums and American Intellectual Life, 1876–1926*. Chicago: University of Chicago Press, 1998.

Cope, Edward D. "On a Gigantic Saurian from the Dakota Epoch of Colorado." *Paleontological Bulletin No. 25* (1877): 5–10.

Cross, C. Whitman. "Description of the Telluride Quadrangle." *U. S. Geological Survey, Geological Atlas, Folio 57* (1899).

———. "Stratigraphic Results of a Reconnaissance in Western Colorado and Eastern Utah." *Journal of Geology* 15 (1907): 634–79.

Daily, William A., and Fay K. Daily. *History of the Indiana Academy of Science, 1885–1984*. Indianapolis: Indiana Academy of Science, 1984.

Davidson, Jane P. *The Bone Sharp: The Life of Edward Drinker Cope*. Philadelphia: Academy of Natural Sciences of Philadelphia, 1997.

Desmond, Adrian. *The Hot-blooded Dinosaurs*. London: Blond & Briggs, 1975.

Dupree, A. Hunter. *Science in the Federal Government: A History of Policies and Activities*. Baltimore: Johns Hopkins University Press, 1986.

Farrington, Oliver C. "Dr. Frederick J. V. Skiff." *Proceedings of the American Association of Museums* 3, nos. 7 and 8 (April–May 1921): 197–98.

———. "The Museum as an Educational Institution." *Education* 17 (1897): 481–89.

Field Columbian Museum. "An Historical and Descriptive Account of the Field Columbian Museum." *Publications of the Field Columbian Museum, Historical Series* 1, no. 1 (December 1894).

———. "Annual Report of the Director to the Board of Trustees for the Year 1898–1899." *Publications of the Field Columbian Museum, Report Series* 1, no. 5 (October 1899).

———. "Annual Report of the Director to the Board of Trustees for the Year 1899–1900." *Publications of the Field Columbian Museum, Report Series* 1, no. 6 (October 1900).

———. "Annual Report of the Director to the Board of Trustees for the Year 1900–1901." *Publications of the Field Columbian Museum, Report Series* 2, no. 1 (October 1901).

———. "Annual Report of the Director to the Board of Trustees for the Year 1901–1902." *Publications of the Field Columbian Museum, Report Series* 2, no. 2 (October 1902).

———. "Annual Report of the Director to the Board of Trustees for the Year 1904–1905." *Publications of the Field Columbian Museum, Report Series* 2, no. 5 (October 1905).

Field Museum of Natural History. "Annual Report of the Director to the Board of

Trustees for the Year 1907." *Publications of the Field Museum of Natural History, Report Series* 3, no. 2 (January 1908).

———. "Annual Report of the Director to the Board of Trustees for the Year 1912." *Publications of the Field Museum of Natural History, Report Series* 4, no. 3 (January 1913).

Foster, John. *Jurassic West: The Dinosaurs of the Morrison Formation and Their World*. Bloomington: Indiana University Press, 2007.

Fuller, Henry B. *With the Procession, a Novel*. New York: Harper & Brothers, 1895.

Gilmore, Charles W. "Osteology of *Apatosaurus*, with Special Reference to Specimens in the Carnegie Museum." *Memoirs of the Carnegie Museum* 11, no. 4 (1936): 175–300.

Gingerich, Philip. "History of Early Cenozoic Vertebrate Paleontology in the Bighorn Basin." In *Early Cenozoic Paleontology and Stratigraphy of the Bighorn Basin, Wyoming*, ed. P. Gingerich. University of Michigan Papers on Paleontology No. 24 (1980): 7–24.

Gould, Stephen J. *Bully for Brontosaurus*. New York: W. W. Norton, 1991.

———. "Dinomania." In S. J. Gould, *Dinosaur in a Haystack: Reflections in Natural History*. 221–37. New York: Crown, 1997.

Grand Junction Daily Sentinel. "Animals of Past Ages: Prof. Briggs of Columbian Museum Lectured Last Night to a Select Audience." July 28, 1900.

———. "Dr. S. M. Bradbury Dies in San Diego Last Night." September 19, 1913.

Grand Junction News. "Academy of Sciences." April 9, 1892.

———. "A Scientific Society." May 30, 1891.

———. "Some Prehistoric Specimens." August 4, 1900.

———. [no name] June 1, 1901.

Granger, Walter. "Memorial to Frederick Brewster Loomis." *Proceedings of the Geological Society of America for 1937* (June 1938): 173–82.

Gregory, Joseph T. "George Reber Wieland, 1865–1953." *Society of Vertebrate Paleontology News Bulletin* 39 (1953): 27–28.

———. "North American Vertebrate Paleontology, 1776–1976." In *Two Hundred Years of Geology in America: Proceedings of the New Hampshire Bicentennial Conference on the History of Geology*, ed. C. J. Schneer, 305–35. Hanover: University of New Hampshire/University Press of New England, 1979.

Hatcher, John B. "*Diplodocus* Marsh, its Osteology, Taxonomy, and Probable Habits, with a Restoration of the Skeleton." *Memoirs of the Carnegie Museum* 1, no. 1 (1901).

———. "Field Work in Vertebrate Paleontology at the Carnegie Museum for 1902." *Science*, n.s., 16, no. 410 (November 7, 1902): 752.

———. "The Jurassic Dinosaur Deposits Near Canyon City, Colorado." *Annals of the Carnegie Museum* 1, no. 11 (1901): 327–341.

———. "Osteology of *Haplocanthosaurus*, with Description of a New Species, and Remarks on the Probable Habits of the Sauropoda and the Age and Origin of the Atlantosaurus Beds." *Memoirs of the Carnegie Museum* 2 (1903).

———. "Vertebral Formula of *Diplodocus* (Marsh)." *Science*, n.s., 12, no. 309 (November 30, 1900): 828–30.

———. "Vertebrate Paleontology at the Carnegie Museum." *Science*, n.s., 18, no. 461 (October 30, 1903): 569–70.

Hermann, Adam. "Modern Laboratory Methods in Vertebrate Paleontology." *Bulletin of the American Museum of Natural History* 26 (1909): 283–31.

———. "Modern Methods of Excavating, Preparing and Mounting Fossil Skeletons." *American Naturalist* 42, no. 493 (1908): 43–47.

Hoagland, Clayton. "They Gave Life to Bones." *Scientific Monthly* 56, no. 2 (February 1943): 114–33.

Holland, William J. "Bone Hunters Starting Well at Their Work." *Pittsburgh Dispatch*, July 25, 1899.

———. "The Carnegie Museum Pittsburgh: Annual Report of the Director for the Year Ending March 31, 1900." *Publications of the Carnegie Museum* Serial No. 7 (1900).

———. "The Carnegie Museum Pittsburgh: Annual Report of the Director for the Year Ending March 31, 1901." *Publications of the Carnegie Museum* Serial No. 10 (1901).

———. "The Carnegie Museum, Pittsburgh: Annual Report of the Director for the Year Ending March 31, 1902." *Publications of the Carnegie Museum* Serial No. 15 (1902).

———. "The Carnegie Museum Pittsburgh: Annual Report of the Director for the Year Ending March 31, 1904." *Publications of the Carnegie Museum* Serial No. 28 (1904).

———. "The Carnegie Museum Pittsburgh: Annual Report of the Director for the Year Ending March 31, 1905." *Publications of the Carnegie Museum* Serial No. 36 (1905).

———. "The Carnegie Museum Pittsburgh: Annual Report of the Director for the Year Ending March 31, 1906." *Publications of the Carnegie Museum* Serial No. 43 (1906).

———. "The Carnegie Museum Pittsburgh: Annual Report of the Director for the Year Ending March 31, 1908." *Publications of the Carnegie Museum* Serial No. 51 (1908).

———. "In Memoriam, John Bell Hatcher." *Annals of the Carnegie Museum* 2 (1904): 597–604.

———. *The Moth Book: A Popular Guide to a Knowledge of the Moths of North America*. New York: Doubleday, Page & Company, 1903.

———. "The Skull of *Diplodocus*." *Memoirs of the Carnegie Museum* 9: 379–403.

———. "The Vertebral Formula in *Diplodocus*, Marsh." *Science*, n.s., 11, no. 282 (May 25, 1900): 816–18.

Horowitz, Helen Lefkowitz. *Culture and the City: Cultural Philanthropy in Chicago from the 1880s to 1917*. Chicago: University of Chicago Press, 1989 [1976].

Howard, Robert W. *The Dawnseekers: The First History of American Paleontology*. New York: Harcourt Brace Jovanovich, 1975.

Illustrated London News. "The Crystal Palace at Sydenham." December 31, 1853.
Indiana Academy of Science. *Proceedings of the Indiana Academy of Science* 1 (1892).
Inter Ocean (Chicago). "Now They Are Saying the Dinosaur Didn't Have Two Brains." March 1, 1903.
Jaffe, Mark. *The Gilded Dinosaur: The Fossil War between E. D. Cope and O. C. Marsh and the Rise of American Science*. New York: Crown, 2000.
Kennedy, John M. "Philanthropy and Science in New York City: The American Museum of Natural History, 1868–1968." PhD diss., Yale University, 1968.
Knight, Wilbur C. "The Wyoming Fossil Fields Expedition of July, 1899." *National Geographic Magazine* 11, no. 12 (December 1900): 444–65.
Kohl, Michael F., Larry D. Martin, and Paul Brinkman, eds. *A Triceratops Hunt in Pioneer Wyoming: The Journals of Barnum Brown and J. P. Sams; The University of Kansas Expedition of 1895*. Glendo, Wyo.: High Plains Press, 2004.
Kohl, Michael F., and John S. McIntosh, eds. *Discovering Dinosaurs in the Old West: The Field Journals of Arthur Lakes*. Washington: Smithsonian Institution Press, 1997.
Kohlstedt, Sally G., and Paul Brinkman. "Framing Nature: The Formative Years of Natural History Museum Development in the United States." *Proceedings of the California Academy of Sciences* 55, supplement 1 (2004): 7–33.
Kuklick, Henrika, and Robert E. Kohler. Introduction to H. Kuklick and R. E. Kohler, eds., *Science in the Field, Osiris* 11, second series (1996): 1–14.
Lanham, Url. *The Bone Hunters*. New York: Columbia University Press, 1973.
Lewis, G. Edward. "Memorial to Barnum Brown (1873–1963)." *Geological Society of America Bulletin* 75, no. 2: 19–28.
———. "Memorial to Charles Whitney Gilmore." *Proceedings Volume of the Geological Society of America Annual Report for 1945* (1946): 235–44.
Lockwood, Frank C. *The Life of Edward E. Ayer*. Chicago: A. C. McClurg & Company, 1929.
Look, Al. *In My Back Yard*. Denver: University of Denver Press, 1951.
Loomis, Frederick B. "On the Jurassic Stratigraphy of Southeastern Wyoming." *Bulletin of the American Museum of Natural History* 14 (June 17, 1901): 189–97.
Lull, Richard S. "Early Fossil Hunting in the Rocky Mountains." *Natural History* 26, no. 5 (1926): 455–61.
Lydekker, Richard. "Some Recent Restorations of Dinosaurs." *Nature* 48, no. 1239 (July 27, 1893): 302–4.
Maier, Gerhard. *African Dinosaurs Unearthed: The Tendaguru Expeditions*. Bloomington: Indiana University Press, 2003.
Manning, Thomas G. *Government in Science: The U.S. Geological Survey, 1867–1894*. Lexington: University of Kentucky Press, 1967.
Marsh, Othniel C. "The Dinosaurs of North America." In *The Sixteenth Annual Report of the U.S. Geological Survey*, 133–244. Washington, D.C.: Government Printing Office, 1896.

———. "Notice of a New and Gigantic Dinosaur." *American Journal of Science* 14 (July 1877): 87–88.

———. "On the Geology of the Eastern Uintah Mountains." *American Journal of Science and Arts* 1, Series 3 (1871): 191–98.

———. "Principal Characters of American Jurassic Dinosaurs, Part IV: Restoration of *Brontosaurus*." *American Journal of Science* 26 (ser. 3) (August 1883): 81–85.

———. "Restoration of *Camptosaurus*." *American Journal of Science* 47 (ser. 3) (March 1894): 245–46.

———. "Restoration of *Stegosaurus*." *American Journal of Science* 42 (ser. 3) (August 1891): 179–81.

———. "Restoration of *Triceratops*." *American Journal of Science* 41 (ser. 3) (April 1891): 339–42.

———. "Restorations of *Claosaurus* and *Ceratosaurus*." *American Journal of Science* 44 (ser. 3) (October 1892): 343–49.

———. *Restorations of Dinosaurian Reptiles, Plate II*. New Haven: Privately printed, 1895.

———. *Restorations of Extinct Animals, Plate I*. New Haven: Privately printed, 1895.

———. "Vertebrate Fossils [of the Denver Basin]." In S. F. Emmons, C. W. Cross, and G. H. Eldridge, *Geology of the Denver Basin in Colorado*, U.S. Geological Survey Monograph No. 27: 473–527. Washington, D.C.: Government Printing Office, 1896.

Matthew, William D. "The Collection of Fossil Vertebrates." Guide Leaflet No. 12, *American Museum Journal* (supplement) 3, no. 5 (October 1903).

———. "Early Days of Fossil Hunting in the High Plains." *American Naturalist* 26, no. 5 (1926): 449–54.

———. "The Mounted Skeleton of *Brontosaurus*." *American Museum Journal* 5, no. 2 (April 1905): 63–70.

McCarren, Mark J. *The Scientific Contributions of Othniel Charles Marsh: Birds, Bones, and Brontotheres*. Peabody Museum of Natural History Special Publication Number 15. New Haven: Peabody Museum of Natural History, 1993.

McCarthy, Kathleen D. *Noblesse Oblige: Charity and Cultural Philanthropy in Chicago, 1849–1929*. Chicago: University of Chicago Press, 1982.

McGinnis, Helen J. *Carnegie's Dinosaurs: A Comprehensive Guide to Dinosaur Hall at Carnegie Museum of Natural History, Carnegie Institute*. Pittsburgh: Carnegie Institute, 1982.

McIntosh, John S. "Annotated Catalogue of the Dinosaurs (Reptilia, Archosauria) in the Collections of Carnegie Museum of Natural History." *Bulletin of Carnegie Museum of Natural History* 18 (1981): 5–67.

———. "The Genus *Barosaurus* Marsh (Sauropoda, Diplodocidae)." In *Thunder-Lizards: The Sauropodomorph Dinosaurs*, ed. V. Tidwell and K. Carpenter, 38–77. Bloomington: Indiana University Press, 2005.

———. "The Saga of a Forgotten Sauropod Dinosaur." In *Dinosaur International:*

Proceedings of a Symposium Held at Arizona State University, ed. D. L. Wolberg et al., 77–12. Philadelphia: Academy of Natural Sciences, 1997.

———. "The Second Jurassic Dinosaur Rush." *Earth Sciences History* 9, no. 1 (1990): 22–27.

Meadows, Anne. *Digging up Butch and Sundance*. New York: St. Martin's Press, 1994.

Mehls, Steven F. *The Valley of Opportunity: A History of West-Central Colorado*. Denver: Bureau of Land Management, Colorado State Office, 1982.

Menke, Harold W. "From a Cabin Window." *Bird-Lore* 1, no. 1 (1899): 14–16.

Miller, Wade E. and Dee A. Hall. "Earliest History of Vertebrate Paleontology in Utah: Last Half of the 19th Century." *Earth Sciences History* 9, no. 1 (1990): 28–33.

Mitchell, W. J. Thomas. *The Last Dinosaur Book: The Life and Times of a Cultural Icon*. Chicago: University of Chicago Press, 1998.

Mook, Charles C. "Notes on *Camarasaurus* Cope." *Annals of the New York Academy of Science* 24 (1914): 19–22.

Morgan, Vincent L., and Spencer G. Lucas. "Walter Granger, 1872–1941, Paleontologist." *New Mexico Museum of Natural History and Science Bulletin* 19 (2002): 1–58.

Museum of Western Colorado. "Dinosaur Hill: An Exhibit on the Tale of a Dinosaur, 90 Years from Beginning to End." Brochure. Grand Junction: Museum of Western Colorado, 1992.

New York Journal and Advertiser. "Most Colossal Animal on Earth Just Found out West." December 11, 1898.

New York Times. "500 to Drink Tea under Big Dinosaur." February 15, 1905.

———. "Dinosaur Skeleton Found." August 14, 1900.

———. "Old and Young Call to See the Dinosaur." February 20, 1905.

Norman, David B. "On the History of the Discovery of Fossils at Bernissart in Belgium." *Archives of Natural History* 14, no. 1 (1987): 59–75.

Nyhart, Lynn K. "Natural History and the New Biology." In *Cultures of Natural History*, ed. N. Jardine, J. A. Secord, and E. C. Spary, 426–43. Cambridge: Cambridge University Press, 1996.

Osborn, Henry F. "Additional Characters of the Great Herbivorous Dinosaur *Camarasaurus*." *Bulletin of the American Museum of Natural History* 10 (June 4, 1898): 219–33.

———. *The American Museum of Natural History: Its Origin, Its History, the Growth of Its Departments to December 31, 1909*. 2nd ed. New York: Irving Press, 1911.

———. *Cope: Master Naturalist*. Princeton: Princeton University Press, 1931.

———. *Creative Education in School, College, University, and Museum: Personal Observation and Experience of the Half-Century, 1877–1927*. New York: Charles Scribner's Sons, 1927.

———. "Fossil Wonders of the West: The Dinosaurs of the Bone-Cabin Quarry, Being the First Description of the Greatest 'Find' of Extinct Animals Ever Made." *Century Magazine* 68, no. 5 (September 1904): 680–694.

———. "J. L. Wortman—A Biographical Sketch." *Natural History* 26, no. 6 (1926): 652–53.

———. "Models of Extinct Vertebrates." *Science*, n.s., 7, no. 182 (June 24, 1898): 841–45.

———. "On the Use of the Sandblast in Cleaning Fossils." *Science*, n.s., 19, no. 476 (February 12, 1904): 256.

———. "*Ornitholestes hermanni*, a New Compsagnathoid Dinosaur from the Upper Jurassic." *Bulletin of the American Museum of Natural History* 19 (July 23, 1903): 459–64.

———. "The Recent Progress of Vertebrate Paleontology in America." *Science*, n.s., 13, no. 315 (January 11, 1901): 45–49.

———. "Recent Zoopaleontology, the Sauropoda." *Science*, n.s., 19, no. 476 (February 12, 1904): 271–72.

———. "Recent Zoopaleontology: Vertebrate Paleontology in the United States Geological Survey." *Science*, n.s., 18, no. 469 (December 25, 1903): 835–37.

———. "A Skeleton of *Diplodocus*." *Memoirs of the American Museum of Natural History* 1, no. 5 (1899): 191–214.

Osborn, Henry F., and Walter Granger. "Fore and Hind Limbs of Sauropoda from the Bone Cabin Quarry." *Bulletin of the American Museum of Natural History* 14 (July 9, 1901): 199–208.

Osborn, Henry F., and Charles C. Mook. "*Camarasaurus, Amphicoelias*, and Other Sauropods of Cope." *Memoirs of the American Museum of Natural History* 3, no. 3 (1921): 247–387.

Ostrom, John H., and John S. McIntosh. *Marsh's Dinosaurs: The Collections from Como Bluff*. New Haven: Yale University Press, 1966.

Parsons, Keith M. *Drawing out Leviathan: Dinosaurs and the Science Wars*. Bloomington: Indiana University Press, 2001.

Peale, Albert C. "Geological Report on the Grand River District." In *Tenth Annual Report of the United States Geological and Geographical Survey of the Territories, Embracing Colorado and Parts of Adjacent Territories, Being a Report of Progress of the Exploration for the Year 1876*, ed. F. V. Hayden, 163–85. Washington, D.C.: Government Printing Office, 1878.

Peterson, Olof A., and Charles W. Gilmore. "*Elosaurus parvus*; a New Genus and Species of the Sauropoda." *Carnegie Museum Annals* 1: 490–99.

Plate, Robert. *The Dinosaur Hunters: Othniel C. Marsh and Edward D. Cope*. New York: David McKay Company, 1964.

Preston, Douglas J. *Dinosaurs in the Attic: An Excursion into the American Museum of Natural History*. New York: St. Martin's Press, 1986.

[Anonymous]. *Progressive Men of Western Colorado*. Chicago: A. W. Bowen & Co., 1905.

Rainger, Ronald. *An Agenda for Antiquity: Henry Fairfield Osborn and Vertebrate Paleontology at the American Museum of Natural History, 1890–1935*. Tuscaloosa: University of Alabama Press, 1991.

———. "Biology, Geology, or Neither, or Both: Vertebrate Paleontology at the

University of Chicago, 1892–1950." *Perspectives on Science* 1, no. 1 (1993): 478–519.

———. "Collectors and Entrepreneurs: Hatcher, Wortman, and the Structure of American Vertebrate Paleontology circa 1900." *Earth Sciences History*, 9, no. 1 (1990): 14–21.

———. "The Continuation of the Morphological Tradition: American Paleontology, 1880–1910." *Journal of the History of Biology* 14, no. 1 (Spring 1981): 129–58.

———. "Paleontology and Philosophy: A Critique." *Journal of the History of Biology* 18, no. 2 (Summer 1985): 267–87.

———. "Vertebrate Paleontology as Biology: Henry Fairfield Osborn and the American Museum of Natural History." In *The American Development of Biology*, ed. R. Rainger, K. R. Benson, and J. Maienschein, 219–56. New Brunswick: Rutgers University Press; 1988.

Rea, Tom. *Bone Wars: The Excavation and Celebrity of Andrew Carnegie's Dinosaur.* Pittsburgh: University of Pittsburgh Press, 2001.

Regal, Brian. *Henry Fairfield Osborn: Race, and the Search for the Origins of Man.* Burlington, Vt.: Ashgate Publishing, 2002.

Reingold, Nathan. *Science in Nineteenth Century America: A Documentary History.* New York: Hill and Wang, 1964.

Riggs, Elmer S. "*Brachiosaurus altithorax*, the Largest Known Dinosaur." *American Journal of Science* 15 (April 1903): 299–306.

———. "The Dinosaur Beds of the Grand River Valley of Colorado." *Field Columbian Museum Publication 60: Geology Series* 1, no. 9 (September 1901): 267–74.

———. "Dinosaur Hunting in Colorado." *Field Museum News* 10, no. 1 (1939): 4–5.

———. "The Discovery of the Use of Plaster of Paris in Bandaging Fossils." *Society of Vertebrate Paleontology News Bulletin* no. 34 (1952): 24–25.

———. "The Fore Leg and Pectoral Girdle of *Morosaurus*, with a Note on the Genus *Camarosaurus* [*sic*]." *Field Columbian Museum Publication 63, Geological Series* 1, no. 10 (October 1901): 275–81.

———. "Fossil-Hunting in Wyoming." *Science*, n.s., 11, no. 267 (February 9, 1900): 233–34.

———. "The Largest Known Dinosaur." *Science*, n.s., 13, no. 327 (1901): 549–50.

———. "Structure and Relationships of Opisthocoelian Dinosaurs, Part I, *Apatosaurus* Marsh." *Field Columbian Museum Publication 82, Geological Series* 2, no. 4 (August 1, 1903): 165–96.

———. "Structure and Relationships of Opisthocoelian Dinosaurs, Part II, The Brachiosauridae." *Field Columbian Museum Publication 94, Geological Series* 2, no. 6 (September 1, 1904): 229–47.

———. "The Use of Pneumatic Tools in the Preparation of Fossils." *Science*, n.s., 17, no. 436 (May 8, 1903): 747–49.

Rogers, Katherine. *The Sternberg Fossil Hunters: A Dinosaur Dynasty.* Missoula, Mont.: Mountain Press Publishing Company, 1991.

Rudwick, Martin J. S. "George Cuvier's Paper Museum of Fossil Bones." *Archives of Natural History* 27, no. 1 (2000): 51–68.

———. *Scenes from Deep Time: Early Pictorial Representations of the Prehistoric World*. Chicago: University of Chicago Press, 1992.

Ryder, Richard C. "Dusting off America's First Dinosaur." *American Heritage* 39, no. 2 (1988): 69–73.

Schuchert, Charles, and Clara M. LeVene. *O. C. Marsh: Pioneer in Paleontology*. New Haven: Yale University Press, 1940.

Scott, William B. "John Bell Hatcher." *Science*, n.s., 20, no. 500 (July 29, 1904): 139–42.

———. *Some Memories of a Palaeontologist*. Princeton: Princeton University Press, 1939.

Secord, James. "Monsters at the Crystal Palace." In *Models: The Third Dimension of Science*, ed. S. de Chadarevian and N. Hopwood, 138–69. Stanford: Stanford University Press, 2004.

Semonin, Paul. *American Monster: How the Nation's First Prehistoric Creature Became a Symbol of National Identity*. New York: New York University Press, 2000.

Shor, Elizabeth N. *The Fossil Feud between E. D. Cope and O. C. Marsh*. Hicksville, N.Y.: Exposition Press, 1974.

———. *Fossils and Flies: The Life of a Compleat Scientist: Samuel Wendell Williston (1851–1918)*. Norman: University of Oklahoma Press, 1971.

Simpson, George G. *Concession to the Improbable: An Unconventional Autobiography*. New Haven: Yale University Press, 1978.

———. *Discoverers of the Lost World: An Account of Some of Those Who Brought Back to Life South American Mammals Long Buried in the Abyss of Time*. New Haven: Yale University Press, 1984.

———. "Memorial to Richard Swann Lull (1867–1957)." *Proceedings Volume of the Geological Society of America Annual Report for 1957* (1958): 127–34.

———. "Memorial to Walter Granger." *Proceedings Volume of the Geological Society of America Annual Report for 1941* (1942): 159–72.

———. "A New Jurassic Mammal." *American Museum Novitates* 943: 1–6.

Skiff, Frederick J. V. "Uses of the Museum." *Chicago Times-Herald*, April 29, 1895.

Sloan, Douglas. "Science in New York City, 1867–1907." *Isis* 71 (1980): 35–76.

Sternberg, Charles H. *The Life of a Fossil Hunter*. Bloomington: Indiana University Press, 1990 [1909].

Sunday Times-Herald (Chicago). "Chicago Has the Largest Land Animal that Ever Lived." October 7, 1900.

Sunday Tribune (Chicago). "Bones of the Largest Known Animal Found." October 7, 1900.

Todd, James E. "The First and Second Biennial Reports on the Geology of South Dakota with Accompanying Papers, 1893–6." *South Dakota Geological Survey Bulletin* 2 (1898): 69–70.

Torrens, Hugh S. "The Dinosaurs and Dinomania over 150 Years." In *Vertebrate*

Fossils and the Evolution of Scientific Concepts: Writings in Tribute to Beverly Halstead, by Some of His Many Friends, ed. W. A. S. Sarjeant. Amsterdam: Gordon and Breach Publishers, 1995.

Vetter, Jeremy. "Science along the Railroad: Expanding Fieldwork in the US Central West." *Annals of Science* 61 (2004): 187–211.

Wallace, David R. *The Bonehunters' Revenge: Dinosaurs, Greed, and the Greatest Scientific Feud of the Gilded Age*. Boston: Houghton Mifflin, 1999.

Warren, Leonard. *Joseph Leidy: The Last Man Who Knew Everything*. New Haven: Yale University Press, 1998.

Wilford, John N. *The Riddle of the Dinosaur*. New York: Vintage Books, 1987.

Wille, Lois. *Forever Open, Clear and Free: The Historic Struggle for Chicago's Lakefront*. Chicago: Henry Regnery Company, 1972.

Williston, Samuel W. "Wilbur Clinton Knight." *American Geologist* 33, no. 1 (January 1904): 1–6.

Wortman, Jacob L. "The New Department of Vertebrate Paleontology of the Carnegie Museum." *Science*, n.s., 11, no. 266 (February 2, 1900): 163–66.

Young, David. "*Brachiosaurus*: The Biggest Dinosaur of Them All." *Field Museum Bulletin* 46, no. 1 (1975): 3-9.

Zangerl, Rainer. "*Brontosaurus*—a Bulky Lump of Ancient Protoplasm." *Chicago Natural History Museum Bulletin* 29, no. 4 (April 1958): 5–6.

Index

Page numbers in italics refer to figures.